This ground-breaking new addition to the series Cambridge Studies in Historical Geography presents some of the first researches into a trove of hitherto inaccessible source material.

A controversial component of Lloyd George's People's Budget of 1909–10 was the 'New Domesday' of landownership and land values. This rich documentation, for long locked away in the Inland Revenue's offices, became available to the public from the late 1970s. For the growing number of scholars of early twentieth-century Britain, Dr Short offers both a coherent overview and a standard source of reference to this valuable scattered archive, and suggests numerous avenues of investigation into an era of increasing interest to historians and sociologists. Part I is concerned with the processes of assembling the material and its style of representation; Part II with suggested themes and locality studies. A final chapter places this new material in the context of discourses of state intervention in landed society prior to the Great War.

Land and society in Edwardian Britain will be of particular use not only to historical geographers, economists and anthropologists, but also to local historians and genealogists.

Cambridge Studies in Historical Geography 25

LAND AND SOCIETY IN EDWARDIAN BRITAIN

Cambridge Studies in Historical Geography

Series editors:
ALAN R. H. BAKER, RICHARD DENNIS, DERYCK HOLDSWORTH

Cambridge Studies in Historical Geography encourages exploration of the philosophies, methodologies and techniques of historical geography and publishes the results of new research within all branches of the subject. It endeavours to secure the marriage of traditional scholarship with innovative approaches to problems and to sources, aiming in this way to provide a focus for the discipline and to contribute towards its development. The series is an international forum for publication in historical geography which also promotes contact with workers in cognate disciplines.

For a full list of titles in the series, please see end of book.

LAND AND SOCIETY IN EDWARDIAN BRITAIN

BRIAN SHORT
Reader in Human Geography, University of Sussex

CAMBRIDGE UNIVERSITY PRESS
Cambridge, New York, Melbourne, Madrid, Cape Town, Singapore, São Paulo

Cambridge University Press
The Edinburgh Building, Cambridge CB2 2RU, UK

Published in the United States of America by Cambridge University Press, New York

www.cambridge.org
Information on this title: www.cambridge.org/9780521570350

First published 1997
This digitally printed first paperback version 2005

A catalogue record for this publication is available from the British Library

Library of Congress Cataloguing in Publication data

Short, Brian, 1944–
The land of Edwardian Britain / Brian Short.
p. cm. – (Cambridge studies in historical geography)
Includes bibliographical references.
ISBN 0 521 57035 2 (hbk)
1. Great Britain – History – Edward VII, 1901–1910. 2. Landscape – Great Britain – History – 20th century. 3. Land use – Great Britain – History – 20th century. 4. Great Britain – Historical geography.
I. Title. II. Series.
DA570.S47 1997
941.082′3–dc20 96-13963 CIP

ISBN-13 978-0-521-57035-0 hardback
ISBN-10 0-521-57035-2 hardback

ISBN-13 978-0-521-02177-7 paperback
ISBN-10 0-521-02177-4 paperback

Contents

Figures

Tables

Preface

This book reports on an excavation of material: material that was kept from public view until the 1970s and which is in process of taking its place amongst the key sources for the study of the geography of the British Isles at the very end of the reign of Edward VII. However, it is not only the geography of that period which can be illustrated – but its sociology and many varieties of history, among them architectural, social and economic, as well as local and genealogical studies. It is a prime source in the reconstruction of a period of British history that has remained of great fascination but for which sources have until now been fragmentary, specialised or spatially incoherent.

In what follows the materiality of the source itself sometimes holds centre stage, sometimes the empirical reconstruction of historical geographies of places scattered throughout the British Isles, sometimes the non-material discourse over land and its significance. The emphasis throughout is necessarily upon land, its ownership, value, use and perhaps, above all, its ideological importance. Intellectual movements, political events of great importance and the development of state machineries of taxation form contexts which come to the fore to greater or lesser extents throughout the volume, but which are never intended to take precedence over the purpose of the text – the demonstration of an archival source. The methodology has of necessity been an interdisciplinary blend of geography, history and possibly many more disciplines as the need to trace the archive and its uses has dictated. Archival material, after all, does not dictate one methodology and one alone – it is open to new conceptual developments and new approaches to understanding and utilisation.

This study has been a long time in progress: so long in fact that my researches have now lasted as long as the original legislation upon which the 1910 survey was based (1910–20)! Indeed the material source is huge, and the temptation remains to continue working on it. Another case study, another search for a different set of documents, another literature search for contemporary views about the political and practical considerations of the great survey initiated finally in 1910. The problem may be perceived readily enough by those who have excavated other comparable sources: how far can one go in making generalisations, when life is too short to cover all the available material? I can only plead that I know full well that other researchers might well discover variations, negations and, hopefully, confirmations of what I have written. But I am confident that I have at least managed to set out the basic outlines of the processes leading to the 1910 survey and the most common forms of documentation that others will encounter.

In undertaking such a long-term piece of work, the debts that one accrues to other scholars and colleagues in the interconnected worlds of Academe and Archives are enormous. Those who have been particularly helpful include friends and colleagues, presently or formerly at the University of Sussex: John Lowerson, Alun Howkins, Susan Rowland (who executed most of the diagrams), Jonathan Rowell, Terry Diffey and especially Mick Reed, without whose enthusiasm, knowledge and skill at the very beginning of the project this text could never have emerged. Past and present postgraduate students have helped too, and I am especially grateful for the shared interest of Charles Rawding, William Caudwell, John Godfrey, Hazel Lintott and Mandy Morris at Sussex, and Olivia Wilson (Durham), David Bell (Birmingham and Staffordshire), Mona Paton and Janet Waymark (Birkbeck). Help has come from universities and archives throughout the United Kingdom and the Republic of Ireland. I am also indebted to all those archivists who responded to the initial questionnaire concerning the location of the 1910 material in 1985–86, and I freely acknowledge help and encouragement received from the following: Dr Geoff Armitage (British Library Map Library), Dr John Andrews, Dr Paul Barnwell (RCHME), Dr Madeleine Beard, Miss Geraldine Beech (PRO), Professor John Beckett, Mr Simon Best (Wiltshire County Council), Mr Michael Bottomley (West Yorkshire Archive Service), Mrs Annette Burton (Glamorgan Record Office), Mr Peter Cassells (Valuation Office Dublin), Mr R. Davies (Department of Pictures and Maps, National Library of Wales), Mr Paul Ferguson (Trinity College Dublin), Mr Albert Fallows (Inland Revenue Chief Valuers Department, latterly the Valuation Office Agency), Mr William Foot (PRO), Mr John Goodchild (Wakefield Library), Mr Hugh Hagan (Scottish Record Office), Professor David Hey, Mr Ian Hill (Scottish Record Office), Professor Roger Kain, Mr Alfred Knightbridge, Dr Peter Laslett, Mr Carl McGee, Mrs Frances McGee (National Archives Dublin), Mr Geoff Mead, Dr Dennis Mills, Ms Michelle Neill (PRO Northern Ireland), Dr Richard Oliver, Professor Charles Phythian-Adams, Dr John Post (PRO), Mr Hugh Prince, Mr Bill Riley, Dr Chris Searle, Dr Keith Snell, Ms Liz Stazicker (Surrey Record Office), Professor Michael Thompson, Mr John Trevelyan (The Ramblers Association), Mr Martin Tyson (Scottish Record Office), Dr Malcom Underwood (St John's College, Cambridge), Dr Sadie Ward (Centre for Rural

History, University of Reading), Mr Christopher Whittick (East Sussex Record Office), Mr Gareth Williams (Gwynedd Archives and Museums) and Professor Charles Withers. I am also grateful for financial help towards the research on which this text is based from the Leverhulme Trust and the University of Sussex, for the support of Dr Alan Baker and his editorial colleagues in the *Cambridge Studies in Historical Geography* Series, Ruth Parr, Richard Fisher and Frances Brown (Cambridge University Press), and two anonymous referees for their helpful and thoughtful comments. The index was prepared by Sharon Hall. Many other individuals and institutions have helped throughout the course of the work, but they are too numerous to mention. I am, nevertheless, truly grateful. As ever, any mistakes are my responsibility, and certainly not those of the people listed above.

My greatest debt, however, remains that to Valerie, David, Matthew and James, with apologies for too many lost weekends.

1

An introduction

So very narrowly did he cause the survey to be made that there was not a single hide nor a rood of land, nor – it is shameful to relate that which he thought no shame to do – was there an ox, or a cow, or a pig passed by, that was not set down in the accounts . . .[1]

Thus were recorded the Anglo-Saxon Chronicle's 1085 complaints of the far-reaching investigations of the original Domesday Book, that William had shamelessly let nothing escape notice in his survey. In an extraordinary throw-back, the Land Union complained in matching rhythm in 1910 that: 'not one acre of land throughout the country, not one house or barn, or fence or wall or pond or ditch, or tree or shrub, but what must be assessed and valued.'[2]

The triple connotations of Domesday – judgement, reckoning (i.e. accounting) and disaster – seem almost to run together in the Anglo-Saxon psyche, and any extensive governmental survey in England has tended to be so branded that, despite any obvious utilitarian function, enormous suspicion is immediately aroused amongst those who own property or other wealth.

Great Britain has no real cadastral history despite its long development of bureaucratic government and relatively early and successful establishment of internal local government. Despite the antiquity of land enquiries in England, from the Domesday Book onwards through an early concern with estate mapping, the central government has been remarkably deficient in establishing land records. This may reflect the fact that Great Britain has traditionally had a lower level of state control in land tax and rent collection.[3] One-off surveys such as the 1086 Domesday Book, or the tithe surveys *c.*1840, were

[1] H. C. Darby, 'The economic geography of England, AD 1000–1250', in H. C. Darby (ed.), *An Historical Geography of England before AD 1800* (Cambridge 1963 edn), 165.

[2] The Land Union, *The Land Union's Reasons for Repeal of the New Land Taxes and Land Valuation* (London 1910), 32. The pro-legislation journal *Land Values* pointed out in a rather po-faced way that, of course, 'the modern enquiry does not concern itself with the individual property or livestock which the holder may have upon the land' (in 'Our old brown mother', *Land Values* (November 1910), 115).

[3] R. J. P. Kain and E. Baigent, *The Cadastral Map in the Service of the State* (Chicago 1992).

complete neither in spatial coverage nor in thematic detail. It has therefore become commonplace to state that we have no cadastral surveys in this country which can compare with those of continental Europe, as in France or Sweden or Iceland. In 1979, a significant year in the public deposition of the Lloyd George records, the Northfield report noted that: 'It is disturbing that so little is known about the pattern of acquisition, ownership, and occupancy of agricultural land and that governments should have to take decisions with far-reaching consequences on the agricultural structure, on the basis of incomplete or non-existent data.'[4] Many types of survey do however exist in Europe for periods before the nineteenth century, mostly for purposes of taxation, and mostly without accompanying maps. But in 1807 Napoleon I instituted the French cadastre with its parcel numbers, area, land use and land values for each owner, based on surveys of each parish, and this format was copied in much of Europe outside England.[5]

Proposals have from time to time been put forward for a more comprehensive land data base to cover Great Britain. Soon after the launching of the Land Valuation procedures outlined in this book, for example, a scheme was proposed for the formation of a 'Domesday Office' which would amalgamate the Land Registry, the Land Values Department and ultimately also the Ordnance Survey. The proposal came to nothing, and although schemes to promote comprehensive surveys of landownership and value have been put forward from time to time since, nothing has yet emerged.[6]

The background to the present author's interest in the records herein described is straightforward and probably replicates the circumstances surrounding initial work on many other records. These particular records were first encountered during research for material in the mid-1980s for a study of a Sussex rural community.[7] Nothing was known about them at the East Sussex Record Office, and little could be gleaned from any secondary writing. Informal soundings were made, casual investigations were implemented at the PRO and at various local repositories whilst the author was involved in other researches, and eventually, as a clearer picture emerged of the possible importance of the material, research grant applications were submitted. In 1985 the Leverhulme Trust agreed to fund a project to investigate the coverage and significance of the material, and during the academic year 1985–86 the author

[4] *The Northfield Report* (HMSO, London 1979), para. 259.

[5] G. Larsson, *Land Registration and Cadastral Systems* (London 1991), 19–25; H. D. Clout and K. Sutton, 'The cadastre as a source for French rural studies', *Agricultural History* 43 (1969), 215–23.

[6] PRO LAR 1/107 Undated scheme for the formation of a Domesday Office in the records of the Land Registry.

[7] B. Short (ed.), *Scarpfoot Parish: Plumpton 1830–1880* (Centre for Continuing Education, University of Sussex 1981). Further plans to investigate the history of the parish of Plumpton between 1880 and 1914 included the meeting of a local studies group at Plumpton which investigated the 1910 material in more detail.

was very fortunate to secure the services of Mick Reed as a Research Fellow at the University of Sussex. Great strides were made in the understanding of the magnitude and complexity of the material, thanks in large measure to Mick Reed's assiduous research, and the attention of fellow academics as well as archivists was in various ways drawn to its potential.

Thus, in one conference paper, the present author compared the amount of information available in the 1910 survey with that available in the 1086 Domesday survey and the 1840s Tithe surveys. The Tithe surveys, with the tithe files, present us with about 1.8 million items of data, but were by no means complete in spatial or information coverage, as demonstrated by Kain and Prince. The original Domesday survey was, of course, less precise, and covered a very much smaller geographical extent. The English translation of the Domesday Book runs to 1.1 million words. The survey of ownership taken in 1873 (the 'New Domesday') was not concerned with the actual delineation of property boundaries, and merely gives an indication of the numbers of owners, the amount of land owned by them in each county, and its value.[8] The 1941–43 National Farm Survey is being evaluated as these words are being written, but it is already clear that this too will revolutionise our thoughts about wartime farming in particular, and about the state of Britain's farmland after years of agricultural depression. That survey has something in the order of 250 pieces of potential information per holding, for about 300,000 holdings over 5 acres in size, or about 7.5 million pieces of information overall. It is unlikely however that any one farm will have information on more than perhaps two-thirds of these items, but this would still leave an estimated 5 million items.[9]

By contrast, the survey under consideration here covered both urban and rural England and Wales. The Lloyd George survey has several different, linked records, each containing many items of information (all of which are explained in the following chapters), for about 13 million hereditaments! Even if each record did nothing more than to name the owner and occupier, this would yield 26 million pieces of information, and the further multiplications of material become almost too daunting to contemplate! Each of the 95,000 Field Books for England and Wales alone contained four pages of information on each hereditament, with up to 100 hereditaments per book! We may well be dealing with something in the order of 40 million items of information. And this is excluding the information for Scotland for which another

[8] R. J. P. Kain and H. C. Prince, *The Tithe Surveys of England and Wales* (Cambridge 1985), Frontispiece map and pp. 112–13; British Agricultural History Society and Historical Geography Research Group of the Institute of British Geographers, Joint Winter Conference, Agricultural Censuses and Statistics (December 1986), B. Short, 'Lloyd George's Domesday: the countryside of England and Wales in 1910'.

[9] Brian Short and Charles Watkins, Economic and Social Research Council Grant No. R000235259 (1994–6).

1.3 million hereditaments were included, and Ireland for which 190,000 forms of return, mostly for urban areas, had been issued by 31 March 1912.[10]

This volume is intended to provide a source of reference, and hopefully a source of stimulus, for work on the 1910 material. In what follows, I have attempted to ascertain its availability, check its accuracy, and suggest avenues of interpretation. It takes the form of two main parts, one dealing with the processes of the survey and the context of political legislation, and one demonstrating the potentials and problems of the source in a series of contrasting thematic and regional settings. In Part I the preliminary chapter deals in the first place with the political and conceptual background to the 1910 survey and attempts to contextualise the development of the legislation by foregrounding it against the broader sweep of the many movements pressing for land reform in Great Britain as a whole at the end of the nineteenth century. The text then proceeds to examine the national structure of the ensuing valuation process which was set in train by Lloyd George's eventual triumph over the Lords in 1910. The survey procedures and the resulting multiplicity of documents are then discussed, and a critical view of the modern archival practices and policies in handling the 1910 material is presented. In Part II the significance of the material is explicitly addressed, together with likely problems that will be encountered by future scholars working with the relatively complex documentation. In order to demonstrate more forcefully what might be possible, four chapters are devoted to case studies which deal successively with urban social area analysis 1909–14, rural society and economy 1909–14, rural industrial communities on the eve of the Great War, and finally contrasts and comparisons which can be drawn with contemporary or near-contemporary primary sources and with secondary accounts of particular localities. And whilst much of the case study material is drawn from England and Wales, it should also be noted that the legislation applied equally to Ireland and Scotland, where its administration was in a modified form; attention is therefore devoted to the context and application of the 1910 material there.

The nature of the volume is that of an overview and an evaluation. It is not intended that any of the case studies should be seen as definitive historical accounts of particular localities, and it is to be hoped that the frustration which will inevitably be felt by specialist historians of those chosen localities will be balanced by the recognition of the splendid potential released by the 1910 records. Both pattern and process are addressed here. The pattern of the records themselves as they exist in varying degrees of completeness in repositories around Great Britain is related to the processes which brought the

[10] *8th Report of Commissioners of HM Inland Revenue*, Year ending 31 March 1912 (BPP 1912–13 (C 6344), XXIX, Chapter 3, 155); *Select Committee on Land Values* 1920, evidence of Mr C. J. Howell Thomas, 13.

records into being, and those which have latterly sought to cope in an archival sense with their huge bulk. And directly related to the pattern of the extant records is the pattern that can be reconstructed of the British Isles at the very end of the Edwardian period: its geography, economy and society as seen through one particular instrument of fiscal and ideological legislation.

It is now clear that the twentieth century will yield enormous riches of documentation to the scholars of the future. At present we are somewhat restricted by codes of confidentiality which ensure that various classes of record are kept locked away for periods of between 30 and 100 years. The 1910 material presented here will become of great significance when the 1911 census enumerators' books become available, as they already have in the Republic of Ireland, and as scholars patiently discover what records from within local authority archives will bear comparison and sit alongside these sources. Historians and historical geographers have not worked extensively on twentieth-century records and there is not the same tradition of scholarship that has been established by workers on Victorian and earlier periods. This will change, and it is to be hoped that the evaluation of the 1910 material presented here will go a little way towards the establishment of a corpus of work on twentieth-century records.[11]

[11] B. Short, 'The twentieth century', in D. Hey (ed.), *The Oxford Companion to Family and Local History* (Oxford 1996), 456–62.

PART I

Processes and representations

2

Lloyd George, the 1909 Budget and the land campaigns

In 1906 John Galsworthy published *The Man of Property*, the first volume in *The Forsyte Saga*, an appropriate book to mark the spirit of an age which set great store on the possession of wealth and influence, appearing in the year which marked the halfway point of the short Edwardian age and the one in which the Liberal government came to power. The Britain of the Edwardian age, seen from the perspective of nearly a century, was outwardly exuberant, hopeful, confident. And yet changes since the height of the Victorian 'Golden Years' had been mostly for the worse, at least in the macro-level terms of society and economy. Urbanisation had now reached a point where 79 per cent of the population were classed as urban in the 1911 Census, bringing pollution, overcrowding and a poor standard of living for huge numbers in the inner cities and a desire to escape the masses which brought undistinguished (yet nevertheless nuanced) suburbia into former rural landscapes. Attracted by the hopes of employment and by a desire to escape the drudgery and intrusiveness of much rural life, many young people had joined friends and relatives in the growing towns, only to find unemployment and squalor beneath the veneer of prosperity. The agricultural depression hit landed incomes and rural jobs alike from the 1870s, as foreign imports of food undercut British producers, with something of an upturn only appearing in the opening decade of the twentieth century.

Thus, many have characterised the period as one of crisis, as mounting social and economic problems joined with darkening foreign relations. A general European war loomed; the threat of civil war in Ireland grew; constitutional crisis brought 'the peers versus the people'; the suffragette issue threatened to break through an assumed patriarchal dominance; and a wave of strikes and industrial unrest started in 1910.[1]

[1] D. Read, 'Introduction: crisis age or golden age?', in D. Read (ed.), *Edwardian England* (Historical Association, London 1982), 14–39; C. M. Law, 'The growth of urban population in England and Wales, 1801–1911', *Transactions of the Institute of British Geographers* 41 (1967), 125–43.

Behind many of these problems was the rise in the cost of living. Up to the end of the previous century real wages had risen but now they were stagnant or falling. Exports had become less competitive although they were still contributing 30 per cent of the United Kingdom's national production, and London remained the world's financial capital, but the 'new' resources of rubber, tin, oil or copper had to come from abroad, and Britain now slipped behind her rivals with newer industrial economies, Germany and the United States, in the production of steel. Much of the UK's industrial strength remained concentrated in sectors with little growth potential but the Empire still gave scope for investment and career paths abroad – to the detriment of investment at home. In fact investment overseas had exceeded net capital formation at home since 1870 and was at double the rate of home investment by the beginning of World War I. Elegant Edwardian stockbroker houses in the Home Counties were thus just the outward sign of financial dealings that depended on what Hobsbawm referred to as a 'parasitic rather than a competitive economy'.[2] Financiers and industrialists came to join the traditional elites, whilst industrial, mining and urban properties increased their importance within their portfolios. At this juncture of two great Kondratieff Cycles in the 1890s, at the transition from manufactures based on machinery to industries managed on scientific principles, the British economy began to look distinctly less healthy than hitherto.

It was immensely significant that this was happening at the same time that an expanded electorate was becoming a more working-class body – perhaps 75 per cent being working-class by 1906 – and local government was becoming more democratic and less dominated by the landed elites. A real vote for change could now be expressed, and if the electorate was now largely composed of the urban working class, suburban dwellers, the rural smallholder and the working men of Scotland, Wales and Ireland, the future for the traditional squirarchy would be limited. As the Earl of Meath had warned in 1899, a large poor and hungry stratum would, 'driven to desperation and beguiled by the honey words of Socialists and Anarchists, endeavour to improve their miserable lot by the general destruction of society'.[3]

As foreign competitors grew stronger, bringing depression to industry and the countryside, many began to turn away from the ideology of free trade towards a greater degree of protection. More specifically, in 1903 Joseph Chamberlain unveiled his ideas on Imperial Preference and tariff reform, causing fundamental splits in Tory and Unionist ranks which led among other

[2] M. Dunford and D. Perrons, *The Arena of Capital* (London 1983), 308–9.

[3] Cited in G. D. Phillips, *The Diehards: Aristocratic Society and Politics in Edwardian England* (Cambridge, Mass. 1979), 113. For the relationship of government to society at this time, but set in a wider context, see P. Thane, 'Government and society in England and Wales, 1750–1914', in F. M. L. Thompson (ed.), *The Cambridge Social History of Britain 1750–1950* vol. III (Cambridge 1990), 1–61.

things to Winston Churchill crossing the floor of the House to join the Liberals. By 1909 a modified version of tariff reform had become Tory opposition policy. The Liberals, by contrast, could unite behind a free trade banner and proceeded to power in December 1905 under Henry Campbell-Bannerman. According to one observer, Campbell-Bannerman offered Lloyd George a seat in the Cabinet at the Board of Trade, saying 'I suppose we ought to include him'.[4]

British society in so many ways was therefore ripe for political change and the Liberals swept to power in the 1906 election, moving soon into a largely uncoordinated sequence of social legislation through to 1911, for which money was to be found by increased taxation of that class which could most easily bear it – the landed elite. In 1894 Harcourt's death duties were a precursor to the swingeing attacks which were to be forthcoming in the Liberal years before the First World War. Cannadine has placed great emphasis upon the period between 1880 and 1914 as marking a catastrophic downturn in landed fortunes as 'the demands of an increasingly hostile, predatory and intrusive state had to be met' and even if his analysis is somewhat overdrawn and the elite moved more successfully through the twentieth century than he has claimed, nevertheless the legislation covered in this text, and the political furore surrounding it, did mark a significant moment in the loss of patrician power.[5]

Approaches to the land reform question

Land reform has a long history of theory but little evidence of practical application in Britain. In the latter half of the nineteenth century proposals came repeatedly from the likes of the Chartists, Cobden, Bright, Mill and Joseph Chamberlain, and ideas were particularly focussed by the Irish Land War and Gladstone's Irish Land Act 1881. Rent strikes and violence similarly characterised the relationships of the crofters of western Scotland with the authorities, and there was Welsh unrest too at this time. Certainly in the 1880s the tackling of the concentration of ownership of large swathes of British countryside by a relative handful of people was seen as an urgent issue, with the concentration revealed in the so-called 'New Domesday' of 1873.[6] The hold on the land by this elite was seen by many to be the root cause of contemporary social problems and of the crisis of capitalism and poverty of the British working people.

[4] P. Rowland, *The Last Liberal Governments: The Promised Land, 1905–1910* (London 1968), 19.

[5] D. Cannadine, *The Decline and Fall of the British Aristocracy* (Yale 1990), 102 and *passim*. For a critique of Cannadine's position see, for example, P. Thompson, *The Edwardians: The Remaking of British Society* (2nd edn, London 1992), 193; and for the resilience of the landed elite at the county level see J. Waymark, 'Landed estates in Dorset since 1870: their survival and influence', University of London, unpubl. Ph.D. thesis, 1995.

[6] Return on the Owners of Land, England and Wales (1872–1873), BPP LXXII, 1874. And see also J. Bateman, *The Great Landowners of Great Britain and Ireland* (London 1883).

David Lloyd George, the young Welsh nonconformist solicitor who entered parliament as a Liberal MP for Caernarfon Boroughs in April 1890, was fully aware of such opinions. He had experienced the Welsh revolt against the payment of tithes and the associated violence beginning about 1886, and in that year his first recorded public appearance contained a call for his Flintshire audience to join a Welsh Land League. He proceeded to carry a loathing for 'landlordism' and an advocacy of some redistribution of wealth with him into government office and into the Chancellorship of the Exchequer (an appointment about which King Edward VII had reservations) in April 1908. The conditions for turning the threat to aristocratic power over the land of Britain into material action could now be effected.[7]

There were three generally recognised solutions to reform of the land, amidst enormous agitation to deal somehow with the land issue at this time. The first, and least popular generally because of its draconian connotations, was the outright nationalisation of land for public or social purposes with compensation, as advocated for example by Lloyd George's fellow Welsh Liberal, A. Williams, MP for Merioneth. Nevertheless, the Land Nationalisation Society, formed in 1881, claimed nearly 130 MPs as supporters at times after the 1906 general election. A system of vice-presidential status accorded to supporting MPs allows some estimate of their strength, and shows, for example that there were 68 such MPs in May 1906, rising to 112 by December 1909.[8] Their campaign was led until his death in 1913 by the distinguished biologist Alfred Russel Wallace, whose interest in the land question was heightened by his association with J. S. Mill (1806–73) and Herbert Spencer (1820–1903), the proponent of social evolution.[9] Their perceived influence on the Liberal government was enough for one practical farmer and landowner from Dorset to write in 1911: 'Though the Liberals have no official land policy . . . they at least have an unofficial one held by an ever-increasing section of the party – the nationalisation of land.'[10]

Secondly there were those who wanted the taxation by the state of land values at a rate of twenty shillings in the pound, thereby taxing away the whole value of any unearned increase in values, with other taxes, both direct and indirect then falling away. These were the views of the 'single-taxers' following the 'Apostle of Plunder', the charismatic American writer Henry George, as particularly exemplified in his extremely influential *Progress and Poverty* (first published in 1880 and available in numerous later editions), and in lecture

[7] C. Wrigley, *David Lloyd George and the British Labour Movement: Peace and War* (Hassocks 1976), 3 and *passim.*

[8] Sadie Ward, 'Land reform in England 1880–1914', University of Reading, unpubl. Ph.D. thesis, 1976, 461–4.

[9] J. D. White, *Land-Value Reform in Theory and Practice* (London 1948), 5.

[10] Christopher Turnor, *Land Problems and National Welfare* (London 1911), 236–7.

tours throughout the United Kingdom. The last of these was in 1889. George was opposed to the outright nationalisation of land. He proposed:

not . . . to purchase or to confiscate private property in land. The first would be unjust; the second, needless. Let the individuals who now hold it still retain, if they want to, possession of what they are pleased to call 'their' land . . . Let them buy and sell, and bequeath and devise it. We may safely leave them the shell, if we take the kernel. It is not necessary to confiscate land; it is only necessary to confiscate rent.[11]

Organisations based on Henry George's teaching included the English and Scottish Leagues for the Restoration of Land ('The land belonged to the people by natural right. The people should reassert their title, taken from them in the past by the robber ancestors of the present landlords'), later renamed the English and Scottish Leagues for the Taxation of Land Values. The actions of these groups, especially the 'Red Van' campaigns between 1891 and 1898, are well known, as are the activities of the Land Nationalisation Society and its 'Yellow Van' campaign of the 1890s. In 1902 the name of the former monthly paper, *The Single Tax*, was changed to *Land Values*, in order, as the editorial of June 1902 put it, to 'meet the requirements of the present political situation'. [12] In March 1907 the leagues of England, Scotland and Ireland finally merged to form the United Committee for the Taxation of Land Values (UCTLV)

Josiah Wedgwood, committed parliamentary leader of the 'single taxers', had been early introduced to the work of Henry George whose convictions on free trade and the taxation of land values had given him his 'anchorage and . . . object in politics'. As he declared in his autobiography: 'Ever since 1905 I have known "that there was a man from God, and his name was Henry George". I had no need henceforth for any other faith.'[13] By 1906, however, the 'single taxers' could claim fewer than ten followers in the new Commons, since many pragmatic Liberal politicians did not wish to be actively associated with Georgite sympathies which might label them as 'confiscators'. But in May 1911 a land tax memorial on lines approved by the UCTLV was signed by 173 MPs.

Thirdly, there were more moderate views which confined themselves to local rating but which advocated using exclusively the value of the site as the basis

[11] H. George, *Progress and Poverty* (London 1881), 364.

[12] For a full account of these activities see A. J. Peacock, 'Land reform 1880–1919', University of Southampton, unpubl. MA thesis, 1961, and Ward, 'Land reform' esp. Chapter 8. The quotation from Henry George is from E. P. Lawrence, *Henry George in the British Isles* (E. Lancing, Michigan 1957), 7–8. The quote from *Land Values* (June 1902), 1, is from Ward, 'Land reform', 387. I am grateful to Dr Ward for her assistance in tracing literature of the land tax movements, much of which is held at the Centre for Rural History, University of Reading. The manuscript records of the English Land Tax societies were mostly destroyed by enemy action in World War II, but Ward has assembled a detailed description of the remaining literature in her Ph.D. thesis, which remains the best source for more information on these movements.

[13] J. C. Wedgwood, *Memoirs of a Fighting Life* (London 1940), 60.

for taxation, rather than the total value of the hereditament which would include land plus buildings and improvements etc. This site value rating argument hoped that idle land, especially urban land, would be forced into development, and as such held some appeal for growing numbers of municipal authorities from the late 1880s, beginning with Glasgow in 1889. In 1894 the London County Council began negotiations with government to attempt to secure such legislation, and in 1895 London was the venue for the first conference on the taxation of land values. By 1897 about 200 assessing bodies were wanting such site value taxation. The Final Report of the Royal Commission on Local Taxation appeared in 1901, and while there was a substantial minority on the Commission in favour of site value rating in towns, the conclusions were insufficiently pointed in any one direction. The debates continued with, by 1906, more than 500 local authorities petitioning Campbell-Bannerman in support of site value rating.[14]

With variations still being actively debated, the taxation of land values by central and local government, but at a rate below the 100 per cent levy, became a solution favoured by many. Within parliament, there were no fewer than fourteen bills and resolutions introduced between 1889 and 1906 on the subject of the taxation of land values, all unsuccessful to varying degrees. Thus, C. P. Trevelyan introduced a bill in 1902 on the subject which was defeated on its second reading, and institutions such as the Land Law Reform Association attracted a group of about 280 MPs under J.H. Whitley. In 1904 and 1905 Bills on the subject were again introduced but predictably failed to get Conservative government backing.[15]

The third solution was also the most favoured by pragmatists within the Liberal party, who were prepared to concede the need for taxation of land values as a redistributive measure and as an alternative to Chamberlain's call for a return to tariff protection. Party policy at this time was 'a collage of separate and often competing proposals', but nevertheless on their coming to power in the landslide victory of 1906 with a majority of 354, they began to implement the idea. The MPs who were Vice-Presidents of the Land Nationalisation Society, and those who were prepared to form a Land Values Parliamentary Group, with an overlapping membership, formed a strong and effective pressure group within parliament. Asquith, as Chancellor of the Exchequer soon after the Liberal victory, was presented with a petition signed by more than 300 local councils asking for the introduction of a land values tax, and some 400 MPs also petitioned Campbell-Bannerman for legislation to tax land values.[16]

[14] Roy Douglas, 'God gave the land to the people', in A. Morris (ed.), *Edwardian Radicalism 1900–1914: Some Aspects of British Radicalism* (London 1974), 148–61; *Final Report of the Commission on Local Taxation* (Cd. 638) 1901 – 'Separate Report on Urban Rating and Site Values', 165–6. Lawrence, *Henry George,* 117–19.

[15] A. J. A. Morris, *C.P. Trevelyan 1870–1948: Portrait of a Radical* (Belfast 1977), 55.

[16] *The Times* 27 February 1906; Ward, 'Land reform', 443, 462.

A first Taxation of Land Values (Scotland) Bill in 1906 passed its second reading by 258 votes and was referred to a Select Committee in April, to be chaired by the single-taxer Alexander Ure. The December 1906 Report of the Select Committee recommended still more radical proposals to provide for a valuation of land in Scotland apart from buildings, and a second Taxation of Land Values (Scotland) Bill appeared in 1907. At public meetings both Campbell-Bannerman and Churchill promised land reform and single-taxers both inside and outside the House of Commons pushed hard for such reform. But in August, following the passage of the third reading of the Bill, it was rejected by the Lords, and a strong counter-attack on 'Socialism by installment' and on those who had 'larger looting designs' was mounted. Critical articles in the *St. James Gazette* and *Quarterly Review* were supplemented by the Conservative Campaign Guide in 1907 which contained a section attacking the taxation of land values as undiluted Socialism, and heavily dependent on the principles of Henry George.[17]

A second Land Values (Scotland) Bill was nevertheless reintroduced in 1908 and passed by a great majority through the Commons. *The Times* reported that the Bill was seen as aiming at 'the complete confiscation of the land values of this country, and at taxing all land values out of existence'. The Lords accordingly wrecked the Bill by amendments to the definition of valuation which, among other things, would have made the valuation optional. The Bill was sent back to the Commons in July 1908, and then withdrawn.[18]

Then in November 1908, Josiah Wedgwood presented a petition signed by 246 Liberal and Labour members associated with the Land Values Parliamentary Committee, urging that taxation of land values be included in the April 1909 Budget, since most constitutional experts agreed that the Lords would not interfere with a Finance Bill. The following month, Lloyd George, as the new Chancellor of the Exchequer in Asquith's Cabinet, put forward some initial ideas on taxing land values in a Cabinet Memorandum entitled 'The imposition of a national tax on land values', written by Edgar Harper, statistical officer of the London County Council 1900–11 and Chief Valuer of the Inland Revenue 1911–25. At the end of January 1909 he put forward a description of land taxation in New York. By this time it was clear that the next budget would indeed be the vehicle for the taxation of land values. Lloyd George referred to the 'thaw in the coming Spring which will release the

[17] Lawrence, *Henry George*, 131–3. Earlier Scottish attempts to introduce similar bills are briefly described in P. Wilson Raffan, *The Policy of the Land Values Group in the House of Commons: An Address Delivered by P. Wilson Raffan at the National Liberal Club Political and Economic Circle* (United Committee for the Taxation of Land Values [UCTLV] 1912).

[18] *The Times* 26 March 1908; J. A. Pease, *The Government's Record, 1906, 1907, 1908 and 1909: Four Years of Liberal Legislation and Administration* (London 1909), 195. Pease was Parliamentary Secretary to the Treasury and Chief Liberal Whip at the time of writing, later to become Chancellor of the Duchy of Lancaster, 1910–11 (and as Baron Gainford, Chairman of the BBC 1922–6). For further detail on the Taxation of Land Values (Scotland) Bill, see Ch.12.

land'.[19] The cause of land value taxation was certainly retained at the forefront of formal and informal politics through the efforts of Wedgwood and his followers inside and outside Parliament. The 'Land Song' was sung at meetings between 1909 and 1914 to the tune of 'Marching through Georgia' (Fig. 2.1).[20]

A strong cry for reform was now ever-present within the Liberal ranks. Many were now becoming impatient at the government's inaction: the November 1908 edition of *Land Values* accused the administration of 'ignorance and cowardice' and of 'betraying the trust' of the people. As many as nine separate societies associated with the Liberal Party featured land reform in their programmes, and there was a particularly vociferous radical group urging the government to divert money from military expenditure to poor-law, educational reform and public works for the unemployed. The Liberal Cabinet argued the matter of the imposition of a land tax of some kind and its possible impact on the property market during the winter of 1908–9. The theory of 'social value' was used to distinguish between wealth that was gained by individuals due to their private ability and enterprise, compared with wealth which accrued due to unearned or fortuitous market circumstances, such as that arising from increases in urban land values around developing towns or the use of sweated labour.[21]

The Liberal concern with the unearned increment as a suitable object of taxation had deep roots, having been proposed by J. S. Mill, who had commented:

> The first step should be a valuation of all the land in the country. The present value of all land should be exempt from the tax; but after an interval had elapsed, during which society had increased in population and capital, a rough estimate might be made of the spontaneous increase which had accrued to it since the valuation was made . . . I see no objection to declaring that the future increment of rent should be liable to special taxation . . . whatever the amount of the tax might be, no injustice would be done to the proprietors (provided the tax did not reduce the market value of the land below the original valuation).[22]

[19] A. K. Russell, *Liberal Landslide: The General Election of 1906* (Newton Abbott 1973), esp. 162–5; Roy Douglas, *Land, People and Politics: A History of the land Question in the United Kingdom, 1878–1952* (London 1976), 142; Wedgwood, *Memoirs*, 66, gives the figure as 241 signatures; J. C. Wedgwood, *Land Values: Why and How They Should be Taxed* (Land Values Publication Department, London 1911), 15; B. B. Gilbert, *David Lloyd George: A Political Life – the Architect of Change 1863–1912* (London 1987), 371, citing PRO CAB 37/96/161; 37/97/9 'The taxation of land values in New York'; Lawrence, *Henry George*, 134.

[20] Wedgwood, *Memoirs*, 68. J. Hyder, *The Land Song* (UCTLV Leaflet No.61 1912). Hyder had been the secretary of the Land Nationalisation Society from 1889 and was a long-time advocate of a complete land valuation (Ward, 'Land reform', 97).

[21] H. V. Emy, 'The land campaign: Lloyd George as a social reformer', in A. J .P. Taylor (ed.), *Lloyd George: Twelve Essays* (London 1971), 35–9; *Land Values* (November 1908), 110–11.

[22] John Stuart Mill, *Principles of Political Economy* (London 1891 edn), 524–5. This passage from Mill was later to be quoted by the Archbishop of York in the Lords Debate on the 1909 Finance Bill in justification of the valuation as a prior necessity for taxation (Pease, *The Government's Record*, 210). F. M. L. Thompson, *Chartered Surveyors: The Growth of a Profession* (London 1968), 325.

Leaflet No. 61.

THE LAND SONG.

Air: "Marching Through Georgia."

| m .,r :d .r | m .s_1 :s_1 .s_1 | l_1 .,t_1 :d r | d:—

1. Sound a blast for Free - dom, boys, and send it far and wide!

| m_1 .,f_1 :s_1 .s_1 | l_1 .s_1 :l_1 .d | r .,d :r .m | r :—

March a - long to vic - to - ry, for God is on our side!

| d .,d :r .m | f .l_1 :l_1 .l_1 | s_1 .,d :d .,r | m:—

While the voice of Na - ture thun - ders o'er the ris - ing tide—

| r :r .,r | r :m .,r | d:— | d:

"God made the Land for the Peo - ple!"

Chorus—

| .m | s :— .m s:— .m | d .,d :d .l_1 | d:—

The Land! The Land! 'Twas God who gave the Land!

.m | s :— .m | s :— | .m | r .,r :r .m | r:—

The Land! The Land! The ground on which we stand!

| d .d :r .m | f .l_1 :l_1 .l_1 | s_1 .,d :d .r | m:—

Why should we be beg - gars, with the Bal-lot in our hand!

| r :r .,r | r :m .,r | d :— | d :

"God gave the Land to the Peo - ple!"

2. Hark! the shout is swelling from the East and from the West·
Why should we beg work and let the Landlords take the best?
Make them pay their taxes for the land—we'll risk the rest:
The Land was meant for the People!

Chorus.

3. The banner has been raised on high, to face the battle din!
The Army now is marching on the struggle to begin:
We'll never cease our efforts till the victory we win,
And the Land is free for the People!

Chorus.

4. Clear the way for Liberty! The Land must all be free!
Britons will not falter in the fight, though stern it be,
Till the flag we love so well shall wave from sea to sea
O'er the Land that's free for the People!

Chorus.

[TURN OVER.

Fig. 2.1 'The Land Song.'
Source: Centre for Rural History, University of Reading 35/10084 (UCTLV Leaflet no. 61).

Clearly many saw that site values, and therefore any increment, were derived from social practice and activity by the community. As Hobhouse put it:

In part it [site value] is due to the general growth of the country to which the increase in town life is to be attributed. In part it depends on the growth of the particular locality, and in part to the direct expenditure of the ratepayers' money in sanitation and other improvements which make the place one where people can live and industry can thrive. Directly and indirectly, the community creates the site value.[23]

As President of the Board of Trade in December 1910, Churchill was prosaic in his campaigning tour of Lancashire, referring to:

the landlord who happens to own a plot of land on the outskirts or at the centre of one of our great cities, who watches the busy population around him making the city larger, richer, more convenient, more famous every day, and all the while sits still and does nothing. Roads are made, streets are made, railway services are improved, electric light turns night into day, electric trams glide swiftly to and fro, water is brought from reservoirs a hundred miles off in the mountains – and all the while the landlord sits still. Every one of these improvements is effected by the labour and at the cost of other people . . . and yet by every one of them the value of his land is sensibly enhanced. He renders no service to the community, he contributes nothing to the general welfare . . . but the land . . . may be what is called 'ripening' – ripening at the expense of the whole city, of the whole country, for the unearned increment of its owner.[24]

Liberal landowners and landed MPs lobbied the Chancellor and the Cabinet Finance Committee at this time to dissuade him from taxing capital values, which they believed would deplete Liberal support in the countryside, destroy relief given by the 1896 Agricultural Rating Act, and overturn leases which placed taxes and rates on the shoulders of the occupiers rather than landlords. Lloyd George replied that he would be willing to see agricultural land excluded from the new capital value taxation, but was otherwise unmoved, and almost certainly underestimated the degree of hostility within his own party at this time, influential though it was. He later claimed that he pushed the Land Clauses through the Finance Committee by 'simple bullying', accusing individuals of concern for 'their friends the Dukes'. But in full Cabinet his proposals for levies on improved land and ground rents were rejected, leaving only his proposals for taxing unimproved land, mineral rights duties, and the Increment Value Duty.[25]

The People's Budget

If the Liberal emphasis upon the obligations rather than privileges of property, as seen in Lloyd George's Limehouse speech, is added to such demands

[23] L. T. Hobhouse, *Liberalism* (reprinted Westport, Conn. 1980), 53.
[24] Winston Churchill, *The People's Rights* (London 1909, repr. 1970), 118–19.
[25] Gilbert, *Lloyd George*, 373–6.

from the Radicals within his own party for direct taxation on unearned income, then the platform for the notorious/celebrated 1909 Finance Bill can be understood. In addition, he was disturbed to hear that his party was losing heart, and the land was an issue which was calculated to rally and enthuse Liberal support. As he said in May 1908, 'It is time we did something that appealed straight to the people. . . to stop the electoral rot.'[26]

By 1909 also, the Liberal government was encountering at first hand the crises noted above, which included rising unemployment, financial deficit, and electoral unpopularity, together with longer-term critical problems of urban overcrowding and impoverishment, as well as rural depopulation and decay. Asquith's government was faced with having to raise about 10 per cent extra revenue to fund various crucial areas of social provision. The need to raise money for labour exchanges, National Insurance, the Development Commission, and the reform of local government finances loomed before the Chancellor, as did the need to find money for the building of the naval Dreadnoughts required for defence against growing German threats. And as Lloyd George was to say in October 1909 at Newcastle, 'A fully-equipped Duke costs as much to keep up as two Dreadnoughts; and Dukes are just as great a terror and they last longer', giving eloquent testimony to the fear of from whose 'broad shoulders' the extra money would come. Taxation of the landed classes now became the Liberals' political symbol and would be used to counteract any move towards tariff reform by opposing taxation of the wealthy as against taxation of food.[27]

Lloyd George's Budget introduced seven entirely new forms of tax: a super-tax of 6d on incomes of £5,000 and over; a graduated tax on motor vehicles; a duty of 3d per gallon on petrol; and four duties on land. These latter were relatively small-scale, and had been watered down because of internal cabinet pressures, Lloyd George himself claiming that they would raise only about £600,000, about 3 or 4 per cent of the additional £16.1 million to be raised through the Budget. The duties were aimed above all at those who made profits through development on land which had increased in value owing to some intrinsic locational factor in the site. Particularly affected, however, would be many Peers who had benefited from the massive spate of urban development over former agricultural land in the previous thirty years. Not for nothing did Lloyd George refer to this as 'a war budget. . . for raising money to wage implacable warfare against poverty and squalidness'. [28]

[26] William George, *My Brother and I* (London 1958), 220.

[27] Emy, 'The land campaign', 43–4; B.K.Murray, 'The politics of the People's Budget', *The Historical Journal* 16 (1973), 555–60; D. Lloyd George, *Slings and Arrows: Sayings Chosen from the Speeches of the Rt. Hon. David Lloyd George, O.M., M.P.* (ed. P. Guedalla, London 1929), 111; N. Blewett, *The Peers, The Parties and the People: the General Elections of 1910* (London 1972), 45–7. This otherwise very full account of the 1910 political scene underestimates the importance of the land issue and Lloyd George's valuation proposals.

[28] *H. C. Deb.* 5s. iv, cc. 472–548, 29 April 1909.

In his Budget speech, apparently an appalling 4½-hour performance, and speaking, as Hilaire Belloc a Liberal backbencher wrote, 'like a man in the last stages of physical and mental decay' after many sleepless nights spent in preparation, Lloyd George spelled this out. He distinguished between 'land whose value is purely agricultural in its character and composition, and land which has a special value attached to it owing either to the fact of its covering marketable mineral deposits or because of its proximity to any concentration of people'. He went on to point out that agricultural land had not appreciated in value over the last twenty or thirty years, but that there had been an enormous increase in the value of urban land and mineral property. And most important, 'the growth in the value, more especially of urban sites, is due to no expenditure of capital or thought on the part of the ground owner, but entirely owing to the energy and the enterprise of the community'.[29] Thus the proposal to levy a tax on 'the increment of value accruing to land from the enterprise of the community or the landowner's neighbours', and with 'a valuation of the land itself – apart from buildings and other improvements – and of this difference, the strictly unearned increment, we propose to take one-fifth . . . for the State'.

His second proposal relating to land was for the imposition of a tax on the capital value of 'unimproved' land on which building was held back for speculative gain, land 'which is not used to the best advantage, an annual duty of 1d [later reduced to ½d] in the £ on the capital value of undeveloped land'. This proposal was later stated by Lloyd George to have been brought forward in order to ensure an honest valuation for the purposes of the Increment Value Duty. Since valuation at the beginning of the process was based on the landlords' own estimates, this unimproved property tax would prevent landowners from overvaluing their land in order to minimise Increment Value. Each tax thus fitted the other – his own analogy likened them to a pair of scissors.[30]

He continued: 'The same principle applies to ungotten minerals, which we propose similarly to tax at ½d in the £, calculated upon the price which the mining rights might be expected to realise if sold in open market at the date of valuation.' A third land measure was for a 10 per cent reversion duty upon any benefit accruing to a lessor at the termination of a lease: 'The reversion at the end of a long building lease having no appreciable market value at the time the lease is granted is, when the lease falls in, of the nature of a windfall, and can be made to bear a reasonable tax without hardship.'[31]

The measures in fact pleased neither the advocates of land values taxation nor of nationalisation. Frederick Verinder, General Secretary of the English League for the Taxation of Land Values, gave a mixed reaction in 1911:

[29] D. Lloyd George, *Better Times: Speeches by the Right Hon. D. Lloyd George, MP* (London 1910), 124–6. [30] *H. C. Deb.* 4s. iv, c. 540, 29 April 1909.

[31] Lloyd George, *Better Times* (1910), 131–5; Peter Rowland, *Lloyd George* (London 1975), 216.

> The key to the complicated provisions of the Land Clauses is to be found in Mr Lloyd George's promise – a great and far reaching promise which he has splendidly redeemed – 'to provide machinery for a complete valuation on a capital basis of the whole of the land in the United Kingdom'. Regarded from this standpoint the meaning of the new 'Land Taxes' becomes clear. They are not the 'Taxation of Land Values', for that implies a universal valuation as its basis. They are not a substitute for, or even an installment of, the Taxation of Land Values. They do not fall upon Land Values as such, or upon all land values . . . The attempt to distinguish between 'land value' and 'increment value', the exemptions of agricultural land and of small holdings of land, have no economic justification. They can only be justified on the grounds of political expediency arising out of the action of the Lords. Partial as they are in their operation, they are so laid as to necessitate a universal valuation, and to make that valuation an inseparable part of the Budget. Their value lies not in their revenue-producing capacity, still less in their economic effects, but in their success as a political expedient for forcing a valuation of land, in all the three Kingdoms, through a hostile House of Lords, even at the cost of a special general election, and of much confusion in the national finances.[32]

Nevertheless all supporters of the legislation supported the land clauses since they required a valuation of all land to be carried out by the Inland Revenue. In the short term this was to enable a differentiation to be made between the Gross Value and the Site Value of the land (between the value of the site covered and the site cleared), but it was widely seen that this would enable a future government to tax land values at whatever higher rate was decided upon, or to compensate owners fairly in the event of nationalisation.

Landowners of course were equally aware of these radical possibilities, and were appalled at the prospect. The Earl of Crawford, noting the difficulties Lloyd George had experienced in dealing with the technical material in his budget speech, nevertheless concluded that 'We shall soon however forget the absurdities of the speech when we realise the vindictiveness of its policy.'[33] Letters and meetings abounded, and subscriptions to charities and sports clubs were cut amid threats of redundancies for employees. Balfour spoke unusually vehemently in debate on the Bill, calling the scheme 'arbitrary and unjust . . . by the very proposals you have made you have given a shock to confidence and credit, which will take a long time to recover'. As Asquith later wrote: 'It was the land taxes and perhaps still more the proposed valuation of land which set the heather on fire'.[34]

[32] F. Verinder, *Form IV: What Next?* (UCTLV, London 1911), 3–4. The extra-parliamentary land tax associations were entirely scornful of the Liberals' social policy. *Land Values* (November 1908, 110–11) noted that while it had 'no word to say against old age pensions' it thought that the idea that unemployment could be relieved by closing public houses was 'a cruel mockery' and Churchill's plan to regularise labour through employment exchanges was a 'gross betrayal' of earlier land tax policies (Ward, 'Land reform', 560, fn. 113).

[33] J. Vincent (ed.), *The Crawford Papers: The Journals of David Lindsay 27th Earl of Crawford and 10th Earl of Balcarres 1871–1940 during the Years 1892 to 1940* (Manchester 1984), 126.

[34] W. Watkin Davies, *Lloyd George 1863–1914* (London 1939), 333; Gilbert, *Lloyd George*, 379.

Lloyd George had effectively smuggled in the notion of a valuation of all land in his Budget, a notion which had been rejected twice recently by the Lords, but which, in the guise of a Finance Bill, would stand a better chance of circumventing their veto. In a Cabinet Memorandum of 13 March 1909 he noted the need to add taxation to valuation to get the package passed.

. . . it must be borne in mind that proposals for valuing land which do not form part of a provision for raising revenue in the financial year for which the Budget is introduced would probably be regarded as being outside the limits of a Finance Bill by the Speaker of the House of Commons. I have consulted Sir Courtenay Ilbert on the subject, and he is distinctly of the opinion that, unless it is contemplated to raise substantial revenue during the year, valuation clauses would be regarded by the authorities of the House as being a fit subject for a separate Bill, and not for a Finance Bill.[35]

Authorship of the Land Clauses is difficult to establish precisely, but Lloyd George, no economic or fiscal expert, received expert help from Harper, and from within the Treasury from Robert Chalmers, and from Charles Masterman, then a junior minister, High Churchman and confidant. The Hobhouse diaries for March 1909 also point to help coming from Hobhouse, Haldane and Harcourt:

George has demanded the preparation by Depts. and Draftsmen of successive schemes of Licences, land taxation, and death duties each more impossible than the other, and every week the Cabinet has thrown out scheme after scheme, very often to have the same scheme brought up in a different language a week later. Neither he nor the Cabinet have had time to consider details of clauses, and it was only owing to my searching scrutiny of some of his land tax proposals, and to my communications with the P.M., Haldane and Harcourt thereafter, that the Govt. failed to adopt *sub silentio* an absolutely unworkable scheme.

The Heads of Dept. are in a state of revolt and insubordination which is quite indescribable . . . [36]

There seems to be no corroborating evidence for such help, except that Lucy Masterman noted in her diary that Hobhouse did help out in at least one late sitting on the Clauses.[37] She recalled in her diary entry for 31 May 1909:

C. [Charles Masterman] has been helping Ll.G. with the Land Clauses of the Budget and has got very fond of him in the process . . . It has been rather amusing to see C. for the first time coming up against first-rate officials, Chalmers, Thring [the government draughtsman] etc., finding them very restful in their single-eyed enthusiasm for the work and their unlimited willingness.[38]

[35] Gilbert, *Lloyd George*, 372, citing PRO CAB 37/98/44.

[36] E. David (ed.), *Inside Asquith's Cabinet: From the Diaries of Charles Hobhouse* (London 1977), 77.

[37] Gilbert, *Lloyd George*, 372–3. Lucy Masterman, *C. F. G. Masterman: A Biography* (London 1939), 144. [38] Masterman, *C. F. G. Masterman*, 129.

Of Lloyd George she wrote:

Of course his weakness is inattention to detail. Very soon it came about that the officials and C. made the Land Clauses without consulting him at all . . . C.'s work has consisted in long conferences with officials going through the Finance Bill in detail . . . 'I can see', C. said one night to me, 'my line is to lie low and not try to speak and get George to rely on me: then I can get things done, and if I get things done I want done I don't care a damn who gets the credit for it'

and she added:

The possibilities of a disaster all the way through have been practically unlimited owing to the insecurity of the Chancellor's knowledge of his own Bill.[39]

Lucy Masterman also described how Lloyd George, Charles Masterman and Sir William Robson, the Attorney-General, handled personally most of the debate on the Land Clauses, with the two latter men dealing patiently with the endless detail but with the Chancellor gaining more of a grasp as the sessions progressed. It would seem that in the drafting and in the arguing of these complex Clauses, perhaps half-a-dozen men were closely involved, and very many others kept busy providing the necessary information and clarification.[40]

In his Limehouse speech of 30 July 1909, after three months of industrious work within Parliament on his Bill, Lloyd George, the 'cottage-born man', returned savagely and provocatively to the theme of the unearned income of the landed proprietors, in the light of the post-budget threats to cut back on benefactions and lay off labour, should the new budget proposals go through:

The ownership of land is not merely an enjoyment, it is a stewardship. It has been reckoned as such in the past, and if the owners cease to discharge their functions in seeing to the security and defence of the country, in looking after the broken in their villages and in their neighbourhoods, the time will come to reconsider the conditions under which land is held in this country. No country, however rich, can permanently afford to have quartered upon its revenue a class which declines to do the duty which it was called upon to perform since the beginning.[41]

Lloyd George himself seems to have had his own private reasons for the legislation, although as an extraordinarily canny politician, it is unclear as to how far along the road to nationalisation of land he really wished to travel. Probably he was less inclined this way than his landed would-be victims believed. A life-long opponent of the landed interest and with the revival of rural Britain through taxation of land values a long-cherished idea and genuine personal committment expressed as far back as 1891 in a speech at Bangor,[42] he expected little revenue to be raised by the new taxes, indeed referring to the Increment Value Duty as 'another little tax' in his Limehouse speech, but necessary constitutionally so that some money could be expected to flow from the legislation in the financial year covered by his budget speech. This tactic laid

[39] *Ibid.*, 129–30, 145. [40] *Ibid.*, 144–5. [41] Lloyd George, *Better Times* (1910), 155–6.
[42] H. Du Parcq, *Life of David Lloyd George* vol. III (London 1911–13), 421.

him open to his opponent's charge of 'tacking' (the annexation of legislation to a finance bill which is not strictly connected to the matter of money).[43]

Despite some initial fears, he appears to have wanted the valuation to ease future reform of the rating system and to facilitate attempts at compulsory purchase or 'communal purchase', where necessary to provide much-needed housing or other items of social provision. Indeed his Limehouse speech hinted broadly at possible land reform, which would require such a survey before its inception. And such a move towards a valuation would bring diverse radical interests within the Liberal and Labour groups together in his support. So, for example, although the 'single-taxers' wanted a straight tax on land values, and Lloyd George eventually delivered a transfer tax on increases in land values at the point of exchange of that land, as Wedgwood later wrote: 'Never mind! It did at least provide for valuation and registration. Armed with that we could at some later date get local taxation raised on the new valuation.'[44] In 1912 Lloyd George told Riddell, 'I knew the land taxes would not produce much . . . I only put them in the Budget because I could not get a valuation without them.' This is confirmed by the notes he made for his case before the Cabinet: 'Enables us to legislate on one of the greatest questions submitted to the country despite the House of Lords.'[45]

The crystallisation of such feelings by the Chancellor of the Exchequer in government legislation and in sensational speeches redolent of class warfare was too much for landowners, Unionists, and indeed for many less radical Liberals. Speeches such as that arranged by the Budget League at Limehouse and again at Newcastle in October, where he asked the famous question as to whether the Lords, '500 men, ordinary men, chosen accidentally from among the unemployed, [should] override the judgment – the deliberate judgment – of millions of people who are engaged in the industry which makes the wealth of the country?',[46] enflamed landed opinion. Vituperative comments rained down on this 'socialistic' and 'German' proposal (an increment value tax had been in use for some years in Frankfurt).[47]

The Bill had faced what Lloyd George had referred to as 'five months' hard labour' of argument (following its second reading on 21 June, it was actually in committee from 21 June to 6 October) and no fewer than 554 divisions, taking up more parliamentary time than almost any other goverment measure

[43] Lloyd George, *Better Times*, 147. For a stout defence against the 'tacking' charge see Churchill, *The People's Rights*, 58. [44] Emy, 'The land campaign', 58; Wedgwood, *Memoirs*, 69.

[45] B. K. Murray, 'Lloyd George and the land: the issue of Site-Value Rating', in J. A. Benyon *et al.* (eds.), *Studies in Local History: Essays in Honour of Professor Winifred Maxwell* (Cape Town 1976), 39, citing Lloyd George Papers C/26/1/2.

[46] Lloyd George, *Slings and Arrows*, 128.

[47] The German tax was charged on the occasion of a sale only, and then on the increment solely with reference to the gross purchase price of land and buildings together. It was thus a far simpler tax theoretically, and easier to administer (B.Mallet, *British Budgets 1887–88 to 1912–13* (London 1913), 306–7, 322).

before or since to comb through its ninety-six sections and six schedules, and with attention particularly focussed on Part I which included forty-two sections setting up the four new duties. Its eventual acceptance came on 4 November after no less than 640 hours of parliamentary discussion. There were two relatively major changes in the Land Clauses: on 13 July it was agreed that agricultural land was to be exempted from Increment Value Duty (though not from the Valuation); and on 9 September that the proposal to tax mining royalties (Mineral Rights Duty) was substituted for the tax on undeveloped ('ungotten') minerals, which was demonstrably impracticable. It was also announced at the latter date that the state would now bear the cost of land valuation and appeals against assessment, and that any land upon which £100 had been spent in the last ten years would be exempt from Undeveloped Land Duty, thereby diminishing still further any income from that source.[48]

After six days' discussion the House of Lords then rejected the Finance Bill on 30 November 1909 by 350 votes to 75, thereby 'refusing supply' (i.e. income from taxation) to the government, and thus precipitating the constitutional crisis of 1909–10. Of the great landed peers, about 60 per cent took the Unionist side, about 4 per cent the Liberal. At this time one-third of the Unionist Party in the Commons were country gentlemen, while in the Lords about two-thirds of those voting for rejection owned at least 5,000 acres of land. The *Daily News* of 28 December 1909, in an article 'The lords who killed the Budget', estimated that the peers voting against the Budget owned some 10.4 million acres. With greater vehemence perhaps than many, the Duke of Beaufort told an audience of his tenants that he would like to see Lloyd George (and Churchill with whom he was frequently coupled) 'in the middle of twenty couple of dog hounds'.[49] As Wedgwood later remarked, 'the Lords threw out the Finance Bill because of the Land Clauses, and gave us a General Election on a good issue if on a bad Bill'. Lloyd George commented: 'Their greed has overcome their craft, and we have got them at last.'[50] But evidence from his private letters seems to point to Lloyd George only conceiving of the issue as being a major one affecting the power of the House of Lords after the Limehouse speech, and certainly not at the time of the framing of the 1909 Budget, although this seems difficult to reconcile with the enormous clamour from the Georgite *Land Values* editions of January and February which certainly did appreciate the forthcoming struggle in its wider dimension.[51]

[48] John Grigg, *Lloyd George: The People's Champion 1902–1911* (London 1978), 219; D. Lloyd George, *The People's Budget* (London 1909),135–90; Mallet, *British Budgets*, 307; *H. C. Deb.* 4s. v, cc. 1878–83, 13 July 1909; Gilbert, *Lloyd George*, 390–1.

[49] Blewett, *Peers*, 76–7; fn. 39, 428; Du Parcq, *Lloyd George*, vol. III, 545.

[50] Wedgwood, *Memoirs*, 69; Davies, *Lloyd George*, 339.

[51] Murray, 'Lloyd George', 37–8. For the alternative view that this was a deliberate attempt to break the Lords' veto, following their wrecking of much of the attempted Liberal legislation on education, temperance, land reform, Welsh disestablishment and Irish Home Rule, see Lawrence, *Henry George*, 13–56.

The Bill actually reached the statute book only after a passionate year-long struggle and a plethora of campaigns by the Liberals' 'Budget League' chaired by Winston Churchill, the 'Budget Protest League', founded by the archetypal landowner Walter Long in June 1909, the 'English League for the Taxation of Land Values' (with associated regional branches such as the Yorkshire and Northern Land Values League, based in Keighley), and the 'Land Defence League' founded by Charles Newton-Robinson which became the 'Land Union' in April 1910. Much of the criticism of the budget was politically naïve: 'Rich men have simply wept in public on rich men's shoulders', and the complaints of the Dukes – 'the dismal dirge of the dilapidated duke', as Churchill put it – provided more ammunition to their enemies and effectively split the Unionist opposition. The Budget League, fed with propaganda by Masterman and Harold Spender, was by contrast particularly successful and attracted funds from magnates such as Joseph Fels, American soap manufacturer from Philadelphia and enthusiastic Georgite 'single-taxer'.[52]

In the general election of 14 January 1910 following the Lords' rejection of the budget, the Liberal Party was returned by a whisker. There were now 275 Liberal MPs to the Unionists' 273, the government losing many seats in the rural south in particular, and the Liberals now became dependent on the votes of the Irish Nationalists (82 seats) and Labour MPs (40 seats). Nevertheless, the rejected Finance Bill of 1909 was pushed through this time, being reintroduced on 20 April 1910 and signed by the King on 29 April, one year to the day since it was first introduced and one week before his death – 'it being alleged in some quarters that the two events were not unconnected'.[53] And later, after the failure of inter-party talks and the second General Election in December 1910, and following the Parliament Act 1911, the power of the House of Lords' veto was also lost, to Lloyd George's delight, thereby striking a political blow at the landed interest to align with the economic impact of the 1909 legislation. The issue of 'Peers vs. the People' had been decided emphatically in favour of the latter, to the delight of the majority of Liberals.[54] In December 1910, on the hustings at Deganwy, Caernarfon, fighting his seventh election, Lloyd George had said:

[52] See the extended critique of the Land Clauses in Charles Newton-Robinson, 'The blight of the Land Taxes: why they must be repealed', *The Nineteenth Century and After* 68 (1910), 389–98; *ibid.*, 'The blight of the Land Taxes: a retrospect and a prospect', *The Nineteenth Century and After* 69 (1911), 1073–85; and *ibid.*, 'The blight of the Land Taxes: why they must be repealed', *The Nineteenth Century and After* 72 (1912), 96–109. For Churchill and derision of the attacks, see Blewett, *Peers*, 74–5, citing *The Times* 29 July 1909 and *The Nation* 31 July 1909. Churchill's attacks were reprinted in Churchill, *The People's Rights*. Fels made a formal offer of support to the cause of land taxation at the Henry George Commemoration Dinner in October 1908. For this and the great Hyde Park pro-budget demonstration of July 1909, which in a final address particularly welcomed the valuation as essential to any policy of land and social reform, see Ward, 'Land reform', 497, 560 fn. 115.

[53] Rowland, *Lloyd George*, 235.

[54] Full studies of this struggle are in Bruce Murray, *The People's Budget 1909–10: Lloyd George and Liberal Politics* (Oxford 1980); and in Gilbert, *Lloyd George*, Lawrence, *Henry George* and

I know something about valuation. I had a good deal of experience of valuation in the days when I was practising down at Portmadoc . . . and nothing struck me more than the inequality of the assessment everywhere, especially for local rating. One thing especially struck me. The man who paid to the full was the owner of a small property. I have never seen a tradesman let off without paying on the full valuation, but I have seen many a mansion let off at a tenth of its value . . . I am more concerned for the poor man at the bottom than the man at the top . . . When we get the complete valuation we shall have a basis then for readjusting the burden of local taxation.[55]

The onset of the valuation

In the meantime, by the Act of 29 April 1910 – the Finance (1909–10) Act – the Commissioners of the Inland Revenue were empowered to:

cause a valuation to be made of all land in the United Kingdom, showing separately the total value and the site value respectively of the land, and in the case of agricultural land the value of the land for agricultural purposes where that value is different from the site value. Each piece of land which is under separate occupation, and, if the owner so requires, any part of any land which is under separate occupation, shall be separately valued, and the value shall be estimated as on the thirtieth day of April nineteen hundred and nine.[56]

Such a valuation would have as its only comparable precedent the Domesday Book itself. And once the land was valued the Act provided also for the levying of several duties on this land. Most significant was the 'Increment Value Duty' to be charged at the rate of 20 per cent of any increase in value accruing after 30 April 1909 on the occasion of the transfer or sale of land, the grant of leases (except those less than fourteen years), on the death of an owner, or as for land held by corporate bodies, every fifteen years (to be first levied in 1914). Increment value was defined as the amount (if any) by which the site value of a plot of land at the occasion of transfer exceeded the original site value as of 30 April 1909 (Section 2).

There were a number of significant exceptions, intended to exclude smaller farmers, owner-occupiers, cottagers etc. For example, agricultural land was exempted from tax (though not from the valuation) if it had no higher value than its current market value for agricultural purposes only (Section 7). Owner-occupiers of less than 50 acres were exempt from the tax (though again

Blewett, *Peers*. The tenor of Lloyd George's defence of his own policies at this time can be caught in D. Lloyd George, 'The issues of the budget', *The Nation* 30 October 1909, 182; and see also D. Spring, 'Land and politics in Edwardian England', *Agricultural History* 58 (1984), 33. Alluding briefly to the issue described by Gilbert as 'old, but not quite dead', It is unlikely that Lloyd George framed his budget with the idea that it would be rejected by the Lords and provoke a crisis between the two houses, but as discussions proceeded over the months, he began to calculate the benefits of this larger class struggle (S. Constantine, *Lloyd George* (Lancaster Pamphlets 1992, 36–7)). [55] *Land Values* (January 1911), 182.

[56] The Finance (1909–10) Act, 1910, 10 Edward 7, Section 26 (1).

not from the valuation), unless their land had an average total value exceeding £75 an acre and as long as the land was not occupied by a house worth more than £30. Owner-occupied dwelling houses were excluded if their annual value did not exceeed £40 in London, £26 in a borough or urban district with a population of over 50,000 at the last census (presumably 1901, since the 1911 Census would not have been available for the framers of the legislation), or £16 elsewhere. Also excluded was land held by corporate or incorporate bodies if used for games and recreation and not yielding profit; and land held by the Crown which was deemed already to have paid duty. The other levies included 'Reversion Duty' (Sections 13 to 15), 'Undeveloped Land Duty' (Sections 16 to 19), and 'Mineral Rights Duty' (Sections 20 to 24). The Budget also included increases in the rate of death duties, the introduction of a graduated income tax, and a supertax. The scene was indeed set for an 'angry drift of unexamined opinion as it swirled through London clubs, regimental messes, and country-house parties'.[57]

The 'single-taxers' now became more powerful within the parliamentary Liberal Party, and many of their leading speakers could now be numbered amongst the Parliamentary Land Values Group. In August 1910 the group asked for a national tax on the full site value determined by the valuation which had been set in train. In May 1911 more than 170 MPs presented a memorial also requesting a speeding up of the valuation process, which was now due for completion by 1915, and asking for the values to be made available for local authority and public use, scrutiny, and as the base for a future budget tax. In the appointment of a Departmental Committee on Local Taxation in 1911 under the chairmanship of Sir John Kempe, Lloyd George acceded to the wishes of this group, and the following year saw a militant campaign for rating reform and for the full implementation of land values taxation. Two by-elections in 1912 returned the radical land-taxers Hemmerde and Outhwaite as Liberals to increase the pressures on Lloyd George. But now the single-taxers and more radical land reformers faced criticisms from moderates within the Party, and it is clear that Lloyd George himself was feeling increasingly uncomfortable with the extremist single-taxers, and with their statements such as: 'The Valuation is a revolution. No revolution was ever so necessary.'[58]

[57] Spring, 'Land and politics', 30.

[58] Emy, 'The land campaign', 38–9; R. L. Outhwaite, *The Rating of Land Values: The Case for Hastings, Harrogate, Glasgow* (London 1912), 15–16. Among the signatories to the 'memorial' were Keir Hardie (with reservations), Philip Snowden and J. C. Wedgwood. The Parliamentary Land Values group's main communication was through *Land Values: Journal of the Movement for the Taxation of Land Values and the Untaxing of Improvements*. It was issued monthly at 1d from their offices at 11 Tothill Street, London SW, and contained notes on the taxation of land values, editorial articles on the situation, extracts from speeches and news of the movement 'at home and abroad'. The 'Revolution' extract is from *Land Values* (November 1910), 115; Murray, *The People's Budget*, 42. For Lloyd George's relations with the growing Labour Party movement at this time see C. J. Wrigley, *David Lloyd George and the British Labour Movement: Peace and War* (Hassocks 1976).

Nevertheless, in June 1912 the Land Enquiry Committee was established from Liberal supporters ('Bitter partisans' in Lord Ailesbury's words)[59] to investigate the condition of the land, as an essential precursor to the Liberals' Land Campaign, and as an attempt to demonstrate responsible and constructive moves towards land reform. Land reform was by now an issue which clearly seized the mind of the electorate and which could not be ignored. Much of the leadership of the enquiry was provided by B. S. Rowntree, with funding partially coming from his father and other sympathetic industrialists such as William Lever. Anxieties about the decline of agriculture and the decay of rural society, low pay, poor housing and rural–urban migration were widely shared, and had featured in the 1909 Budget proposal to establish Lloyd George's 'pet child', the Development Commission, to stimulate rural industry and afforestation.

The Enquiry reported in October 1913.[60] In the same month Lloyd George officially opened the land campaign with an eloquent speech at Bedford, and a few days later he announced the government's intention to create a new Ministry of Lands and Forests (an idea which had been put up originally in 1910 as a 'Domesday Office') to assume all responsibility for land valuation, small-holdings, land purchase and the development of rural industries.[61] The valuation, now anticipated for completion in 1915, would form the basis for land acquisition by local authorities in anticipation of development. Such a Ministry would also deal with matters such as wages, rents and tenure. The idea was enthusiastically endorsed by most Liberals, including Asquith, since it not only unified and energised the party but took concern for the countryside interest away from its traditional champions, the Conservative Party, who attacked the proposals as a step on the downward slope towards land nationalisation. The proposals were certainly ingenious politically as well as socially, and included a minimum wage for farm workers to enable them to pay an economic rent for housing, rather than continue to rely on tied cottages. The increased wage bill for farmers would be met by pro-rata rent remissions from the landowners. There were other aspects of this policy, but essentially the bill would be picked up again by the landowners, not the tenant farmers. A complete survey was also to be made of all the land in the country, and provision

59 Lord Ailesbury, 'The rural problem: some reflections on the Land Enquiry', *The Nineteenth Century and After* 74 (1913), 1087.

60 The Report of the Land Enquiry Committee, *The Land* vol. I (London 1913) and vol. II (London 1914). The Committee was chaired by A. H. Dyke Acland, and comprised E. R. Cross, E. Davies MP, Baron de Forest MP (who submitted a minority memorandum on land nationalisation), E. Hemmerde MP, J. I. Macpherson MP, B. Seebohm Rowntree and R. Winfrey MP. Acland in particular was a long-standing friend of Lloyd George. See also B. Gilbert, 'David Lloyd George: the reform of British landholding and the Budget of 1914', *The Historical Journal* 21 (1978), 117–41.

61 *The Times* 13 October 1913; 23 October 1913; 10 November 1913; *Daily News* 23 October 1913. National Library of Wales, Lloyd George Papers C/1/1/18. Progress was slow with this proposal, and 1915 became the anticipated start-date.

made for rural housing. Much of 1914 was to have been devoted to this programme, and by the end of May all seemed to be going very well, with great enthusiasm for the Liberals' rural regeneration campaign being expressed from all parts of rural England, although it was noticeably less well received in urban areas. The second and third reports of the Committee, dealing with urban land and with Scotland respectively, appeared in 1914, but the War intervened before the Ministry could begin, and also before any real progress could be made on measures to improve urban conditions.[62] Not all land-taxers by any means were appeased and instead of Lloyd George approaching land reform as they expected, 'like a Daniel come to judgement', *Land Values* could only bemoan that they had not had 'the knife of land value taxation to cut out the cancer but the plaster of state control'.[63] In fact many land-taxers were outraged that Lloyd George did not concentrate more on the single tax issue within the towns. Trevelyan wrote on the eve of Lloyd George's Land Campaign that: 'His main appeal for power must be to the towns . . . I can guarantee that the mass of land-taxers will swallow a good deal which they think heresy if rating of land values gets a real place. Otherwise I feel certain his campaign will go off in smoke.'[64]

Landed resistance

Whilst Lloyd George faced pressure from within Liberal ranks to declare himself in favour of land values as the basis for local rating, the defeat of the Lords in 1910 had by no means ended landed resistance to the 'New Domesday' either, since it was perceived by the Unionist Party as 'vindictive and socialistic'.[65]

Thus: 'There can be no manner of doubt that the institution of private property is seriously menaced at the present time – more seriously menaced perhaps in Great Britain than anywhere else in the world.'[66]

Two illustrations demonstrate the landowners' mood at this time. First, it was reported that in August 1910, between the passing of the Bill by the Commons and its rejection by the Lords, that about 150 West Riding property owners met at Leeds under the presidency of Lord Mowbray and Stourton, and formed the West Riding Land and Property Owners' Defence Association, 'for the mutual defence of its members against the taxation imposed by the Finance Act, and for the purpose of securing a repeal of such taxation'. Furthermore, according to the *Manchester Guardian* report of 31 August, 'Each must judge for himself as to whether he should put a high value or a low value on his property, but it was unfair to put the property-

[62] A. Offer, *Property and Politics 1870–1914: Landownership, Law, Ideology and Urban Development in England* (Cambridge 1981), 371–83; Douglas, 'God gave the land', 158.
[63] Ward, 'Land reform', 535. [64] Murray, *Lloyd George* (1976), 43.
[65] Mallet, *British Budgets*, 307. [66] Lord Milner, cited in Offer, *Property and Politics*, 380.

owner in this dilemma.' And secondly, two years later, when the valuations were taking place, when a buyer for the Duke of Bedford's Covent Garden property appeared in 1913, Rowland Prothero, his chief estate agent, trusted Conservative spokesman on land policy and author of *English Farming Past and Present* (1912), advised him to sell, fearing confiscation without even gaining compensation. The London properties of the Duke of Bedford were also the scene of a fiscal struggle between the land agent and the Inland Revenue valuers which was fought individually over each of the 1,664 hereditaments. And with the valuers being frequently inexperienced, the tax assessments proved less onerous than the Duke had originally feared.[67]

Moreover, as parts of the original Act were modified after 1910, landowners banded together in organisations such as the above-mentioned 'Land Defence League', which shortly afterwards became the 'Land Union' to combat the valuation and the levying of duties, with their activities sympathetically reported in the *Land Union Journal*. The Land Union was formed 'to combine all interested in land industry for their own defence', and included members from the land-connected professions as well as landowners and farmers of all degrees of wealth. Supposedly a non-party political organisation, but heavily backed by the Dukes of Westminster, Portland and Bedford, the Union attacked the legislation at all points. In 1910 a pamphlet described the 'ulterior motives for the Land Valuation' and noted 'Mr Lloyd George's and Mr Winston Churchill's inclination to Land Nationalisation and State Control, are well known, and it is necessary for the country to realise that these forms of Socialism must be fought in their earliest stages . . .' thereby precluding 'the admitted policy of the Land Taxers to exterminate the land owner, by taxing him out of existence' and creating 'partial Land Nationalisation without compensation'. In Lloyd George's words it worked 'day and night in every part of the country to find faults with the valuation; stirring up suspicions; working up complaints; and picking holes'.[68]

Their champion was Ernest G. Pretyman MP who mastered the details of the valuation better than anyone else. At every opportunity he attacked the measures in Parliament, quoting extensive research, for example in the debate on the 1912 Budget to demonstrate the high cost of the valuation exercise in relation to its returns. His main strategy, however, was through the courts, where more than fifty cases concerning the legislation were decided. Here the Inland Revenue experienced a number of humiliating reverses, beginning with the Lumsden Case (1911), sponsored by the Land Union, which established

67 Spring, 'Land and politics', 27–32; *Land Values* (October 1910), 91; Greater London Council Record Office, Bedford Mss E/BER/C9/E12/8 'Brief account of the principal steps taken in the settlement of the Provisional Valuations on the Bedford (London) Estates, 1910–17'.

68 *The Times* 12 March 1910 and 14 April 1910; *The Land Union's Reasons for Repeal of the New Land Taxes and Land Valuation* (Land Union, London 1910), 9, 11, 21. Lloyd George's opinion is cited in the *Daily News* 21 June 1912.

that speculative builders' profits were liable for Increment Value Duty, thereby rendering the government responsible for a forecast housing famine despite Lloyd George's previous promise of immunity for the builders back in 1910. The culmination of these court actions was the Norton Malreward Case, sponsored by the Land Union on behalf of Lady Emily Smyth of the 385-acre holding known as 'The Model Farm', Norton Malreward, Somerset, when the judgement of 28 February 1914 by Mr Justice Scrutton in the Court of King's Bench found, *inter alia*, that the whole basis upon which the Inland Revenue had hitherto valued agricultural land was invalid. New methods were ordered which proved impossible to apply in practice. The implication of the Norton Malreward Case (and the Chells Farm Case heard on the same day by Mr Justice Scrutton), for example, was that the exact position reached on 30 April 1909 by tenant farmers throughout the country regarding their tillage, manurings, sowings and ordinary farming operations would have to be taken into consideration by the Inland Revenue when arriving at a valuation. This, together with other reconsiderations arising from the Lumsden Case meant that the Undeveloped Land Duty had to be discontinued, and the whole valuation procedure was put in doubt.[69]

During 1913 Lloyd George had been under constant pressure from the land-taxers. The Land Tax Group within parliament presented a memorandum to him on 17 April signed by 176 MPs proposing that a Revenue Bill be introduced incorporating those amendments to the 1911 Finance Act which had been ruled out of order in August 1912 (particularly those relating to the valuation of agricultural land), and also that the modified valuation be used as the basis for both a uniform national land tax and local rating. In July Lloyd George did bring forward a Revenue Bill which was chiefly concerned with revisions to the Increment and Reversion Duties, and also with deductions for improvements over a 30-year period prior to the valuation of agricultural land. The Opposition threatened to block the passage of the legislation, and Lloyd George dropped the measures relating to agricultural land in an agreement with them, much to the fury of the land-taxers. In the end the entire Bill was dropped.[70]

In October 1913, battling against Unionist concentration on Ulster and Home Rule issues, Lloyd George began to draw up schemes for site value rating for legislation in 1914, to tackle the issue of rating reform and to appease the more radical site value rating enthusiasts. By this time he had probably been persuaded by Masterman that the Valuation might best be used not for Treasury taxes, but as the basis for local taxes. By February 1914 some measure of agreement had been reached in the Cabinet, and he was confident that the differential rating of land and structures on the land could be undertaken. The Kempe Committee reported at the beginning of March, but as with

[69] Offer, *Property and Politics*, 368–9. [70] Ward, 'Land reform', 528–9.

the report of the Royal Commission on Local Taxation in 1901, there was again a substantial minority on the Commission in favour of site value rating but the voting had been seven against and six for.

Lloyd George ignored this finding, and the Revenue Bill in May 1914 was again used to further the necessary legislation, as it had been in 1909. A national system of valuation for local taxation was proposed for site value rating, together with large grants to Local Authorities from the Exchequer. He stated that 'We intend that the taxation of site values shall henceforth form an integral part of the system of local taxation.' It was also framed to attempt to remedy the legal difficulties presented by landed interests, but this legislation never passed into law, since with the outbreak of war the valuations were suspended. This had been partly at the request of Pretyman on behalf of the Land Union, who argued in August 1914 that the men holding the necessary information in the countryside were being mobilised. In fact the 1914 Budget was very muddled, and and in June the Cabinet actually decided to retract and postpone much of the proposed reform, since Lloyd George had evidently underestimated the difficulties in the practical application of his ideas. Rating reform henceforth became lost amid preparations for war, and funding from its main benefactor, Joseph Fels, ceased on the latter's death in February. Nevertheless, the process of valuation did continue, although owing to the legal complications noted above, no valuations were actually served, pending new legislation. However, another 20 million acres of agricultural land were valued after 1914. Procedures were by now, however, open to doubt owing to the legal attacks.

The decline of the valuation

The taxation of land values never really recovered after 1914. Its political influence shrank and membership of the various land leagues fell away. By February 1916 *Land Values* noted that the 1909 Budget was a 'dead letter as far as concerns valuation and the collection of the land values duties', and by 1917 the journal was sending out urgent appeals for funds to maintain its own publication.

Following the Armistice and the election of a coalition government with Lloyd George at its head in 1918, a Select Committee was appointed in July 1919 to investigate the current position of those various duties which had been initiated in 1910, and to make recommendations as to their retention, alteration or repeal. They were also to investigate the current position of land valuation, 'regard being had to the desirability of state valuations of land being available for public purposes'. Ernest Pretyman, not surprisingly, was a member of the Select Committee. Sir Thomas Whittaker chaired the first two of the meetings in August and October 1919 until his death, when Pretyman took the chair in November, to be followed in the chair for the last two meetings by Mr Beck. The fourteen original members who had attended the first (August) meeting of the Commission had shrunk to seven by its last meeting just before Christmas. A very short report ensued, noting that 'owing to difficulties arising from different interpretations of the order of reference and divergent views as to the scope of their enquiry, they had been unable to consider the matters to them referred'.[71]

More helpfully perhaps, the evidence given to the Committee was also published, consisting primarily of statements from officials from the Inland Revenue Department. Mr Alexander Blair, Chief Valuer (Scotland), pointed to the problems of implementation owing to differences in Scottish law. Charles H. Gott, Superintending Valuer (West Riding Division), concentrated on the valuation of minerals in his evidence. C. J. Howell Thomas, Deputy Chief Valuer, described the establishment of the valuation procedures and the implications of the different legal decisions since 1910 on their implementation. Percy Thompson, a Commissioner of the Inland Revenue, looked at the position in 1919 and the legislation which would be necessary if the valuation and duties were to be continued. And Edgar Harper, Chief Valuer, gave evidence on the difficulties, progress and costs of the valuation, its soundness and reliability, and on the utility of the Valuation Office for purposes other than those connected with land value duties.[72]

In addition the Commission also received memoranda from the Council of the Surveyors' Institution declaring that the duties should not be retained in their present form and that 'the machinery of the Act cannot be satisfactorily amended'. They also heard from the President of the Law Society, Mr W. A. Sharpe, who wrote of the expense, delay and uncertainties surrounding Increment Value Duty, and whose 'vices infect all the others', leaving a 'wide door open to argument and delay', and urging their repeal, since 'No tax . . . has cost so much to work in proportion to its result.'[73]

The Commission also heard from the Land Union secretary and assistant secretaries who dwelt particularly on the adverse impact of the duties on the building and land development industry, and on the disutility of the original survey, since: 'not only is a universal valuation as on one certain date a useless expense to the State, but even if it could be completed – of which we have doubts – it would be of no value except for collecting artificially-defined taxes'. Such a valuation, they claimed, rapidly became obsolete because of new legislation (the Rent Restriction Acts, Housing and Town Planning, and the Land Resettlement Act were cited), variations in the value of money and other changes. For these and many other reasons, they therefore advocated repeal of the legislation. Appendices submitted by the Land Union dwelt at great length on the deleterious effect that the legislation was having on the building trade

[71] Report of the Select Committee on Land Values, BPP, 1920 (Cmd. 556), XIX, 2–4.
[72] *Ibid.*, 5–63. [73] *Ibid.*, 64–9.

and the consequent housing shortages, with opinions drawn across the board from Sir John Tudor Walter, from Urban and Rural District corporations and councils, professional and voluntary associations, industrialists such as the managing director of the London Brick Company, and from many auctioneers, building societies and estate agents.[74]

Evidence also came in from practising barristers, surveyors and land agents, such as Sir Francis Elliot Walker of the Land Agents' Society and estate manager for the Duke of Northumberland. He gave adverse evidence against the legislation, much as his colleagues had always done since its inception, and quoted the first memorandum of the Land Agents Society in 1910 on the subject of the theoretical nature of many of the valuations having to be made: 'The Bill indeed introduces a general atmosphere of make-believe and pretence which, however ingenious it may be in theory, is very out of place in a taxing measure.' Sir Francis was worried by the thought that the real object of the initial 1909 valuation might be to introduce periodical revisions at some future dates for new taxation and rating purposes. And furthermore since valuations of agricultural land had been made on a basis which had since been ruled upon adversely by the Court of Appeal, it would follow that practically all the completed valuations were incorrect and would require costly revision to give proper values as at 30 April 1909, values which were now over ten years out of date.[75] James Dundas White, longstanding critic of tariff reform, now a barrister but originally one of the 'single-taxers', Secretary of State for Scotland by 1912, and now one who was in the process of leaving the Liberals with Wedgwood for the Independent Labour Party, noted the difficulties into which the legislation had fallen. He noted the rejection in 1910 of a simple single tax and continued to advocate a direct tax on land values rather than the Incremental Value Duty, Reversion Duty and Undeveloped Land Duty. His was the only evidence to speak even partially in favour of the retention of the 1909 valuation as a valid part of taxation procedure.[76]

Although the Commission's report was once more inconclusive, the land clauses of the Act were finally repealed by Section 57.1 of the Finance Act 1920, 'the obligation . . . to cause a valuation to be made of all land in the United Kingdom shall cease', and the duties, with the exception of those on minerals, were removed. The remaining land-taxers marched to defeat into the division lobby singing the 'Land Song' on 14 July 1920.

No more collections were to be paid, and repayment of the duties collected as Increment Value Duty, Reversion Duty and Undeveloped Land Duty was

[74] *Ibid.*, 69–91.

[75] *Ibid.*, 91–4. The Duke of Northumberland had been singled out by Lloyd George in his Limehouse speech in 1910 in a well-known case where the Duke had demanded £900 per acre for land for a school which had been worth 30s an acre in terms of his contribution to the local rates (Lloyd George, *Better Times* (1910), 148).

[76] Report of the Select Committee on Land Values, 94–6.

made possible on application within six months. The complex legislation had stumbled from one legal difficulty to the next, and many of the original intentions had never, in fact, materialised, such as the periodical duty payment by corporate (and unincorporate) bodies. Lloyd George abandoned the legislation as part of the price for keeping the Unionists inside his Coalition government after the war.[77] Symbolically, *Land Values* became *Land and Liberty* from June 1919.[78]

In the War Cabinet proceedings of 30 July 1919 Lloyd George as Prime Minister indicated that the information from the survey should be available to local authorities in connection with compulsory acquisition of land, a sentiment echoed by Chamberlain as Chancellor of the Exchequer in 1921. Less ideologically but more practically, the valuation had certainly helped departments such as the Admiralty in the requisition of land at the outbreak of the war, as it did the Ministry of Health and Ministry of Agriculture after the war with particulars of land for local authority housing and for land resettlement respectively.[79]

Henry George had once declared: 'We would simply take for the community what belongs to the community – the value that attaches to land by the growth of the community; leave sacredly to the individual all that belongs to the individual.'[80]

But the radical Liberal land proposals of the period 1906–20, perhaps the closest thing to approach genuine land reform in the modern period, could not be simple in a country so entrenched in its attitudes to land as the United Kingdom. Lord Justice Scrutton remarked in 1920, looking back over his earlier decisions:

> the habits you are trained in, the people with whom you mix, lead to your having a certain class of ideas of such a nature that, when you have to deal with other ideas, you do not give as sound and accurate judgements as you would wish . . . It is very difficult sometimes to be sure that you have put yourself into a thoroughly impartial position between two disputants, one of your own class and one not of your class.[81]

Lloyd George himself, as head of the postwar coalition government, was forced to end the legislation he had formerly nourished amid schisms and dissent within the Liberal Party, from which it never recovered. Thereafter, no Labour, coalition or Conservative government had the combination of political strength and ideological will to resuscitate the legislation. Something similar was included in Philip Snowden's budget of 1931, but three months later Ramsay MacDonald's Labour government was swept away and the new

[77] Finance Act (10 and 11 Geo. 5) 1920, 57 (1); Murray, *Lloyd George*, 47.
[78] Lawrence, *Henry George*, 166.
[79] *The Valuation Office 1910–85: Establishing a Tradition* (Inland Revenue, London 1985).
[80] Quoted in *Land Values* (November 1911), 136.
[81] T. E. Scrutton, 'The work of the commercial courts', *Cambridge Law Journal* 1 (1921), 8.

Chancellor of the National Government, Neville Chamberlain, suspended the valuation and the valuation staff as an economy measure. The tax was repealed in the 1934 Budget, and there were no further attempts to impose a national tax on land values.[82] In 1975 the Labour Government's Community Land Act proposals were introduced in words that sounded familiar, for one declared aim was 'to restore to the community the increase in value of land arising from its efforts'.[83] But the onset of the Thatcher administration again demolished all such ventures.

[82] Lloyd George came under bitter Liberal attack for his move to annul the taxes. For his answers see the *Lloyd George Liberal Magazine* 1 (October 1920), 6; and for the background to the Liberal Party problems at this time see Roy Douglas, *The History of the Liberal Party* (London 1971), 150–1. For developments in the land values taxation debate after 1920 see also Lawrence, *Henry George*, 170–92.

[83] Department of Environment *Circular 12175. Community Land: Circular 1*: General Introduction and Priorities (HMSO, London, 3 December 1975), 1.

3

The national structure of the valuation process

In his 1909 Budget speech Lloyd George had little to say about the mechanisms by which his new taxes and the valuation they ushered in would be established, leaving such detail for the Committee stages.[1] However, the Inland Revenue had begun to set up the machinery to implement the valuation even before the Finance Bill became law, and on the assumption that the 1909 Budget would become law by about September 1909.[2] An interim report was drawn up from a committee, chaired by Sir Robert Thompson, to formulate regulations for carrying the provisions of the Finance Bill 1909 dealing with duties on land values into effect.[3] The organisation in Scotland, although similar to that in England and Wales, was to be controlled separately through the Scottish Secretariat of the Inland Revenue in Edinburgh, while that in Ireland was under the aegis of the existing General Valuation Office of Ireland

[1] The Bill was remodelled between the Committee Stage and Report to include two new definitions of values: gross value and full site value, probably at the behest of what Yardley in 1930 referred to as 'theorists' (R. B. Yardley, *Land Value Taxation and Rating: A Critical Survey of the Aims and Proposals with a History of the Movement* (printed for the Land Union by W. H. and L. Collingridge, London 1930), 641–2).

[2] For legal guidance see W. H. Aggs, *The Finance (1909–10) Act 1910, with Full Notes on the Land Duties* (London 1910); C. E. Davies, *Reports on Land Valuation Appeals: Finance (1909–10) Act 1910*, 3 vols. (London 1913–14); The Land Union, *The New Land Taxes and Mineral Rights Duty: The Land Union's Handbook on Provisional Valuations* (London n.d. [1911]); C. E. Davies, *Land Valuation under the Finance (1909–10) Act 1910: The New Land Duties, Licensing Duties, Stamp Duties, and Alteration in Death Duties* (Estates Gazette, London 1910).

[3] Chief Valuer's Library, Inland Revenue Head Office, Carey Street, London, 'Duties on Land Values: Interim Report of the Committee appointed to Draw up Regulations for Carrying into Effect the provisions of the Finance Bill, 1909, dealing with the duties on Land Values' [nd]; see also PRO IR 74/146 M3.11, H. Thomas, 'Draft notes on the organisation of the land valuation department' (June 1910). A large amount of other supplementary material concerned with the establishment and subsequent procedures of the Valuation Office has survived. The proposals for the taxes themselves and discussions prior to the 1909 budget constitute one set of papers in the PRO. The organisation of the Department and the progress of the valuation are also well documented (PRO IR 73/6). Some files of the Valuation Office itself are in IR 40. Organisation and progress are recorded in IR 74/146–8; IR 40/2878.

under Sir John Barton, Commissioner of Valuation, and thereby under the immediate control of the Treasury.

The creation of the structures

The first task was to arrange for the presentation of instruments under Section 4.3 of the Act, which provided that transfers and leases of land for periods exceeding fourteen years had to carry an Increment Value Duty stamp. On 29 April 1910, the Bill received the royal assent and that same day a circular was sent to all solicitors in the United Kingdom, calling their attention to Section 4 relating to collection and recovery of duty in cases of transfers and leases, and enclosing copies of the regulations.

These initial steps completed, the Inland Revenue Department began to create the structure necessary for carrying out the valuation and administering the new taxes. Two main kinds of staff, valuation on the one hand, and administrative and secretarial on the other, were needed: the former to value the land, the latter to assess and collect the duties and to administer the Act generally – so far as it related to the land duties – under the jurisdiction of the Secretaries Office of the Inland Revenue. Thus the actual assessing and collection of the Land Values Duties remained with a branch of the Secretariat of the Inland Revenue, but the valuation itself required a staff of experts whose sole duty should be that of the valuation. It was still the case, however, that the assessment and collection of the Land Value Duties had to proceed simultaneously with the process of conducting the valuation, and that the former relied to a great extent on having a valuation. Where 'occasions' arose requiring taxation, a valuation was needed urgently, and staff would be diverted from their valuation duties – a source of some internal friction at times.[4]

At this date, the Valuation Office, which had originated in 1909 as a branch of the Estate Duty Office, comprised only sixty-one persons and it therefore had to be augmented rapidly with permanent and temporary staff to begin the survey and to deal with *ad hoc* valuations for Increment Value Duty and Reversion Duty as such cases came in. Recruitment of the necessary staff began immediately around the nucleus of valuers who had previously dealt mostly with death duty valuations. Advertisements for valuers to join the new department were then published:

VALUATION SURVEYORS WANTED in Government Department; age 35 to 45 years; must have had good experience in the valuation of real property and qualified

[4] 54th Report of Commissioners of HM Inland Revenue, Year ended 31 March 1911, BPP 1911 (Cd. 5833), XXIX, 157; Report from the Select Committee on Land Values, BPP 1920 (Cmd. 556), XIX, Evidence of Mr Percy Thompson, a Commissioner of Inland Revenue, 16. For internal friction see Chapter 11.

as Fellow of the Surveyors' Institution. Salary to commence at £550 per annum. – Address, with copies of three testimonials. – Valuation (118), 'Estates Gazette' office, 34 and 35 Kirby Street, E.C.[5]

The bulk of the new valuers who were recruited were members of professional bodies such as the Surveyors' Institution. Attitudes to the legislation from the valuers as a body were predominantly hostile, in line with most other landed interest groups. Not only were the taxes themselves thought to be inequitable, but also the machinery for their collection was unworkable. The Surveyors' Institution did provide a memorandum to Lloyd George which confined itself to the machinery of valuation, but the political neutrality of the Institution as such was severely tested. Its President, Sir Alexander Rose Stenning, and many senior members, for example, were conspicuously present at protest meetings such as that of the great Land Taxes Protest Committee which alleged that the Budget 'smacked of Star Chamber methods and outright confiscation unknown in any civilised country'. Nevertheless, many new jobs were created, and a scale of charges for valuation work undertaken under the Finance Act was published by the Institution.[6]

The first Chief Valuer appointed was Sir Robert Thompson (1909–11), soon to be succeeded by Sir Edgar Harper, who resigned from the London County Council in May 1911 to advise the government's departmental committee on local taxation, which was investigating developments since the 1901 Royal Commission. However, he was appointed to head the Valuation Department in September 1911. And he it was who also oversaw the demise of the Increment Value Duty legislation in 1920 but the continuation of the Valuation Office as a central source of advice for government departments. He retired in 1925, and the office still continues today. Indeed, almost from the moment of its inception in 1909 the Valuation Office began to receive requests for assistance from other departments, and by 1918 it had almost become another government department, although much of its work at this time centred around the valuing of land for war purposes and reconstruction. By 1920 the land values work had become only a small part of its overall remit, but nevertheless, the repeal of the duties (except minerals) threatened its existence. A review by the Select Committee on National Expenditure was initiated, and reported that the office should have a continuing role in valuation work for other government departments. Its new role in government service was set out in a Minute to the Board of Inland Revenue from the Chancellor of the Exchequer.[7]

[5] Advertisement reprinted in F. M. L. Thompson, *Chartered Surveyors: The Growth of a Profession* (London 1968), Plate 48, between pp. 208–9.

[6] Thompson, *Chartered Surveyors*, 302, 320–1.

[7] Select Committee on Land Values, Evidence of Mr C. J. Howell Thomas, F.S.I., Deputy Chief Valuer to the Board of Inland Revenue, 9–15; A. Offer, *Property and Politics 1870–1914: Landownership, Law, Ideology and Urban Development in England* (Cambridge 1981), 366.

The spatial organisation of the valuation

Regional decentralisation was an important initial consideration, and England and Wales were initially divided into twelve Divisions in 1910, although this was very soon afterwards increased to fifteen (including Scotland). This was achieved by the splitting of the large Home Counties Division into Home Counties (North) and Home Counties (South), with the original Superintending Valuer of the Home Counties Division, J. Cawter, taking over the northern half, and C. W. Hayley Mason, originally a Superintending Valuer working at Somerset House with C. J. Howell Thomas and Sir Robert Thompson, taking on the southern half. The Central Division was split into Central (East) and Central (West), and there was also the later addition of a Cumbrian Division. Each Division was placed under the control of a Superintending Valuer, who returned annually a report of valuation progress within the Division. From such reports one may glimpse something of the differences in average values of hereditaments, e.g. by 1915 the figure for Home Counties (South) was £557 compared with S.Lancashire £481 and Wales £436. Scotland, treated as a separate Division, had a Chief Valuer (Scotland) who took instructions from the Board through the secretariat located in Scotland (Fig. 3.1 and Table 3.1).[8]

These large regional Divisions were in turn subdivided into Valuation Districts, the key spatial unit for the purpose of the valuation, and from which valuations were issued and at which the records were kept (Table 3.2). Each was headed by a District Valuer, with a number of Assistant Valuers, as well as clerical and technical assistants. By the end of July 1910 nearly 100 offices had been opened, although staffing and equipment lagged behind. In turn the Valuation Districts were made up from convenient groupings of Income Tax Divisions, each in the charge of a Valuer assisted by a junior or valuation assistant. In some cases there was but one Income Tax Division to a Valuation District – as for example in the cases of Kensington, Manchester or Leeds, with large and busy urban populations. Elsewhere the number of Income Tax Divisions could increase in rural areas to as many as eighteen at Welshpool or twenty at Ipswich. In Scotland the Income Tax Divisions were the counties:

Offer notes of Harper that 'in the nature of his talents, he only succeeded in making confusion worse confounded. Robert Chalmers' lucid and terse advice gave way to opaque disquisitions of mounting desperation.' For an authoritative description of the Inland Revenue, with sections on the Valuation Office and its continuing work after 1920, see Sir Alexander Johnson, *The Inland Revenue* (The New Whitehall Series 13, London 1965), 20–9, 170–85.

[8] PRO IR 74/148. Report by the Chief Valuer and Deputy Chief Valuer upon the Progress of the Original Valuation under the Finance (1909–10) Act 1910, together with copies of the Board's order thereon, 1912–13. Valuation Department, *List of Valuation Districts* (HMSO, London, October 1910); *ibid.*, *Return Showing the Names and Qualifications of the Chief Valuer, Superintending Valuers . . . 31st Day of January 1911* (for Mr Austen Chamberlain, Return to an Order of the House of Commons, 14 March 1911, HMSO, London, March 1911).

Fig. 3.1 The Valuation Divisions of England and Wales by the end of 1910.
Source: PRO IR 74/148.

Perth, Stirling, Kinross and Clackmannan were administered from Perth; and the huge area of Inverness, Ross and Cromarty, Sutherland, Caithness, Orkney and Shetland was administered from Inverness. In some areas, Mineral Valuers were appointed and located in the Divisional Offices, and

Table 3.1 *The initial regional structure of the Valuation Department 1910*

Divisions	Superintending Valuer	Valuation Districts	Income Tax Divisions
N.London	R. A. Dash (Strand, WC)	8	23
S.London	S. Martin (Somerset House)	9	13
Home Counties	J. Crawter (Somerset House)	15	105
Western	J. E. Tory (Bristol)	12	119
Eastern	C. G. Eve (Nottingham)	11	130
Central#	T. G. Fisher (Birmingham)	9	38
S.Lancs.	J. W. Marsden (Manchester)	6	14
N.Lancs.+	F. W. Thompson (Bolton)	6	21
W.Riding	C. H. Gott (Bradford)	9	20
Northern	W. Townend (Newcastle)	10	43
Wales	D. T. Davies (Cardiff)	15	122
Scotland	J. W. Fyfe (Edinburgh)	9*	33
Totals Divisions 12		Valuation Districts 119	Income Tax Divisions 681

Notes:
#Later divided into Central (East) and Central (West).
+Part of this division ultimately became the Cumbrian Division.
*Scotland was eventually divided into twelve Valuation Districts.
Source: 54th Report of Commissioners of Inland Revenue, *BPP* 1911, XXIX.

Table 3.2 *The spatial hierarchy of Valuation in England and Wales*

Head Office (Chief Valuer) Somerset House, London
12 (later 14) Valuation Divisions (Superintending Valuers)
119 Valuation Districts (District Valuers)
681 Income Tax Divisions (Land Valuation Officers)
7,000 Income Tax Parishes
10.7 million hereditaments

were usually directly responsible to the Superintending Valuer. Special valuers with knowledge of the licensed trade dealt with premises selling intoxicating liquor. Most of the appointments at the Valuation District level were temporary and non-pensionable

Upon completion of the organisational structure during 1910, the now fourteen Divisions of England and Wales had been subdivided into 111 Districts, increased to 118 by 1914. The valuation was conducted under the control and general supervision of the Board of the Inland Revenue, who met periodically – on average, not less than three mornings a week in the first years – to discuss cases which raised questions of principle. Insofar as their decisions were generally applicable they were embodied in the Instructions for Valuers which were issued from time to time.

Within the Income Tax Divisions, the pre-existing 'Income Tax Parish' (ITP) was adopted as the basic unit for valuation purposes.[9] Sometimes several civil parishes were united to form one ITP, and in these cases the first parish in the group was used to identify the hereditament. For example, the ITP of Ashley Walk, in the Hampshire New Forest, included the civil parish of that name as well as the civil parishes of Breamore, North and South Charford, Fordingbridge, Hale, Rockbourne, and Wood Green. A hereditament in Hale, for example, would be identified as 'Ashley Walk' followed by a number.

Organisation at the local level

One documented example of the structure being set up at the local level is provided by the Dorchester District Valuation Office, which was opened in June 1910 at 28, High West Street, Dorchester by the Superintending Valuer for the Western Division, Mr J. E.Tory. The office was initially run with a staff of two until November when two Temporary Valuation Assistants were appointed. Mr E. Dale was the appointed District Valuer for the Dorchester Valuation District which administered the valuations for ten Income Tax Divisions, including Dorchester itself and Blandford, Poole, Shaftesbury, Sherborne and Wareham. Mr H. Dean, the only clerk (Temporary Clerk at £5 per annum) until February 1911, recalled working frequently until midnight to cope with the enormous workload, and not receiving the official instruction book as to procedures until December 1910. No typewriter arrived until October!

In 1911 new staff were appointed: a draughtsman, a male shorthand typist, two more Temporary Clerks, and two more Temporary Valuation Assistants. In 1912 Mr Dean enticed three more Temporary Clerks away from local auctioneers' offices with offers of 15s per week. By late 1913 the staff had grown to thirteen, but with the outbreak of war the men were called up, leaving the District Valuer with one man of 63 and subsequently three sisters with no office experience who worked until the return of two Temporary Clerks in April 1919 when they resigned.[10]

[9] Income Tax Parishes may consist of a single poor law parish, a union of several poor law parishes, or in some cases (principally in large towns) a part of a poor law parish. *Instructions to Valuers* Part I (HMSO, London 1910), 23. Income Tax Parishes were authorised under the Taxes Management Act (43 & 44 Vict. c. 19) 1880, Section 37. They were proposed by local Commissioners of Taxes as convenient units for administrative purposes and had to be approved by the Board of the Inland Revenue. I am grateful to John Goodchild, Principal Local Studies Officer and Archivist, Wakefield Library, for this information.

[10] *The Valuation Office 1910–85: Establishing a Tradition* (Inland Revenue, London 1985), Part 5 [np].

The appointment of Land Valuation Officers

Across the whole country Temporary Land Valuation Officers (LVOs) were appointed to each ITP. The Assessors of Income Tax (many of whom were also Assistant Overseers) were considered suitable to take on the task, and were already appointed to each parish by the various bodies of District Commissioners of Taxes. The Inland Revenue made provisional enquiries before the Bill received the Royal Assent, as to the willingness of Assessors to undertake the duties as LVOs, and requested the 600–700 separate Boards of Commissioners to allow them to be so used. Only in a relatively small number of cases were individual Assessors unable or unwilling to act in this capacity, whereupon other 'suitable persons', nominated usually by the Surveyor of Taxes in each District, were appointed. These appointees remained as Assessors of Income Tax, and were not appointed full-time to the service of the Inland Revenue. In Scotland, Surveyors of Taxes acted as LVOs for counties and some burghs, whilst local Land Valuation Assessors acted for other burghs.

This scheme enabled the Inland Revenue to have people with the necessary local knowledge undertaking the crucial task of identifying each hereditament. A Declaration of Secrecy (Form 26 – Land) was required, but this did not prevent the criticism that confidentiality could be breached, especially in small communities, and that the disclosure of private affairs to the local tax-collector (sometimes a rival in business) could result in their becoming 'the public property of the village taproom'. This was a charge which Lloyd George was quick to refute, and to point out that landowners did have the option of making returns to the District Valuer directly, bypassing the LVO. Indeed this problem had been foreseen in 1909 by Sir Robert Thompson's initial committee to determine structures and procedures, who noted that the local officer 'is often a shopkeeper or a person of no position whatever'. Gentlemen also complained of gross incivility from local officials, complaints which hold overtones of class conflict between larger landowners and their petit bourgeois tax-collector neighbours. There were doubts expressed also about the competence of those surveyors who would 'abandon lucrative professional careers to enter into the service of the State' while Newton-Robinson, noting that the officials were being put forward by Lloyd George as being eager to help landowners complete the forms accurately, wrote that they were 'such persons as local tax-collectors or assessors, and very often small tradesmen; experienced enough in their own way, no doubt, but for the most part utterly incapable of interpreting the Finance Act, even to themselves, much less to other people'. Nevertheless, the co-operation of the Boards of Commissioners of Taxes did enable the Inland Revenue to recruit some 7,000 qualified LVOs across the country, and their practical knowledge of values in their respective areas was to prove very important in the valuation process. [11]

They were urged in 1910: 'to devote their time and energy whenever practicable to the necessary outdoor work of valuation, with a view to completion at the earliest possible moment of the Original Valuation throughout each particular district, and so thoughout the whole country'.[12]

A satirical book, *Alice in Plunderland*, supposedly written by a certain Loris Carllew in 1910, caricatured a landowner as the March Hare who reads his 'orange-yellow-backed booklet with a conspicuous black title, "LAND UNION GUIDE"', and Lloyd George as the Welsh Rabbit who informs Alice of his schemes (Fig. 3.2):

. . . he clapped his hands.

'Let Valuation Officers appear!'

Alice was reminded of the story of Aladdin and the wonderful lamp; for all of a sudden the whole park seemed to be populated with an immense host of shabby, elderly men, with an expression of deep disgust and astonishment on their pinched and worn faces. The Welsh Rabbit appeared startled, as if he had expected something quite different. 'Step forward number 1184,' he said in a quavering falsetto. They all called out, 'No.1184,' at the top of their voices, and after several minutes' wait, a seedy-looking person gradually emerged from the edge of the crowd, and came forward, wearing a rusty-black frock-coat and a greasy chimney-pot hat of an obsolete pattern, which he politely took off to Alice, apparently not perceiving the Welsh Rabbit at all.

'Let me present to you,' said the latter to Alice, 'a good average specimen of the Valuation Officer. He is a creation of my own, sparely and frugally maintained out of the expected gross proceeds of my new land taxes. As such, he was only born a few months ago, but is already a most precocious infant – and you can't ask him any question he can't answer offhand! Now, 1184, please go and explain to the March Hare, first how to fill up his Form properly and then how to leave it to me.'

The Valuation Officer's face wore a look of the blankest amazement. 'Fill up forms,' he said, 'why I'm an old schoolmaster and have been trying hard to keep my forms filled all my life – and never quite succeeded. Nothing in the wide world would ever induce me to give up a single one!'

The Welsh Rabbit turned a trifle yellower than usual . . . [13]

[11] 54th Report of Commissioners of HM Inland Revenue (1911), 158; Lloyd George received in conference several gentlemen in the land professions and dealt (not altogether convincingly) among other issues with 'Local Inquisitiveness' (*Land Values* (October 1910), 101); for the 'village taproom' charge see 'The Land Inquisition', in *The Spectator* 105 (3 September 1910), 337; Chief Valuer's Library, Interim Report 1909, 4; W. H. Aggs, *The Finance (1909–10) Act 1910* (London 1910), 8–9; Sir Edgar Harper, 'Lloyd George's Land Taxes' (paper to International Conference on Land-Value Taxation, Edinburgh 1929), reprinted in *Land and Liberty* [nd]; Charles Newton-Robinson, 'The blight of the Land Taxes: why they must be repealed', *The Nineteenth Century and After* 68 (1910), 397.

[12] Inland Revenue Valuation Department, *Instructions to Valuers* Vol. I, 1910, 33, para. 172.

[13] Loris Carllew (pseud.), *Alice in Plunderland* (London, 1910).

THE WELSH RABBIT INTRODUCES THE VALUATION OFFICER TO ALICE

Fig. 3.2 'The Welsh Rabbit introduces the Valuation Officer to Alice.'
Source: Carllew, *Alice in Plunderland* (1910), facing p. 26.

Progress of the valuation

However, the unusual nature of the values to be ascertained, and the prolonged opposition from landed pressure groups, together with internal inconsistencies in the legislation which required lengthy legal proceedings, all meant that the valuation process inexorably fell behind schedule. The basis for the Undeveloped Land Duty and Increment Value Duty was the Valuation, but since the collection and assessment of the duties ran concurrently with the Valuation, administration was further hampered. With the Scrutton Judgement of 1914 came the further problem that from that date no valuations of farmland or which included agricultural land could be served to owners. Furthermore the onset of war brought a concession from the Chancellor of the Exchequer that the time for raising objections should be extended to sixty days after a date to be set for landowners served with a provisional valuation within sixty days prior to, or at any time after, mobilisation. In fact that date was never set.[14]

By November 1910 all Division and District Offices were fully staffed and in working order, and in September 1911 the staff of the Valuation Office totalled 2,015 in Great Britain (of whom 152 were in Scotland) which comprised:

199 Valuers on the Permanent Staff
504 Temporary Valuers
346 Temporary Valuation Assistants
126 Temporary Draughtsmen
2 Established Clerks
838 Temporary Clerks

By 31 July 1914, numbers had attained a maximum as follows:

1 Chief Valuer (£1,200 pa)
1 Deputy Chief Valuer
1 Chief Valuer (Scotland) (both £850 rising to £1000 pa)
1 Assistant Chief Valuer (Scotland) (£800 to £850)
13 Superintending Valuers (£800 to £850)
1 Acting Superintending Valuer (£700)
120 Valuers (First Class) (£550 to £700)
120 Valuers (Second Class)(£350 to £500)
200 Valuers and Valuation Assistants (Third Class)(£100 to £350)
1 Minor Staff Clerk (£250 to £350)
2 Second Division Clerks (£70 to £300)
120 Clerks (£45 to £150)

[14] Select Committee on Land Values, Evidence of Mr Percy Thompson, 16, and citing letter from H. P. Hamilton on behalf of Lloyd George to C. H. Kenderdine, secretary to the Land Union (who had written on behalf of E. G. Pretyman), 17 August 1914, 21.

There were also the following temporary staff (all at various personal salaries),

753 Valuers
1,581 Valuation Assistants
138 Draughtsmen
2,171 Clerks

Most of the main work on valuation was completed by the Autumn of 1915, six months beyond the date originally promised by Lloyd George, but final completion was rendered impossible because of the war. All valuations had been made by the end of July 1915 with the exception of three Districts, in one of which a fire had destroyed all records, necessitating much of the work having to be repeated. In summer 1915 many staff were transferred from completed survey areas to those areas still to be completed. A departmental committee established by Lloyd George in July 1915 expected that little more could be undertaken under wartime conditions, and that the smaller staff would be able to cope with arrears that had accumulated by that date. Therefore the staff was slimmed down and at the end of September 1915 the total for Great Britain was 2,425, falling rapidly at the completion of Valuation to 1,146 (including 79 for Scotland), comprising:

457 Valuers on the Permanent Staff
123 Established Clerks
566 Temporary Staff

The progress made annually from 1910 to 1915 is shown in Table 3.3. The period to March 1911 was much taken up by training of staff, although their valuations were still classed as 'cheap and nasty' by opponents of the legislation. But progress moved ahead, especially in the year 1912–13 when the valuation tackled larger agricultural holdings. By September 1915 nearly 7.2 million valuations had been issued, and a further 611,522 made but not issued, having been made after the Scrutton Judgement of 1914. The number of valuations covered more than 10.5 million hereditaments, and over 56 million acres representing 98.9 per cent of the total acreage. By the beginning of the War about 60 per cent of the total Valuations had become final, and 88.3 per cent of those which had been served, although there was some regional variation between the Divisions, with North London being notably slower than average, and with Cumbria, West Riding, and Central (East) being well above average in completion rates (Table 3.6).[15]

The position for England and Wales was very similar in terms of the timing of progress and the costs (Table 3.4).

[15] 54th Report of Commissioners of HM Inland Revenue (1911); Evidence of Mr Edgar Harper, Chief Valuer to the Board of Inland Revenue, 42–50. The 'cheap and nasty' allegation was by Charles Newton-Robinson, 'The blight of the Land Taxes: a retrospect and a prospect', *The Nineteenth Century and After* 69 (1911), 1075.

Table 3.3 *Yearly progress of the Finance (1909–10) Act 1910 (excluding mineral valuations) in Great Britain*

Period	No. of valuations (1000s)	Hereditaments valued (1000s)	Acres valued (1000s)	Staff nos.	Aggregate salaries (£1000s)
August 1910– 31 March 1911	298.2	370.0	256.3	1,428	130.7
Year ending March 1912	1,501.2	1,848.4	3,381.6	2,829	309.2
Year ending March 1913	1,703.2	2,263.1	10,639.6	4,096	444.6
Year ending March 1914	2,113.9	3,038.7	21,189.4	4,611	529.5
Year ending March 1915	1,719.0*	2,462.2	18,227.9	3,735#	565.0~
1 April 1915– 30 September 1915	449.0**	603.2	2,449.5	2,425#	251.3~
Totals	7,784.4	10,585.6	56,144.3		2,230.4

Notes:
* Of which 321.9 were made but not issued.
** Of which 289.6 were made but not issued.
# Excluding men in HM forces.
~ Including pay of men in HM forces.
Source: Select Committee on Land Values, Table A, 49.

The position in Scotland can be summarised as in Table 3.5, and again the same trend can be seen between 1910 and 1915.

The Unionist press made much of the high cost of collection of the three taxes (excluding the rather different Mineral Duties which did not depend on the Valuation). Lloyd George himself estimated originally that the yield of the three taxes would be £500,000 in the year following April 1909, but then reduced his estimate to £425,000 later in 1909. In his second Budget in July 1910 the Chancellor revised the figure downward again to £250,000. In his third Budget of May 1911 he put the yield at £300,000 for 1911–12. An estimate of £550,000 for the two years 1910–11 and 1911–12 was thus put forward. This was compared unfavourably with an actual yield of £60,273.

However, the cost of the valuation and collection of the duties down to 31 March 1912 was £680,000, excluding not insubstantial sums for stationery and legal expenses, but final estimates put the total cost of mounting the

Table 3.4 *Yearly progress of the Finance (1909–10) Act 1910 (excluding mineral valuations) in England and Wales*

Period	No. of valuations (1000s)	Hereditaments valued (1000s)	Acres valued (1000s)	Percent total acreage
August 1910–31 March 1911	291.8	335.5	195.9	0.5
Year ending March 1912	1,464.6	1,672.5	2,899.1	7.8
Year ending March 1913	1,631.7	1,987.0	8,423.5	22.6
Year ending March 1914	1,977.2	2,554.8	13,087.5	35.1
Year ending March 1915	1,597.7*	2,152.9	10,270.4	27.5
1 April 1915–30 September 1915	425.4**	549.8	1,859.1	5.0
Totals	7,388.3	9,252.4	36,735.6	98.4

Notes:
* Of which 276.7 were made but not issued.
** Of which 276.6 were made but not issued.
Source: Select Committee on Land Values, Table A, 49.

valuation and collection of duties at more than £2 million (Table 3.3), and bringing in only a quarter of that in revenue.[16]

It was this ominous gap that was soon perceived between income and expenditure which further fuelled the controversy about the real nature and intentions of the Valuation. Many contemporaries, however, mistakenly assumed that the cost of establishing the Valuation Office was the same as the cost of the original Valuation. But, as Edgar Harper explained in 1920, the Valuation Office also performed many other, quite unrelated, tasks connected with such items as valuations in connection with old age pensions and

[16] 'The collapse of the Land Taxes', *The Spectator* 110 (19 April 1913), 646–7. PRO IR 63/35, N22.1, E. J. Harper, 'Cost of Valuation' (2 May 1914) f. 168. Exact estimates of income and expenditure are difficult to determine and contemporary, or near contemporary, estimates differed wildly. Yardley put the overall cost down to 1920 at £5 million and the income at £1.3 million (excluding the minerals duty). The former sum, which included rents of offices, stationery and travelling costs as well as salaries and the issuing of forms, was taken from an estimate by the Chancellor in April 1920 (*Hansard*, 26 April 1920, col. 860) (Yardley, *Land Value Taxation*, 651–2).

Table 3.5 *Yearly progress of the Finance (1909–10) Act 1910 (excluding mineral valuations) in Scotland*

Period	No. of valuations (1000s)	Hereditaments valued (1000s)	Acres valued (1000s)	Percent total acreage
August 1910–31 March 1911	6.5	34.4	0.4	0.3
Year ending March 1912	36.6	175.9	482.6	2.5
Year ending March 1913	71.5	276.2	2,216.1	11.4
Year ending March 1914	136.6	483.8	8,101.9	41.6
Year ending March 1915	121.3*	309.4	7,957.5	40.9
1 April 1915–30 September 1915	23.6**	53.4	590.4	3.0
Totals	396.1	1,333.2	19,408.7	99.7

Notes:
* Of which 45.2 were made but not issued.
** Of which 13.0 were made but not issued.
Source: Select Committee on Land Values, Table A, 49.

licences, determining monopoly values, and work for other government departments.[17]

Finally, a Reference Committee comprising a panel of fourteen men was established for England and Wales before the end of the 1910–11 Financial Year to hear appeals against the valuation procedures. The Committee selected the referee, and clear procedures were laid down, and in December 1910 they published their Land Values (Reference) Rules 1910 to guide appeals, as required under Section 33 of the Finance (1909–10) Act 1910. Nevertheless, many found the appeals procedures difficult and expensive in the early years of the legislation.[18] An appeal to the High Court was then possi-

[17] 54th Report of Commissioners of HM Inland Revenue (1911), Evidence of Mr Edgar Harper, Chief Valuer to the Board of Inland Revenue, 46–8. Details of work for other government departments is given on these pages, and includes work for the Post Office, Admiralty, Air Ministry, War Office, Ministry of Munitions, Board of Agriculture (for help with the plough-up campaign and Land Settlement), War Agricultural Committees, Ministry of Health (information on housing sites for Local Authority development) and much else.

[18] The Land Values (Reference) Rules, dated 5 December 1910, made by the Reference Committee for England under Section 33 of the Finance (1909–10) Act, 1910. A version of this is reprinted in *The New Land Taxes and Mineral Rights Duty: The Land Unions Handbook on Provisional Valuations* (London nd [1911]). Procedures prior to December were printed in

Table 3.6 *Number of Provisional and Final Valuations completed by 30 June 1914*

	No. of Provisional Valuations (1000s)	No. of valuations made final (1000s)	Percentage
N.London	423.7	334.1	78.9
S.London	538.2	445.3	82.7
Home Counties (North)	431.5	392.1	90.9
Home Counties (South)	423.4	371.5	87.7
Western	376.6	339.1	90.0
Eastern	450.8	414.0	91.8
Central (East)	458.9	428.5	93.4
Central (West)	230.8	205.3	89.0
S.Lancs.	516.9	447.5	86.6
N.Lancs.	335.4	306.8	91.5
Cumbria	90.1	84.6	93.9
W.Riding	367.4	350.8	95.5
Northern	461.3	413.3	89.6
Wales	385.4	323.5	83.9
Scotland	262.1	224.4	85.6
Totals			
England and Wales	5,490.6	4,856.5	88.5
Great Britain	5,752.7	5,080.9	88.3

Source: Select Committee on Land Values, Table B, 50.

ble on either side on points of law, a procedure entailing great delay in final settlement of the valuations. Harper estimated the number of appeals as 1,662, of which 736 were still outstanding in September 1919; 832 were withdrawn or had a compromise agreed; and out of the ninety-three which went to a hearing, forty-nine went in favour of the Inland Revenue, twenty-one went against and twenty-three were drawn.[19]

By 1919 the Valuation was virtually at an end, although any land with an element of agricultural value which had been valued after 1914 still had not been subject to the serving of any value upon the owner. In all other cases the issue to owners was practically complete, but there were many appeals which had not been settled.

C. E. Davies, *Land Valuation under the Finance (1909–10) Act, 1910* (Estates Gazette, London 1910), 248–53. Registers of appeals and a few sample case files, mostly relating to minerals and including copies of Form 76–Land (mineral rights) and Form 77–Land (notice of appeal), are to be found in PRO LT 5.

[19] *Ibid.*, Appendix II (C), 173; Harper, 'Cost of Valuation' (1929); Select Committee on Land Values, Evidence of Mr Edgar Harper, 46.

4

The survey procedures and documents

First stages: the Valuation Book

Once the staff were in place, a Valuation Book was sent to the Land Valuation Officer (LVO) in each Income Tax Parish (ITP), who arranged for the local assessor of taxes (if somebody other than themselves) to copy into it, from the Rate Book or 1910–11 'Schedule A' registers (themselves transcripts from the Rate Book), the description of each property, together with names of owners and occupiers, and the figures given in the Rate Book or registers for the extent of the property and its rateable value.[1] For this they were paid 1d per two lines. Any unrated properties did not of course appear in the Rate Book, and these had to be separately identified and entered into the Valuation Book, where they can now be found at the end of the other entries, often in a different hand, and normally with less information attached. The LVOs were instructed:

> To ascertain by careful observation and local enquiry, whether any land in your parish, particularly large or small plots of 'waste', 'vacant' or 'building' land, has been omitted from the Poor Rate, and if you discover any such omissions, you are to enter particulars of the lands in the Valuation Book, at the end of the copy of the Poor Rate . . . You are also to enter in the Valuation Book particulars of land forming the sites of churches, chapels and similar buildings . . . [2]

Every hereditament was then attributed a unique number, the number running consecutively throughout the ITP. The Books thus contain a 'record of valuations made by the Commissioners of Inland Revenue, in accordance with the provisions of Part I of the Finance (1909–10) Act, 1910'.

Known within the Valuation Department, and subsequently in many County Record Offices and local repositories, as the Domesday Books (unfortunately thereby ensuring a degree of confusion with the 'New Domesday' of

[1] The information was to be entered directly from the Rate Book, or if this was not possible, then from the Schedule A Income Tax Assessment Book for 1910–11, which was itself a copy of the Rate Book, *General Instructions to Land Valuation Officers*, 2.
[2] *Ibid.*, 7 para. 23.

Table 4.1 *Information in the Valuation Books*

Column	Information
1	Assessment (Hereditament) no.
2	Poor rate no.
3	Name(s) of occupier(s)
4	Name(s) and address of owner
5	Description of property
6	Address of property
7	Estimated extent
8	Gross Annual Value
9	Rateable Value
10	Map reference
11–14 (inc.)	Extent as determined by the valuer in acres, roods, perches and yards
15–39 (inc.)	Calculated figures for the various values
40	Comments

1873), the Valuation Books were the first major document to be compiled. The book actually comprised three Land Forms: the first was Form 20 Land, 'Duties on Land Values: Record of Valuations made by the Commissioners of the Inland Revenue', and this actually formed the title page of the Valuation Book. Secondly, Form 24 Land was a record of the numbers of 'Forms of Return' sent out by the LVO, and was usually signed by him. This comprised page 2 of the Valuation Book. Thirdly, the bulk of the book was made up of pages officially known as Form 21 – Land. Onto these pages the LVO transcribed the details from the Rate Book, prior to sending out the forms of return to the various owners. Valuation Books survive for most areas so far as is known, and are to be found in local repositories. The only exceptions are those Books covering the City of London and Westminster (actually covering Paddington), which are located in the Public Record Office.[3]

The information that should be in the Valuation Books is set out in Table 4.1. Not all records attain perfect compilation standards by any means! So Table 4.1 (and all similar tables below) demonstrate what should theoretically be present in the document, but actual research may prove the LVO's clerical abilities to have fallen far short of the ideal. The books are of large format, measuring 43 cm by 34 cm, bound in cloth, with leather corners and with the name of the Valuation District on the spine. The size precludes the reproduction of a typical page.

Columns 2 to 9 (inc.) are the transcriptions of the 1909–10 Poor Rate Book

[3] The Valuation Books for the City of London and for Westminster (Paddington) form the Class PRO IR 91.

or the Schedule A Assessment Book. These were entered up by the LVO who then numbered the hereditaments in the ITP consecutively. These hereditament numbers are found in column 1, and any research using the documentation should note the relevant numbers, since they are of crucial importance for using the other categories of document.

Columns 10 to 40 were entered up by staff in the District Valuation Offices following valuation. Columns 11 to 14 are also of great importance as they give area extents taken from the Ordnance Survey plans, after the boundaries of the hereditament had been established. Also of great potential importance are the comments found in the final column, which may provide vital evidence for interpreting and analysing the material. In some cases the Valuation Books also contain street indexes: in Derbyshire, for example, the entries for many of the larger settlements have such indexes as at Buxton or Glossop in the High Peak Valuation Office.[4]

Alterations, especially to the names of occupiers and owners in columns 3 and 4, may often be found, although the status of the alterations is not always clear. It may indicate that the owner or occupier had changed between the compilation of the Rate Book and the issuing of the Form of Return in August 1910, or that the change occurred later, but before the hereditament was valued often several years later. Later subdivisions of property were also to be squeezed against the appropriate entry if possible. Since the valuation books continued to be actively consulted long after the repeal of the land clauses in 1920, it is also possible that the changes might be more recent.

The great Form 4 outcry

The 'Forms of Return' were those delivered (personally for preference unless the owner lived at a distance, when it was by registered post) by Land Valuation Officers to owners of land or other resources and intended for them to complete and return with the information necessary for the levying of the various duties. The methodology for distributing the huge volume of forms gave rise to considerable anxiety and it was clear that people with local knowledge were needed, as noted in Chapter 3. It could still, however, take several weeks to deliver the forms in some of the larger ITPs, and there were claims from opponents that 'the elaborate and prolix government forms . . . are being dribbled out by degrees, but . . . often the forms have been served on the wrong person. It is not always easy to find the actual owner.'[5]

The most common 'Form of Return' was Form 4–Land, of which 12 million copies were printed. This was supposed to be sent to landowners of each here-

[4] Derbyshire RO D595R/3/1/7a; 11–14.

[5] C. Newton-Robinson, 'The blight of the Land Taxes: why they must be repealed', *The Nineteenth Century and After* 68 (September 1910), 393.

ditament by August 1910 by the LVOs, accompanied by Form 1–Land, giving the notification to make the return and by Form 2–Land, giving the instructions for making the returns on Form 4 'framed with a view to making it clear to the owner what details he is required to return; and to giving him general information which will assist him in filling in the particulars required on Form 4'.[6] Before dispatching the forms, the hereditament number and the particulars of the rateable value etc. appearing in the Valuation Book were inserted into columns 1 to 9. The completed document was then deposited in the District Valuation Office, where some may still remain.

The exceptions to this related to land owned by 'statutory companies', defined as 'any railway company, canal company, dock company, water company, or other company who are for the time being authorised under any special Act to construct, work, or carry on any railway, canal, dock, water, or other public undertaking . . . '.[7] These were required to complete Form 3–Land. Form 5–Land was completed where Mineral Rights Duty was likely to be levied. Form 6–Land was sent to owners of unworked minerals, and Form 6A–Land was a revised version of this. But where the minerals consisted of clay, brickearth, sand, chalk, limestone or gravel, they were to be treated as part of the land from whence they were obtained, and thus dealt with through Form 4.

Thus, during August 1910, landowners across the country received a 'Notice to make Returns' which required them to provide detailed information on every hereditament – or unit of occupation – in their ownership. The Form 4–Land was soon to become notorious amongst landowners as an imposition upon their individual rights and freedoms. Where the owner of a hereditament was unknown, the officer was enabled, under Section 31 of the Act, to require every person who paid rent in respect of any land, and every person who as an agent for another person received rent in respect of any land, to furnish the name and address of the person to whom he paid rent or on behalf of whom he received rent, as the case may be. The demand for information was made on Form 8–Land, and a fine of £50 was payable if no return was made.

The forms were distributed to owners, one for each hereditament owned by them, and as far as possible within one envelope. The reality could differ from central government intentions to some greater or lesser degree. In the parish of Pewsey (Wiltshire), for example, 563 forms had to be distributed. Of these 552 were delivered on 24 August, seventy-one in registered packets and the rest personally. The others were sent during September and as late as 11 October by registered post. The Valuation Book was handed over finally on 26 November 1910.[8]

[6] Inland Revenue Chief Valuer's Library: 'Duties on Land Values: "Interim" Report of the Committee appointed to draw up Regulations for carrying into effect the provisions of the Finance Bill, 1909, dealing with the duties on Land Values' [nd, np].

[7] Finance (1909–10) Act 1910 Section 38.4.

[8] Wiltshire Record Office, Form 24–Land (Inside Valuation Book for Pewsey 1910, Inland Revenue register no. 57. Pewsey).

Table 4.2 *Information contained on Form 4–Land*

1. Parish(es) in which land was situated.
2. Name of occupier.
3. Full name and address of person making return.
4. Whether interest was freehold, copyhold, or leasehold:
   (a) name of manor if copyhold,
   (b) term and date of commencement of lease if leasehold, and information on renewal covenants. The name and address of the lessor was also required.
5. Name and precise situation of the land.
6. Description of land and particulars of any buildings and other structures thereon, and the purposes for which the property was used.
7. Extent of the land if known.
8. If the land was let by the person making the return:
   (a) whether let under lease or agreement,
   (b) if neither, then period of let,
   (c) if let under lease/agreement:
      (i) term for which granted,
      (ii) date of commencement of term,
      (iii) whether there was a money consideration involved,
      (iv) whether tenant was responsible for laying out money for improvements etc.
9. If owner-occupied, the person was to give the annual rent obtainable if let to a yearly tenant keeping it in repair.
10. Amount of any Land Tax and by whom borne.
11. Amount of any Tithe Rent-charge, or commutation payment, and by whom borne.
12. Amount of Drainage or similar rate, and by whom borne.
13. Who bore all usual tenant's rates and taxes.
14. Who bore cost of repairs, insurance, and other necessary maintenance costs.
15. Whether land was subject to any:
   (a) fixed charges (exc. Tithe) and amount thereof,
   (b) public rights of way,
   (c) public rights of user,
   (d) right of common,
   (e) easements affecting the land,
   (f) restrictive covenants and their date.
16. Details of last sale (if any) within 20 years before 30 April 1909, and of subsequent expenditure:
   (a) date of sale,
   (b) amount of purchase money and any other considerations,
   (c) capital expenditure upon land since sale.
17. Observations with full details of any part of the land that the owner required to be separately valued.
18. Name and address of any agent or solicitor that information should be sent to.
19. Various questions about possible mineral rights.
20. Owners could also provide estimates of the different categories of value, if they so wished.

See also Appendix 1 for the original layout of Form 4.

Since Form 4 was by far the most common of the Forms of Return, this description is confined to the information that should be found on this document (Table 4.2). A copy of the form is to be found in Appendix 1.

Owners were required to provide the information as far as possible with reference to the circumstances existing on 30 April 1909. They frequently also kept 'unofficial' copies of these forms for their own records, and these are sometimes found today in estate, solicitors' and similar collections in local repositories and in the PRO. Some have been deposited locally by DVOs, but it would seem that most have been destroyed.

The reactions to Form 4 were extraordinarily vociferous. Lloyd George had to spend a few days in London quieting this storm that had blown up over this early stage in the process of land valuation. As a result of receiving Form 4 'there was loud complaint in the newspapers, references to the Spanish Inquisition, and cartoons of honest John Bull spending hours which were needed for the harvest torturing himself to provide numbers for blockheaded civil servants'. Even C. F. Brickdale, the Registrar, had commented in an earlier memo of July 1910 that the forms were 'formidable in appearance to a layman', although he went on to say that despite this they were in reality 'really nothing more than sketches of abstract'.[9]

The cartoon of the 'Free-born Briton who, within the period usually allotted to his holidays, is required, under threat of a penalty of £50, to answer a mass of obscure conundrums relating to land values, in order to facilitate his future taxation', published in *Punch,* 24 August 1910, sums up this attitude rather well (Fig. 4.1).

While Lloyd George was thought to be on a motoring holiday, possibly with golf as an important item, in Germany, Switzerland and Italy, *Punch* also ran an eleven-verse poem by 'O.S.' on 24 August 1910 under the heading of 'The two holidays', one of many such items of doggerel verse engendered.[10] It began:

> So you are off to see the sights,
> To taste an unofficial beano
> Full of the keen and pure delights
> Which none but those with conscience free know;
> And Heaven, you hope, is sure
> To bless your German–Swiss–Italian tour!

The eighth verse ran:

> We, too, had hoped to take our ease
> In spots renowned for natural beauties,

[9] A. Offer, *Property and Politics 1870–1914: Landownership, Law, Ideology and Urban Development in England* (Cambridge 1981), 364; B. Gilbert, *David Lloyd George: A Political Life – The Architect of Change 1863–1912* (London 1987), 421; PRO LAR 1/24 (File no. 005/9).

[10] *Punch* 24 August 1910, 128.

Fig. 4.1 'The Holiday Task.'
Source: Punch 24 August 1910, 129 (reprinted, with text reset, from Offer, *Property and Politics*, 365).

> But have, instead, to grind at these
> Condemnable Land-Value-Duties;
> Yes, while you romp about
> We've got to work your silly puzzles out

and it continued in similar vein:

By flowery routes you lightly bound,
But we, our holidays all rotted,
Await a fine of fifty pound
In case the answers can't be spotted;
And how to find the clue
We have no notion more than you

Within a space of thirty days
(In this the month for gathering roses)
We've got to solve the sinuous maze
Or pay your minions through our noses;
While you at your sweet will
Go round and gambol with a rubber pill.

In *The Financial Times*, F.W. wrote a four-verse poem[11] that ended:

We've got no heart for shooting;
For walks we dare not go;
We neither bathe nor paddle;
We neither golf nor row.

Our brows are sadly puckered;
To think, it seems a bore;
To tell the truth about it,
We're filling up Form IV.

Lord Rosebery was apoplectic:

There is a cry of anguish throughout the land, and with all connected with the land. I do not know how many millions of my fellow-creatures – everyone who has a foothold on the land, either the temporary foothold of a lease, or what was once considered, but I think is no longer so, the more durable occupation of a freehold – every one of those suffering fellow-countrymen is at this moment exposed to an inquisition unknown since the middle ages, but it has tortured them almost to extinction. The boot and the thumbscrew have not yet come, but the inquisition in every other form is complete.[12]

Watkin Davies noted that income tax payers had actually been answering similar questions for years:

not indeed with joy, but at least without proclaiming themselves tortured martyrs; but it was something new for the owners of land to have to fill in forms of any kind, or to render to the community a return for their stewardship. The fact that so vocal and widespread an agitation rose from so trivial an irritation is proof of the sensitiveness of landowners . . . and their determination to keep the nation from prying into what they regarded as their rights.[13]

Land Values in September 1910 noted that:

[11] *Financial Times* 17 September 1910.
[12] W. Watkin Davies, *Lloyd George 1863–1914* (London 1939), 370. [13] *Ibid.*

We are only at the beginning of the valuation and already it is engaging more serious attention than any other subject in the world of politics. This is a reason for satisfaction . . . the most prominent incidents connected with its progress are the fierce and angry protests of its opponents.[14]

There was, of course, more moderate landed opinion to be found. The Earl of Crawford, writing in his private journal on 'the new Land Tax', noted on 28 September 1910 that 'I fancy we are squeaking too loudly about the hardships of Form IV', and on 14 November 1910 that 'before long the unpopularity of these new burdens will grow'.[15] But the great 'Form 4' outcry certainly enlivened the Tory press during what may have otherwise been rather a dull 'silly season' and it even also fleetingly provided material for the stage! As Lloyd George himself noted on 14 September, in conference with members of land-related professions to discuss the difficulties being encountered (Fig. 4.2):

I was not altogether responsible for the forms having been sent out in August, I agree that it is very unfortunate that they were sent out in the holiday season, but the press were rather short of copy, and it had therefore been a boon to them that they should have the land taxes to think about, and in fact, as far as I can see, they have divided the attention between that and 'Dr' Crippen.[16]

The timing was governed by the very short timescale allowed from the April Budget for the appointment of officials, the printing of forms and other preparations. From all over the country came reports of meetings and statements. A meeting of Scottish landowners – the Scottish Land and Property Federation – on 22 September in Edinburgh noted the need for a memorandum setting out a guide to Form 4. And very many smaller owners consulted the LVO for advice, again delaying the progress of the survey.

Nevertheless, by late November, as in the case of Pewsey (Wiltshire) cited above, the District Valuers were able to take the Valuation Books, together with the completed Forms of Return, from the LVOs, whose task was now effectively finished. As Mr Dean, a temporary clerk in the Dorchester Office, remembered, the handover was complex:

There must have been between 20 and 30 LVs [LVOs]. This took some arranging, appointments made at the rate of 8 LVOs per day. All had to be checked, and only 4 of the staff to do this:– (1) the total of entries in Domesday Book. (2) No of Form IVs issued and returned. Forms signifying these entries and duly signed. They were paid 2d for each Form IV issued, and ½d for each Form IV returned.[17]

[14] *Land Values* (September 1910), 67.

[15] J.Vincent (ed.), *The Crawford Papers: The Journals of David Lindsay 27th Earl of Crawford and 10th Earl of Balcarres 1871–1940 during the years 1892 to 1940* (Manchester 1984), 164, 167.

[16] H. Du Parcq, *Life of David Lloyd George* (London 1911–13) vol. III, 564–5; *Land Values* (October 1910), 100.

[17] *The Valuation Office 1910–1985: Establishing a Tradition* (Inland Revenue, London 1985), Part 5 [np]. According to the Instructions to Land Valuation Officers, Mr Dean should have been getting rather more than the quoted sums for his work! All claims for remuneration by LVOs

(*By kind permission of the* Westminster Gazette.)

"THE FOURTH FORM BOY."

Head Master: *You'll find your fourth form work quite easy, my boy, if you'll only give your mind to it, and not lose your temper. If there is anything you don't know, say so!*

Fig. 4.2 'The Fourth Form Boy.'
Source: Originally published in the *Westminster Gazette*, and reprinted in *Land Values* (October 1910), 101.

In some of the larger ITPs, and in urban and leasehold areas, the handover was delayed, but by the end of the financial year, some 10,700,000 forms had been sent out, and 9,600,000 returned completed. The services of the LVOs had cost the state £174,342 during the financial year 1910–11.

Upon receipt of the Valuation Books and Forms of Return from the LVOs, the District Valuation Office staff sorted out and filed numerically all Forms of Return relating to each hereditament. Upon examination of the returns, it was often found that the information given needed verification or amendment, although the Commissioners of the Inland Revenue found that the returns, from large landowners especially, had been 'fully and accurately rendered'. All Valuation Books and Returns were to be carefully preserved at the District Valuer's Office, although some larger landowners had

were to be made on Form 31 which was certified as correct by the DVO before transmission upwards to the Chief Valuer. Supplemental claims could be made on Form 32! LVOs were paid 2d for each copy of Form 4 sent out (*H. C. Deb.* 5s. xxii, c. 927, 6 March 1911).

their own copies of Form 4 made out, which were boldly stamped unofficial at the top.

As a concession to landowners of smaller properties Section 5 of the Revenue Act 1911 subsequently enacted that hereditaments tenanted by different people might be valued as one unit, provided that they were spatially contiguous and that they did not exceed 100 acres in total, and that the Inland Revenue was satisfied that this was an equitable procedure. Form V.O.58 was completed to enact this. The form contains the name, signature and full address, of the owner of the property concerned, with the parish, hereditament numbers, postal description and area of the hereditaments to be aggregated. For information, a definition from Section 41 of the Finance Act of 'the owner' was appended at the foot of the page.[18] The government originally proposed no spatial limit to the Commissioners' powers to value contiguous pieces of land. Wedgwood wished these powers restricted to 10 acres, and since this was unacceptable, Chiozza-Money's compromise of a 100-acre upper limit was accepted. Many were worried about the 'averaging-out' of high- and low-value land in the vicinity of towns which might follow from this principle, and the journal *Land Values* raged at this weakening of 'the fabric of valuation'.[19]

The Field Books

The necessary information from the owner's return and Valuation Book was in each case transcribed into the valuer's Field Book in which the valuation was actually made. Temporary Valuation Assistants were then posted to different parts of the Districts, armed with the Field Books, Ordnance Survey maps, and an authority to inspect land which had to be produced on demand. Valuations were made following inspection of the property, and the various types of value assessed as at 30 April 1909. The steps in the assessment were to be entered into the Field Book, as a result of both the owners' own returns and those of the valuer, following his inspection (Table 4.3). So far as practicable, valuers inspected and valued neighbouring properties at the same time.[20]

The Field Books number some 95,000 volumes and are now available for consultation in the PRO as Class IR 58. They were used by the valuers for

[18] Section 5 of the Revenue Act 1911 replaced Section 26 (1) of the Finance (1909–10) Act 1910. The form was entitled 'APPLICATION for the aggregation of hereditaments under the provisions of Section 5 of the Revenue Act 1911'. For internal discussion on the basis on which aggregation should be allowed, following the Revenue Act, see Chief Valuer's memo on grouping of hereditaments 8/2/12 in The Chief Valuer's Library Box 3/4. I am grateful to Mr A. Fallows for allowing privileged access to the Chief Valuer's Library.

[19] *Land Values* (May 1911), 268, 'averaging the valuation'. As many as sixty-nine Liberal and Labour MPs voted against the government to support Wedgwood's 10 acre upper limit for grouping land for valuation purposes.

[20] This section concentrates on the English and Welsh Field Book format only. For variations in the cases of Scotland and Ireland, see Chapters 11 and 12.

Table 4.3 *Information in the Field Books*

Page 1

1. Reference [hereditament] no. [also at top of pages 2–4] and Ordnance Survey Map no.
2. Situation
3. Description
4. Extent
5. Gross Value (i) Land (ii) Buildings
6. Rateable Value (i) Land (ii) Buildings
7. Gross Annual Value, Schedule A
8. Occupier
9. Owner
10. Interest of owner
11. Superior interests
12. Subordinate interests
13. Occupier's tenancy (i) Term (ii) Date of beginning of tenancy
14. How determinable
15. Actual (or estimated) rent
16. Any other consideration paid
17. Outgoings (i) Land Tax (ii) Paid by (iii) Tithe (iv) Paid by (v) Other outgoings
18. Who pays (a) Rates and taxes (b) Insurance
19. Who is liable for repairs
20. Fixed charges, Easements, Common rights and Restrictions
21. Former sales (i) Dates (ii) Interest (iii) Consideration (iv) Subsequent expenditure
22. Owner's estimate (i) Gross value (ii) Full site value (iii) Total value (iv) Assessable site value
23. Site value deductions claimed
24. Roads and sewers (i) Dates of expenditure (ii) Amounts

Page 2

25. Particulars, description, and notes made on inspection
26. Charges, Easements and Restrictions affecting market value of Fee Simple
27. Valuation [and continuing as nos. 30 to 33 on p. 4]
    Determination of GROSS VALUE
    (i) Market Value of Fee Simple in possession of whole property in its present condition
    (ii) Deduct Market Value of site under similar circumstances, but if divested of structures, timber, fruit trees, and other things growing on the land
    (iii) Difference Balance, being portion of market value attributable to structures, timber etc.
    Divided as follows: Buildings and structures
    Machinery
    Timber
    Fruit trees
    other things growing on land
    (iv) Market Value of Fee Simple of Whole in its present condition (as before)

Table 4.3 (*cont.*)

(v) Add for Additional Value represented by any of the following for which any deduction may have been made when arriving at Market Value
Charges (excluding Land Tax) Restrictions
(vi) GROSS VALUE (derived from i to v above)

Page 3
28. Table showing Description of Buildings, with columns for:
(i) Index letter
(ii) Description of buildings
(iii) Dimensions: Frontage, Depth, Height
(iv) Cubical contents
(v) Condition
(vi) Remarks
29. Space for sketch map of buildings etc.

Page 4.
30. Gross Value (brought over from foot of p. 2)
31. Valuation: determination of FULL SITE VALUE
(i) Gross value less Value attributable to Structures, timber etc. (as before)
(ii) Yields FULL SITE VALUE
32. Valuation: determination of TOTAL VALUE
(i) Gross value (as above)
(ii) Less deductions in respect of
Fixed charges (including Fee Farm Rent, rent seek, quit rent, chief rents, rent of Assize)
Any other perpetual rent or Annuity
Tithe or Tithe Rent Charge
Other burden or Charge arising by operation of law or under any Act of Parliament
If Copyhold, estimated Cost of Enfranchisement Public Rights of Way or User
Rights of Common
Easements
Restrictions
(iii) Yields TOTAL VALUE
33. Valuation: determination of ASSESSABLE SITE VALUE
(i) Total value less value attributable to Structures, timber etc. (as above)
(ii) And less Value directly attributable to:
Works executed
Capital expenditure
Appropriation of Land
Redemption of Land Tax
Redemption of other charges
Enfranchisement of Copyhold, if enfranchised Release of Restrictions
Goodwill or personal element

Table 4.3 (*cont.*)

(iii) And less Expense of Clearing Site
(iv) Yields ASSESSABLE SITE VALUE
34. If agricultural land, the value for agricultural purposes including or excluding (specify) Sporting Rights
35. Value of Sporting Rights
36. If Licensed Property, the annual licence value
37. Liable to Undeveloped Land Duty as from (date)
38. For further reference as to apportionments etc. see . . .

Notes:
All punctuation and capitalisation as on the original Form.

actually making the valuation, and are thought to survive for most areas, although there are gaps, the extent of which is not yet known.

The Field Books should contain the fullest information of any category of document. The entry for each hereditament comprised four pages in the book and includes of course the hereditament number, being the same number as on the Forms of Return and in the Valuation Books. Each book contains information on up to 100 hereditaments, all to be within the same ITP, and with the first Book having all the information on hereditaments 1 to 100, the second on 101 to 200 etc. All of the information entered upon the Forms of Return detailed above was transcribed onto the first page of the Field Book entry for the hereditament by clerical staff in the DVO prior to an inspection by the valuer. As Mr Dean of Dorchester put it:

> All particulars on Form IV had to be copied in FIELD BOOKS 100 pages at a time, of course separate for each parish.
>
> So urgent it became that the two TVAs [Temporary Valuation Assistants – appointed in November 1910] had only two days leave at Xmas (Xmas Day and Boxing Day). They had to come back as Mr Dean [the writer, he writes in the third person] had been allocated 4 days leave, and all had to be taken before 31 December. No such thing as overtime.
>
> Each district was divided and TVAs were posted to different parts of the County, taking with them Field Books and 25″ Ordnance Maps. Made their own arrangements as to inspection of properties, marking boundaries on 25″ Map and Ref No. in Field Books.[21]

The second page of the entry (Table 4.3) included a description of the property and notes made on inspection. The description was to be as full as necessary in order to make a correct valuation, and in order to defend any future litigation. In practice, descriptions frequently included details of building

[21] *The Valuation Office 1910–1985: Establishing a Tradition, passim.*

materials, numbers and use of rooms, comments on repair and condition as well as suitability for the purpose used, facilities, ancillary buildings and their condition, water supply and sanitary facilities and so on. For agricultural holdings, comments on state of cultivation, drainage, land use, etc., are common. Sketch plans of farmsteads are also common, being traced from the Ordnance Survey sheets with details of each building and its use noted. This is usually found on p. 3 of the entry which was used to provide additional information of a descriptive kind. As the work began to fall behind schedule in 1912, the sketch plans were dispensed with, and these are therefore only found for those hereditaments valued before this date. Exceptionally, a photograph might be included, as on p. 2 of the Field Book for Penshurst Place, Kent (Fig. 4.3). The lower half of the second page of the entry, and the final page, include figures for four various values, and calculations made in this respect. These values were the highly controversial, and partly theoretical (and ultimately indefensible), 'gross value', 'full site value', 'total value' and 'assessable site value'. These values are described more fully below.

In some cases the information on hereditaments was too bulky for the designated spaces in the Field Book. In such cases an additional Field Notebook, with two pages and one inset sheet for each hereditament, was to be used, with any further records, calculations or notes required for future reference. Thus we are enjoined to 'see notes' in the Field Book for the 437-acre farm of Ernest Gardner MP at Maidenhead, Berkshire, but unfortunately such notes have in this case, as in so many others, been lost.[22]

The Field Books were to be kept 'neatly and carefully preserved, as it may be required for reference many years hence, and may in some cases have to be produced in evidence'.[23] Entries made in the field were to be in pencil, with the Book being written up in ink at a later stage. In some cases, as in Suffolk for example, many Field Books were systematically updated through to the 1930s and are thus doubly valuable.[24]

Form 37: the Provisional Valuation

Form 37–Land was the statement of the Provisional Valuation. It was retained in the DVO, and a copy sent on Form 36–Land to the owner(s) of each hereditament and other interested parties. If amendment to the provisional valuation was made, then this was entered onto Form 39–Land. This was identical with Form 37 apart from colour, title, and the amended values, and is likely to be found with the latter form. Copies of amended provisional valuations were sent to the owner(s) on Form 38–Land. Similar forms for minerals were

[22] Inland Revenue Valuation Department, *Instructions to Valuers* Part I (HMSO, London 1910), 30. The Field Book for Spencers Farm, Maidenhead is in PRO IR 58/52286.

[23] *Instructions to Valuers*, 31.

[24] I am grateful to Mr William Foot, Public Record Office (Kew) for this information.

from 33 (Continuation) Reference No. 34

Particulars, description, and notes made on inspection

Carpenters shop harness room loosebox & 2 stall stable
3 mens rooms over
Johnston's rooms
Dining room 3 bedrooms over & wc

Range of store rooms & office brick & slate
2 Vineries (1 lean to 1 span) lily pond with fountain

Charges, Easements, and Restrictions affecting market value of Fee Simple

33 · 4 6 0
34 19 3 9
part 64 kept 5 11 3 informally app^d

Valuation.—Market Value of Fee Simple in possession of whole property in its present condition

PENSHURST CASTLE
(The Seat of LORD De L'ISLE and DUDLEY).

(as before) .... £

Add for Additional Value represented by any of the following for which any deduction may have been made when arriving at Market Value :—

Charges (excluding Land Tax) ....£

Restrictions ....£ £

GROSS VALUE...£

Fig. 4.3 Penshurst Place, Kent: the Field Book entry.
Source: PRO IR58/85804.

Table 4.4 *Information on Form 37–Land*

1. Description of property
2. Situation (County, Parish, No. of hereditament)
3. Name of occupier
4. Extent
5. Original Gross Value
6. Deductions made to arrive at Original Full Site Value
7. Deductions made to arrive at Original Total Value
8. Deductions from Total Value to arrive at Assessable Site Value
9. Original Assessable Site Value
10. Value of agricultural land for agricultural purposes where different from Assessable Site Value

On the reverse

11. Name and address of person(s) on whom the foregoing Provisional Valuation was served

Form 47–Land (Provisional Valuation of Minerals); Form 48–Land (Copy of Form 47); Form 50–Land (Amended Provisional Valuation of Minerals); and Form 51–Land (Copy of Form 50).

The Form 37 contains the information from the valuation arrived at following the visit of the Valuation Officer (Table 4.4).

We therefore once again have the hereditament number, the address and description of the property; the name of the occupier and the extent of the property in acres, roods, perches and yards. Form 36–Land was identical, except for a box for a signature of the Clerk to the Valuer to signify that this was a true copy (of Form 37).

The significance of Form 37 lies in its particular place in the valuation process. Because it followed on after the valuation had been made, the information is likely to be more accurate and up-to-date than that contained in the Valuation Book or on Forms 4, which were based on Poor Rate records which might have been outdated by 1910, or, in the case of Form 4, on information to hand given by owners in August 1910. On Form 37 the area of the hereditament is given down to the nearest yard (theoretically), the full address of the owner is normally given, rather than an abbreviated address as in the Valuation Book, and in some areas, such as Gloucestershire, the forms were quite consistently updated to about 1945 with changes in ownership.

The forms take on an added importance in cases where schedules detailing individual properties have been attached to them. In some cases such as that for Corsley (Wiltshire) descriptive material was included on schedules attached to Form 37 and this is referred to in the relevant Field Books. But as the Forms were not accepted by the local record office, the information is now unfortunately lost (see Chapter 5). The same problem of the omission of information from the Field Books because it was included with the Forms 37 has been found at Lockinge and Ardington (Berkshire). In Ashley Walk, on the fringes of the New Forest in Hampshire, details of land use on a field-by-field basis are often to be found on the surviving schedules in the Hampshire Record Office.[25]

Moreover, Form 37 is the only document that can always be linked directly and unambiguously to the hereditament numbers on the Record Plans discussed below. Individual hereditaments in the Valuation and Field Books were frequently merged into larger units, following the amending legislation in 1911 to make the owners' tasks more straightforward (as discussed above), so that the hereditament number given in these simply may not appear on the Record Plan.[26] The practical significance of this for research is also discussed below.

Appeals against the Provisional Valuation could be made within sixty days of its issue, and were usually dealt with through negotiation at District level, although the panel of referees could adjudicate if local negotiation failed. District Valuers dealt with all but six cases during the financial year 1910–11. Notification of an amended Provisional Valuation was made to the landowner on Form 39–Land. After settling an appeal – or after sixty days if no appeal was made – the Original Valuation was established, and should have been entered into the Valuation Book.

The Ordnance Survey Plans

Whilst the work of preparing the Field Books was progressing, staff were also busy working on the relevant plans of the properties. Each District Office requisitioned for, and received, two sets of the largest-available-scale Ordnance Survey sheets for its area of jurisdiction, and something in the order of 200,000 plans were distributed throughout the country. These were usually at the scale 1:2500, although other scales were also used (Tables 4.5 to 4.7), especially the 1:10560 in areas where the 1:2500 had not at that time been published.

The most recent available editions were used and, where appropriate, updating was carried out to the extent of adding new roads and buildings, although not other details such as railway tracks. In so doing, an intermediate revised edition of the large-scale series was created for Great Britain which, though only partial in revision detail, does nevertheless provide useful dating clues for histories of urban development etc. Some sheets were published entitled

[25] The Corsley (Wilts) Field Books are in PRO IR 58/73332–5; see also Brian Short and Mick Reed, *Landownership and Society in Edwardian England: The Finance (1909–10) Act 1910 Records* (University of Sussex 1987), 35–9 and 85.

[26] Brian Short and Mick Reed, 'An Edwardian land survey: the Finance (1909–10) Act 1910 records', *Journal of the Society of Archivists* 8(2) (1986), 99–100.

Table 4.5 *Ordnance Survey Map scales used for the Valuation*

1:2500 (25 inches to the mile)	the standard scale used
1:10560 (6 inches to the mile)	for moorland and upland areas of low population density
1:1250 (about 50 inches to the mile)	used for urban areas
1:1056 (60 inches to the mile)	used for Central and Greater London
1:500 (about 127 inches to the mile)	used for densely populated town and city centres
1:528 (120 inches to the mile)	used as the 1:500 series for a few small areas only as an older survey, some sheets dating to the 1850s

Source: Public Record Office, *How to Find and Use Valuation Office Maps and Field Books* (1992), 1.

Table 4.6 *Versions of large-scale maps employed by the Valuation Office*

1:500	late nineteenth century, not subsequently revised
1:528	late nineteenth century, not subsequently revised
1:1056	late nineteenth century (? London only)
1:1250	enlargements from 1:2500
1:2500	ordinary sales copies, unrevised
1:2500	advance copies, prior to publication
1:2500	partial revisions, the SPECIAL EDITION 1912
1:2500	partial revisions, printed, and with note in upper margin
1:2500	ordinary sales copies with MSS additions and alterations in dark blue
1:2500	ordinary sales copies with MSS revisions, probably VO copies of OS manuscript mapping
1:2500	post-1912 (e.g. 1940s)
1:10560	ordinary sales copies, unrevised

Source: Partly based upon Oliver's addendum to G. Armitage, 'Ordnance Survey Land Valuation Plans', *Sheetlines* 35 (1993), 8–9.

'Special Edition 1912', but many remained in manuscript versions only.[27] Where the ground had recently been resurveyed, but revised maps had not yet been published, the Ordnance Survey supplied advance proof sheets containing the latest revisions. The main series of sheets date mostly between 1880 and 1915, but a few of the 1:528 sheets date from the 1850s and some from the

[27] Richard Oliver, *Ordnance Survey Maps: A Concise Guide for Historians* (Charles Close Society for the Study of Ordnance Survey Maps, British Library, London 1993), 25; G.Armitage, 'Ordnance Survey Land Valuation Plans', *Sheetlines* 35 (1993), 8–9; and H. Winterbotham, *The National Plans* (Ordnance Survey Professional Papers, NS No.16, HMSO, London 1934), 80.

Table 4.7 *Valuation Office Maps: counties and scales of coverage*

County	1:500	1:528	1:1056	1:1250	1:2500	1:10560
Anglesey	*			*	*	
Bedfordshire				*	*	
Berkshire				*	*	
Brecknockshire				*	*	
Buckinghamshire	*			*	*	
Cambridgeshire				*	*	
Cardiganshire	*			*	*	
Caernarfonshire	*			*	*	*
Carmarthenshire	*			*	*	
Cheshire	*	*		*	*	*
Cornwall					*	
Cumberland	*			*	*	*
Denbighshire				*	*	*
Derbyshire	*			*	*	
Devon	*			*	*	
Dorset				*	*	
Durham	*			*	*	*
Essex				*	*	
Flintshire				*	*	
Glamorganshire	*			*	*	
Gloucestershire	*			*	*	
Hampshire				*	*	*
Herefordshire	*			*	*	
Hertfordshire	*			*	*	
Huntingdonshire				*	*	
Kent				*	*	
Lancashire	*			*	*	*
Leicestershire	*	*		*	*	
Lincolnshire	*			*	*	
London			*	*		
Merionethshire				*	*	
Middlesex				*	*	
Monmouthshire				*	*	
Montgomeryshire				*	*	
Norfolk	*			*	*	
Northamptonshire				*	*	
Northumberland	*	*		*	*	*
Nottinghamshire	*			*	*	
Oxfordshire	*			*	*	
Pembrokeshire				*	*	
Radnorshire				*	*	
Rutland				*	*	

Table 4.7 (*cont.*)

County	1:500	1:528	1:1056	1:1250	1:2500	1:10560
Shropshire	*			*	*	
Somerset	*			*	*	
Staffordshire	*			*	*	
Suffolk	*			*	*	
Surrey				*	*	
Sussex				*	*	*
Warwickshire	*			*	*	
Westmorland				*	*	*
Wiltshire	*			*	*	
Worcestershire	*			*	*	
Yorkshire (ER)	*			*	*	
Yorkshire (NR)	*			*	*	*
Yorkshire (WR)	*			*	*	*

Source: PRO, *How to Find and Use Valuation Office Maps and Field Books* (1992), Appendix 2.

1870s. The Valuation Office used the old series maps for the entire county of Northumberland, for example. In some cases the 1:2500 sheets were complemented by 1:1250 or 1:500 scale maps for the same area, especially in urban locations. The 1:1250 series was introduced in 1911 as a specific reaction to requests from local authorities and the Land Valuation department, although many are simply photographic enlargements of the existing 1:2500 plans, and contain no new information.[28]

A great variety of large-scale mapping was thus employed by the Valuation Offices (Table 4.6). The coverage to be found within different counties at different scales, as deposited in the PRO, is given in Table 4.7.[29]

The Ordnance Survey (through the concurrence of the Board of Agriculture) cooperated with the valuation procedures throughout. They agreed to make large-scale revisions where necessary for the valuation, and these special editions may also feature in deposits of these plans. The state of revision of the 1:2500 plans as at 31 March 1912 is shown in Fig. 4.4. The revisions concentrate around the areas which had seen the most rapid urban, suburban and industrial development over the thirty to forty years prior to 1912. Thus the revisions are for areas such as suburban London, South Wales, South Yorkshire, Birmingham and the East Midlands, and the Northeastern coalfield district.

[28] Report from the Select Committee on Land Values, BPP 1920 (Cmd. 556), XIX, 12, evidence of C. J. Howell Thomas, Deputy Chief Valuer; Public Record Office, *How to Find and Use Valuation Office Maps and Field Books* (1992), 2, 10.

[29] Details of holdings of these maps in the PRO are amplified in Chapter 5.

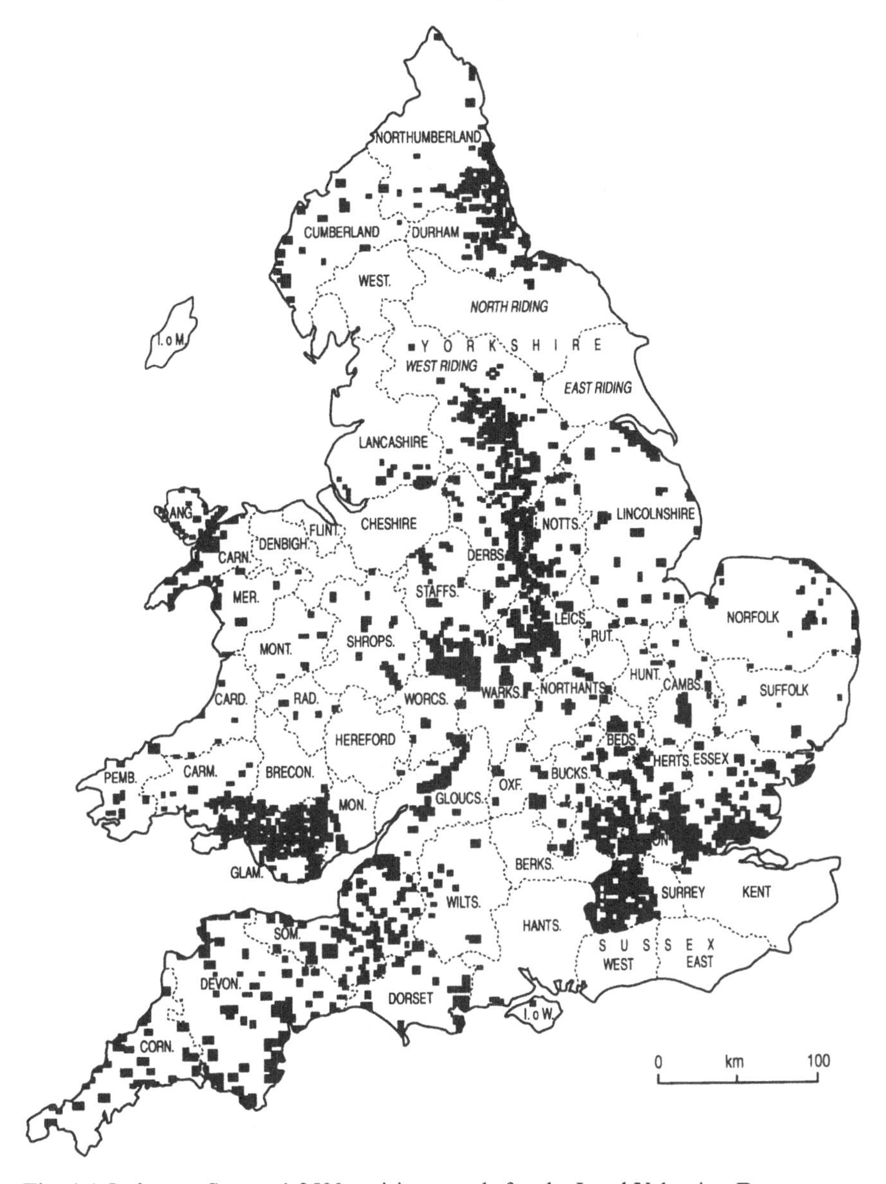

Fig. 4.4 Ordnance Survey 1:2500 revisions made for the Land Valuation Department as at 31 March 1912.
Source: Ordnance Survey Annual Report, 1911–12; BPP 1912–13 (Cd. 63723), XLII, 641.

There were also revisions for substantial rural areas in Devon, Cornwall and Somerset, Essex, the coastal strip north of Grimsby, the Cambridge area, Caernarfonshire and Anglesey. The number of sheets involved is extensive, involving nearly all the OS revisions, and amounting probably to about 1,000 sheets by 1912. Much of Kent, Sussex and Hampshire had been revised shortly before this and so these counties do not feature strongly on the map of revisions. In addition the Valuation Department requested 8,000 enlargements to 1:1250, that is 2,000 ordinary 1:2500 sheets or 3,000 sq. miles, at a cost of £80,000 – partly defrayed through sales to the general public.[30]

One set of the maps was used as a working copy, the other, the Record Sheet Plans, as a permanent record. The latter were mounted on linen and the boundaries of each hereditament were marked out and its identification number was entered once the valuation figure was finalised. A complete set of these plans was to be completed as soon as possible, and they were to be filed with an index and were not to leave the office.

The working sheets were precisely that, and the information they contain is uneven, but they still have value in that whilst they probably contain little more information than on the Record Sheet Plans themselves, and are frequently incomplete, they were often used by valuers in the field and may contain pencilled comments relating to particular hereditaments (Fig. 4.5). A separate set of plans was kept for minerals as appropriate, and each individual unit of valuation was to be recorded by a blue border, to differentiate such plans from the others.[31]

The Record Sheet Plans are documents of crucial importance. They should show the boundaries of the ITP as existing 'within the Metropolis on the 6th April 1909, and elsewhere on the 6th April 1910' in yellow, and the boundaries of each unit of valuation in pink or otherwise in green.[32] In practice however, unit boundaries are frequently shown in other colours, especially red, as different Valuation Offices adopted slightly different local practices, and often the entire unit was shaded in with a colour wash. In any event, each unit of valuation is demarcated and its number entered onto the plan in red (Figs. 4.6 and 4.7). Detached portions should be braced together, and the various parts of the hereditament given a suffix to the main number, e.g. Hereditament No. 263/1; 263/2 etc.

The original instructions to valuers also included instructions (instruction no.151) on how to record subsequent subdivisions of hereditaments on the plans. Further pink, green or blue boundaries were to be drawn in, with the necessary numbers added. The fact of the subdivision was also to be recorded in the Valuation Book and Forms. However amended instruction no. 504 replaced the original with information that subdivisions were not to be shown

[30] I am most grateful to Richard Oliver for this information, and see also the Ordnance Survey Annual Report 1911–12, BPP 1912–13, XLII, 641.

[31] *Instructions to Valuers* I, 29.

[32] *Ibid.*, 28.

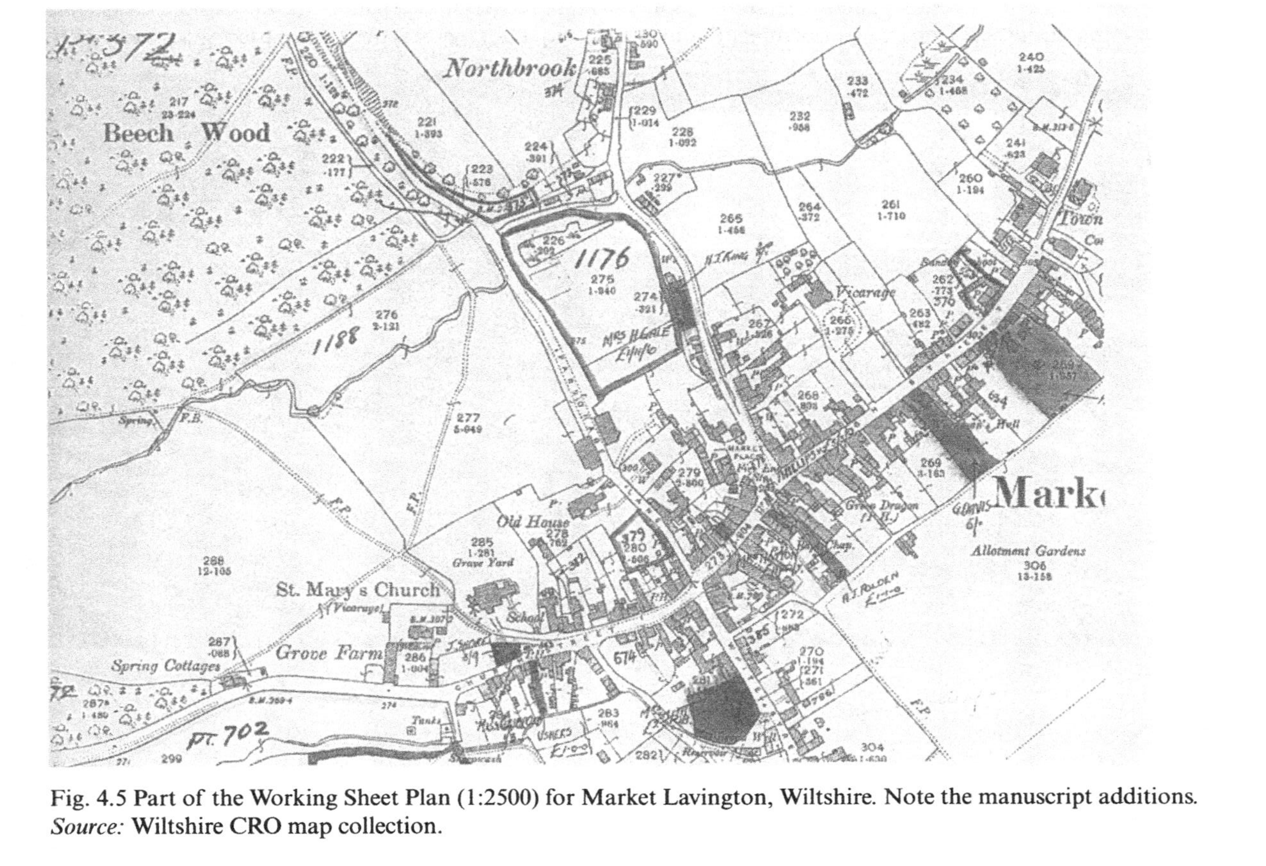

Fig. 4.5 Part of the Working Sheet Plan (1:2500) for Market Lavington, Wiltshire. Note the manuscript additions.
Source: Wiltshire CRO map collection.

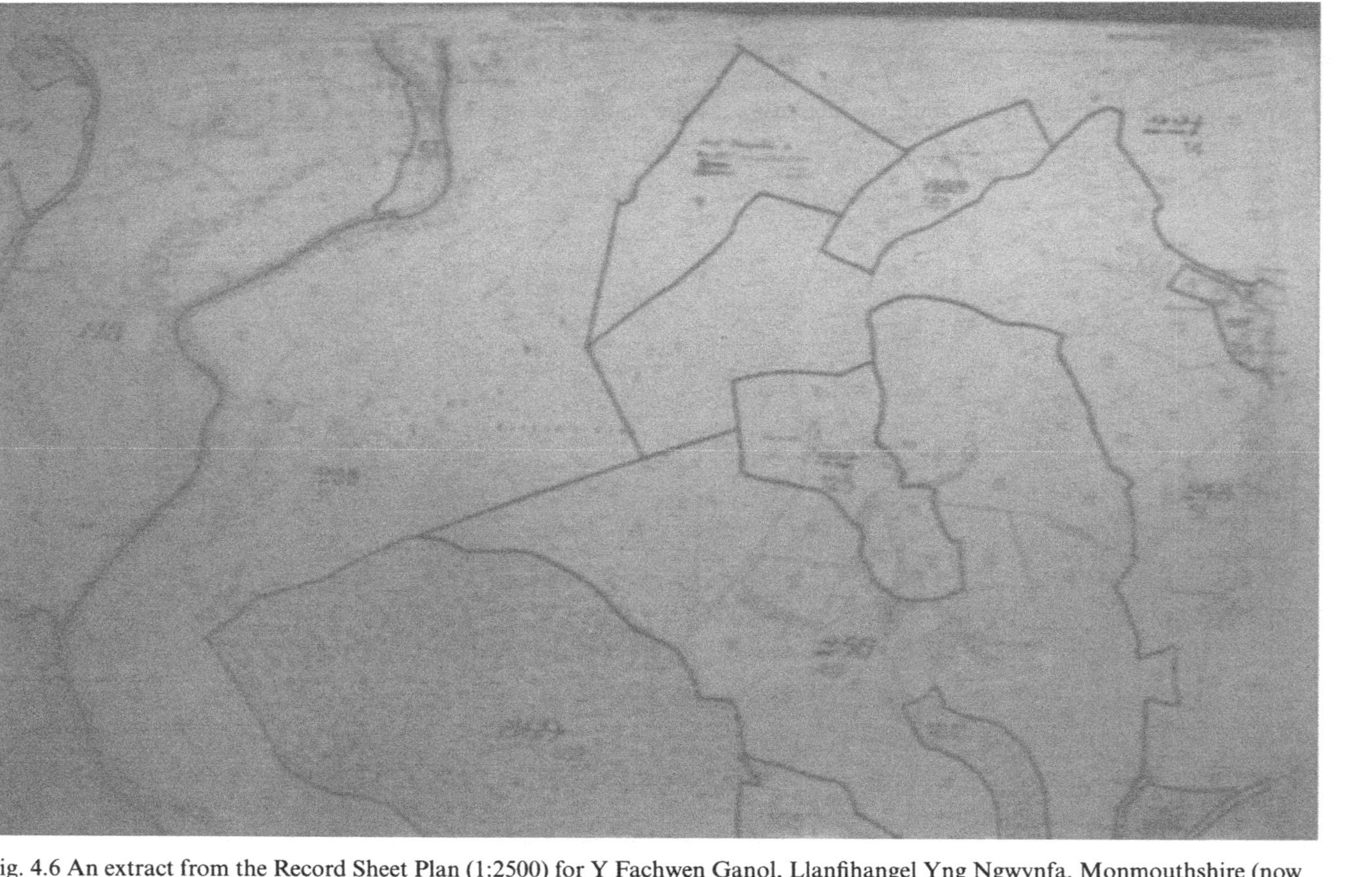

Fig. 4.6 An extract from the Record Sheet Plan (1:2500) for Y Fachwen Ganol, Llanfihangel Yng Ngwynfa, Monmouthshire (now Powys). Common sheep pastures are annotated. Coloured outlining denotes hereditament boundaries.
Source: District Valuation Office, Welshpool (prior to removal to PRO).

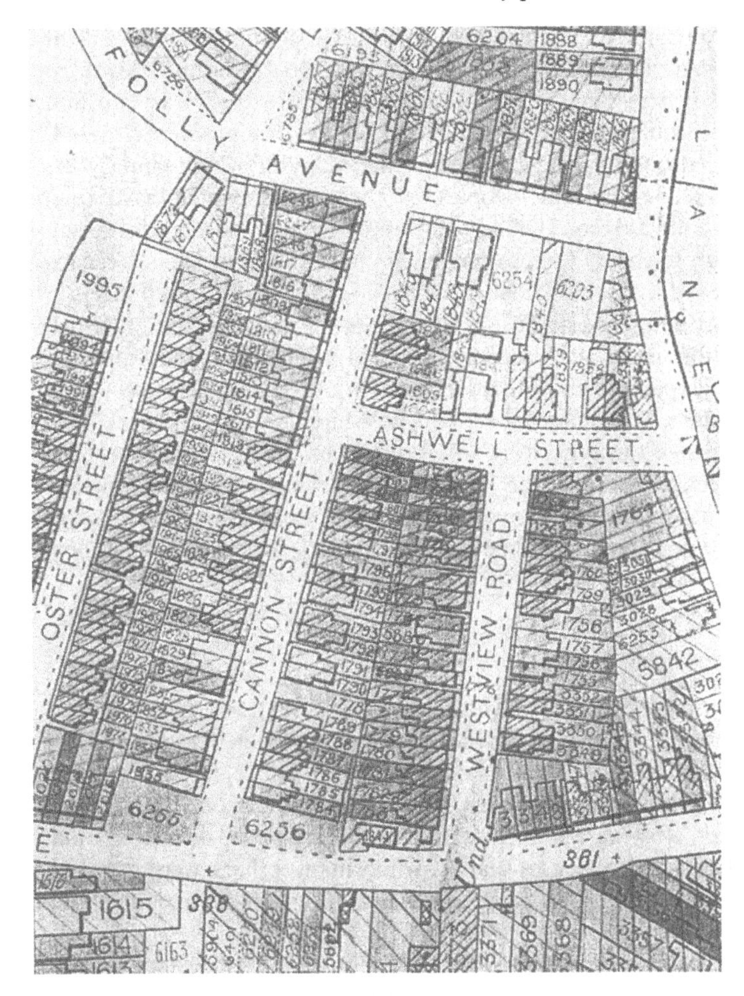

Fig. 4.7 The Record Sheet Plan (1:1250) for an area of St Albans, Hertfordshire. Colour washes in green, pink, blue, mauve, yellow, and buff denote individual hereditaments.
Source: PRO IR 126/8/257.

on the Record Plans. Instead a revised plan was to be prepared showing the changes, and this was to be filed with the papers and was to show both the original layout and the amended one. Thus the Record Plans were to 'therefore show for all time the boundaries and identification numbers of the several units of valuation, as originally valued'.[33]

[33] *Ibid.*, I, 29–30. The change is noted in a handwritten addendum numbered 4803/1910LV. This copy is from the library of the Chief Valuer.

Other duplicate copies of plans were also used for submission to landowners, especially where large estates were involved, to enable boundaries to be drawn on by the owners themselves, or their agents. Therefore at least two, and often three (where minerals were involved) or more sets of plans were required.

There are variations in practice to be found, many of which amplify the information available. In many copies of these plans which have been deposited at the Public Record Office, some updating in manuscript form is to be found through to the 1970s, presumably up until the time when the maps were being removed from the Valuation Offices. And on some 1:2500 sheets in the PRO there are details of land use inserted onto the maps, something which was not called for under the generally accepted practices of the Valuation Office. But arable, pasture and other land uses are shown on sheets covering parts of Lancashire, Yorkshire, Surrey and Wiltshire.[34]

The significance of these maps is clear. At no other time do we have reliable, contemporary maps covering the entire country which delineate property boundaries in both town and country. And given the wealth of supplementary information outlined above, it will be clear that this is a resource of the utmost importance.

Miscellaneous documents

Two main categories of forms were created by the Valuation Office to collect and record information required for the purposes of the act. These were what became known as 'Land' forms and 'V.O.' [Valuation Office] forms.[35] At least 183 forms were promulgated by the Office during the period that the land clauses operated! Most were simply internal office forms such as memoranda, stores requisitions and expenses claims, while many others were standard letters of one sort or another. Appendix 2 lists the most significant of the 'Land' forms.[36]

A great variety of material is also likely to be found in various archives, and this is noted in the following chapter. The interest of the material will depend upon the project being undertaken but apart from the forms already mentioned,

[34] Information from Mr William Foot, Public Record Office. Large numbers of such updated maps are to be found at the PRO, and a separate class is being contemplated for them (as at April 1994). For the use of the 25 inch plans in Valuation Offices in 1921, following the repeal of the Finance Act, see PRO LAR 1/133. OS sheets containing land use information include Lancashire CIII.2 (IR 133/7/49); XCV.11 (IR 133/7/40); XCV.14 (IR 133/7/43); Yorkshire CLXIX.4 (IR 134/4/1002); CLXX.5 (IR 134/4/1017); CCXXVII.13 (IR 134/5/548); CCXXVII.5 (IR 134/5/541); CCXXVI.4 (IR 134/5/526); CCXXVI.8 (IR 134/5/530); Surrey X.9 (IR 125/3/4); Wiltshire XXXV.7 (IR 125/11/402).

[35] A specimen set of 'Land' and VO forms can be consulted at PRO IR 9/62–4, Various Specimen Forms.

[36] An annotated list of those forms in common use by 1910 is given in C. J. Lake, 'Finance (1909–10) Act, 1910: a list of the forms in common use and an abstract of exemptions', Surveyors' Institution, September 1910, 275–83.

correspondence between landowners and valuation officials, between landowners and oppositional organisations and so on, are likely to be found, as well as copies of the various forms which were retained in local solicitors' collections or in various estate archives. Much of this will clearly serve to contextualise the procedures outlined above in terms of their local political impact.

The values

The ultimate aim of the Chancellor, at least in the published version of the Act without the underlying political connotations of future taxation or even land nationalisation trumpeted by his opponents, was to arrive at values for all property subject to the various taxes on land values put forward in the Budgets of 1909 and 1910. The entire programme of land reform envisaged by Lloyd George, however notional, hinged around the values. But they were surrounded by great controversy from the start: their derivations and meanings were criticised from the very beginning, and opinions from all sides of the landed professions were offered. When, in the Autumn of 1908 Lloyd George had begun work on his first Budget, he had envisaged a complete valuation of all land showing Total Values (including improvements) and Site Values (minus improvements).[37] In the event, the values became far more complex, and none of the values to be ascertained in the 1909 Budget was strictly a true exchange, or market value, as generally understood in land-related professions, and indeed as used in Section 60 of the same Act when dealing with Estate Duty. The latter uses the simple definition of 'the market price at the time of the death of the deceased' to define the Principal Value of Landed Estate for Estate Duty,[38] in marked contrast to the lengthy definitions embodied in the formal descriptions, as given in Section 25 of the Finance (1909–10) Act 1910, which are given below.

GROSS VALUE

means the amount which the fee simple of the land, if sold at the time in the open market by a willing seller in its then condition, free from incumbrances, and from any burden, charge, or restriction (other than rates or taxes) might be expected to realise.

[i.e. The market value of the land in its existing condition, all burdens and restrictions being ignored (other than rates and taxes). The phrase 'by a willing seller' puzzled many.][39]

[37] Bruce K. Murray, 'Lloyd George and the land: the issue of Site-Value Rating', in J. A. Benyon *et al.* (eds.), *Studies in Local History: Essays in Honour of Professor Winifred Maxwell* (Cape Town 1976), 42.

[38] Sir Edgar Harper, 'Lloyd George's Land Taxes' (A paper delivered at the International Conference on Land-Value Taxation in Edinburgh 1929, and printed in *Land and Liberty* – the 'monthly Journal for Land-Value Taxation, Free Trade and Personal Freedom').

[39] See, for example, C. Newton-Robinson, 'The blight of the land taxes: why they must be repealed', *The Nineteenth Century and After* 72 (1912), 104.

FULL SITE VALUE

means the amount which remains after deducting from the gross value of the land the difference (if any) between that value and the value which the fee simple of the land, if sold at the time in the open market by a willing seller, might be expected to realise if the land were divested of any buildings and of any other structures (including fixed or attached machinery) on, in, or under the surface, which are appurtenant to or used in connection with any such buildings, and of all growing timber, fruit trees, fruit bushes, and other things growing thereon.

[i.e. The Gross Value less the difference between the Gross Value and the value of the cleared site. Its primary purpose was to obtain a difference between two other values, Gross Value and Full Site Value, to be used to obtain the Assessable Site Value.]

TOTAL VALUE

means the gross value after deducting the amount by which the gross value would be diminished if the land were sold subject to any fixed charges and to any public rights of way or any public rights of user, and to any right of common and to any easements affecting the land, and to any covenant or agreement restricting the use of the land entered into or made before the thirtieth day of April nineteen hundred and nine, and to any covenant or agreement restricting the use of the land entered into or made on or after that date, if, in the opinion of the Commissioners, the restraint imposed by the covenant or agreement so entered into or made on or after that date was when imposed desirable in the interests of the public, or in view of the character and surroundings of the neighbourhood, and the opinion of the Commissioners shall in this case be subject to an appeal to the referee, whose decision shall be final.

[i.e. Gross Value less the depreciation due to any burdens or restrictions which permanently diminish the value of the land. This is probably the nearest value to true market value.]

ASSESSABLE SITE VALUE

means the total value after deducting–

(a) The same amount as is deducted for the purpose of arriving at full site value from gross value; and

(b) Any part of the total value which is proved to the Commissioners to be directly attributable to works executed, or expenditure of a capital nature (including any expenses of advertisement) incurred bona fide by or on behalf of or solely in the interests of any person interested in the land for the purpose of improving the value of the land as building land, or for the purpose of any business, trade, or industry other than agriculture; and

(c) Any part of the total value which is proved to the Commissioners to be directly attributable to the appropriation of any land or to the gift of any land by any person interested in the land for the purpose of streets, roads, paths, squares, gardens, or other open spaces for the use of the public; and

(d) Any part of the total value which is proved to the Commissioners to be directly attributable to the expenditure of money on the redemption of any land tax, or any fixed charge, or on the enfranchisement of copyhold land or customary freeholds, or on effecting the release of any covenant or agreement restricting the use of land which may be taken into account in ascertaining the total value of the land, or to goodwill or any other matter which is personal to the owner, occupier, or other person interested for the time being in the land; and

(e) Any sums which, in the opinion of the Commissioners, it would be necessary to expend in order to divest the land of buildings, timber, trees, or other things of which it is to be taken to be divested for the purpose of arriving at the full site value from the gross value of the land and of which it would be necessary to divest the land for the purpose of realising the full site value.

[i.e. Total Value less the difference between Gross Value and Full Site Value; less any enhanced value due to expenditure on development etc. on the part of persons with an interest in the land; and less the expense of clearing the site where necessary for the purpose of realising the Full Site Value.]

Thus Gross Value and Full Site Value refer to the value of the site covered and site cleared respectively without reference to burdens or restrictions. Total Value corresponds approximately to market value, and Assessable Site Value represents the price the cleared site would fetch if the permanent burdens remained. To put the equation as simply as possible:

Assessable Site Value=Total Value minus the difference between Gross Value and the original Site Value added to those deductions allowed.

In cases where properties were burdened by heavy rent charges, as in the North and West of England and in Scotland where ground annuals and feu duties were to be paid, the assessable Site Value could be nil, or even a minus quantity. Increment Value Duty would still be assessable from this negative quantity, but the concept of 'Minus Site Values' certainly helped to fuel the criticisms against the whole valuation process. Ridicule was not slow in coming from *Punch*. A six-verse doggerel poem supposedly from the owner of a small and worthless plot 'or hereditament (I love that word)':

And more, I yearn, I really yearn, to see
With how much justice valuers hold the scales;
What worth, in their opinion, there may be
In these few yards of dirt and shattered rails, A holding which entails
Upon its owner (as I've said, it's mine)
An average annual loss of 3s.9d.[40]

[40] *Punch* 25 October 1911.

The legality of a Minus Site Value was raised in an Appeal case in the Court of Session in Edinburgh on 9 March 1912 where it was argued that in such cases the value should be nil, rather than a negative. In the particular case the Gross Value was taken to be £4,828, the difference between Gross and Full Site Value £4,320, the Total Value £3,775 and the amount to be deducted for Feu Duty £1,053, giving a Site Value of minus £545. The judges unanimously upheld the view that a site could not have a negative value, and that in this case the Assessable Site Value should be taken as nil. The significance of this was, of course, to lessen any Incremental Value Duty due on any future 'occasion'.[41] A government appeal to the House of Lords was then announced.

These, then, are the four values to be obtained from the survey. In the course of the passage of the original Finance Bill through Parliament in 1909 considerable alterations were imposed after the committee stage. On 19 October the Attorney-General, Sir William Robson, proposed alterations to the wording of Sections 2 and 14 (later 25) which were carried. The effect was to include 'Gross Value' for the first time.

In the case of Statutory Companies only the cost of the land was to be ascertained, as a substitute for the original Site Value of the land. On the occasion when Increment Value Duty was payable the Original Site Value and the Site Value at the time of the 'occasion' (thus 'Occasion Site Value') were compared and the difference would then be liable to tax, after the first 10 per cent of increment. That the Original Total Value and Original Site Value related to one moment in time was also seized upon by opponents who railed at the use of the Budget of 1909 as an instrument of social policy, rather than as a purely accounting exercise as budgets had been under his Gladstonian predecessors.[42] Furthermore, the device of taxing future increments in values necessarily would be expected to influence present land prices, since those prices would normally be influenced by expectations as to future profits to be made. Lloyd George was probably not aware of this possibility, at least in his reply to one critic of this aspect of his scheme, James Hope, who raised just this point on the third day of the Committee stage of the 1909 Finance Bill. Many predicted dire slumps in land markets as a result of the assessments. As one economist wrote in 1912: 'Land is bought with the knowledge that its future value is problematic. The purchaser gives a price for the probability. If you announce that all the chances that turn out well will be taxed, is it not exactly the same thing as knocking something directly off the present value of all the chances?'[43]

[41] Land Union, *The New Land Taxes and Mineral Rights Duty: The Land Union's Handbook on Provisional Valuations* (London nd [*c.*1913]), 41, 158.

[42] H. V. Emy, 'The Land Campaign: Lloyd George as a social reformer, 1909–14', in A. J. P. Taylor (ed.), *Lloyd George: Twelve Essays* (London 1971), 66–7.

[43] C. F. Bickerdike, 'The principle of land value taxation', *The Economic Journal* 22 (March 1912), 3.

There were many who initially wished, as did one writer in the *Agricultural Gazette* on 16 May 1910, that the assessment of the Gross Value should be a secret between the Commissioners and the owner, to prevent would-be compulsory land purchasers in the future from knowing the values of land.[44] An undertaking to that effect was subsequently given. There were also changes in subsequent legislation, such as that of March 1910 which allowed owners to substitute the site value which their land had on sale during their lifetime rather than in the twenty years prior to April 1909 as had previously been the case.

One question was repeatedly asked. Did the valuations represent real facts? The Land Union certainly thought not, regarding them as extremely technical and 'of a more or less hypothetical nature' in their very informative 1911 handbook on the valuations.[45] In order to extract a direct contribution from the landowners whose properties were enhanced in value 'by the action of the community at large'[46] a distinction was drawn for the first time between the Gross Value and the Site Value of land – that is between the value of the site covered and the site cleared. The Increment Value Duty was to be imposed on Site Value, but was Site Value to include trade interest, trade profits, owners' improvements etc. or was it the true unimproved market value of the land divested of the owner's improvements? In his Budget speech of 29 April 1909 Lloyd George had declared in describing the Increment Value Duty that:

> We begin, therefore, with a valuation of all land at the price which it may be expected to realise at the present time, and we propose to charge the duty only upon the additional value which the land may acquire. The valuations upon the difference between which the tax will be chargeable will be valuations of the land itself, apart from buildings and other improvements.[47]

However, as the first valuations began to arrive, many felt uneasy that the letter of the law being followed by the Commissioners did not match up with the pronouncements that buildings and improvements would not be taken into account when arriving at the provisional valuations. In a conference with representatives of the building and allied trades Lloyd George said: 'The builders were under a misapprehension that their profits were going to be taxed as increment. Nothing which is due to their expenditure, brains, or capital is to be taxed at all, and if anyone tells me that the Act does not carry that out, I am perfectly willing to put in words that will carry that out.'[48] Certainly in a memo of July 1910 from C. F. Brickdale, the Registrar, to the Lord Chancellor, which

44 *Land Values* (July 1910), 6.

45 Land Union, *The New Land Taxes*, Chapter VIII, 'The various "values" under the Act', 35–47.

46 54th Report of the Commissioners of HM Inland Revenue: Year ended 31 March 1911, BPP XXIX (Cd. 5833), 1911, 149.

47 Cited in R. B. Yardley, *Land Value Taxation and Rating: A Critical Survey of the Aims and Proposals with a History of the Movement* (printed for the Land Union by W. H. and L. Collingridge, London 1930), 260.

48 *Ibid.*, 262.

was duly forwarded to the Prime Minister, the definition of Increment Value was given as: 'The IV of any land shall be deemed to be the amount (if any) by which the Site Value of the land on the occasion on which IVD is to be collected . . . exceeds the original Site Value of the land . . .'[49] No mention here, then, of market value on the occasion when IVD [Increment Value Duty] was to be collected. And again on 7 July 1909 we find the Attorney-General in a Commons Committee stating that: 'We adhere to the plain principle of our tax, which is that we are taxing a site, and all we have to do is to ascertain the value of that site.'[50] Nevertheless, a series of cases was brought forward by the Land Union to demonstrate that alleged increment values on small building plots with houses were being calculated merely on the basis of the difference between the original Total Value as on 30 April 1909 and the purchase money on subsequent sale. Duty was being collected in cases where there was no evidence of a rise in the value of the site *per se*. And when questioned about the validity of such valuations on builders' legitimate profits on bricks and mortar in Parliament, ministers upheld the Inland Revenue. Thus Masterman, Secretary of the Treasury, asserted that the 'windfall which is above the market value of the buildings has always been interpreted as a fitting subject for Increment Value Duty'.[51]

The actual procedures for ascertaining the Site Values on the necessary 'occasions' were set out in a White Paper of 21 January 1911, published in July. The Somerset House officials instructed that Increment Value Duty should be collected in all cases where either: (a) there has been an increase in the value of the site as compared with the original site value; or (b) the unit of valuation (or an interest therein) has actually been sold for more than it is worth at the time.[52]

In the words of the Land Union: 'This is one of the most remarkable and most important documents that have so far been issued in connection with the working of the valuation clauses of the Finance Act. It came as a great shock to those who had studied the Act and it is bound to give rise to much controversy.'[53]

This fallback to market value at the time of the sale meant effectively that duty would be payable whenever any speculative builder sold a house (i.e. bricks and mortar rather than land) at a price higher than the theoretical original Total Value. Such actions resulted in a series of test cases, such as the Palmer's Green Case, the Richmond Case and then the Lumsden Case in the House of Lords which upheld the government decision at the expense of the speculative building profession, thereby inhibiting housebuilding, taxing 'brains, energy and enterprise' and exacerbating housing shortages as profits

[49] PRO LAR 1/24. File no.005/9. [50] *Land Union, The New Land Taxes*, 159.
[51] *H. C. Deb.* 5s. xxxvii, c. 1538, 29 April 1912.
[52] Yardley, *Land Value Taxation*, Appendix VII, 642–3. Appendix VII includes a thorough review of the difficulties of the Commissioners in arriving at the values necessary for derivation of the taxes under the aegis of the Act. [53] *Land Union, The New Land Taxes*, 161–2.

on the development of land were seen to be taxed.[54] Now the tax was seen to be leviable even when the value of the site had not increased, and the tax was seen to be levied from the profits of the builder's trade. And since many felt that the imposition of the new duties had caused a depreciation in land and house property, the Land Union stressed the importance of ascertaining correctly the value of land as at 30 April 1909, and not the values at the time when prices were unduly depressed 'whether by political or investment conditions'.[55] The Revenue Bill of 14 May 1914 was intended to remedy the defects in the original legislation by substituting a valuation on the 'occasion' for the market price where land had been developed for building as well as other considerations, but owing to the war this Bill never proceeded, although from 1914 until the repeal of the legislation in 1920, all such analogous cases to that dealt with in the Lumsden issue were held in abeyance if there was no genuine increase in the value of the site.[56]

Another unconsidered side effect of the Act was decided upon in the Norton Malreward Case, or 'Scrutton Judgement'. The Court decided that in arriving at the Full Site Value, and subsequently at the Assessable Site Value, of agricultural land the valuations had been arrived at using an erroneous principle concerning tenants' unexhausted improvements and with grass as a 'growing thing'. The main question was whether the Gross Value of agricultural land should or should not include the tenants' crops and unexhausted improvements, and whether the Full Site Value should or should not include grass as a 'growing thing'.[57] The effect of the Norton Malreward decision was to oblige the Inland Revenue to ascertain the value of all growing crops, tillages, tenant right on feeding stuffs, and manure as at 30 April 1909 after a space of nearly five years. These were in nearly all cases the property of the tenant under the provisions of the Agricultural Holdings Act 1908, and they were not obliged to divulge such information to the valuers, even if they could reasonably be expected to do so after that lapse of time.

No further progress in taxing agricultural land could be made until this was sorted out and again the Revenue Bill of 1914 (Clause 2) was intended to do so, but as noted above, the War frustrated that intention. From 28 February

[54] PRO T 171/40 Builders' deputation to Lloyd George 17 April 1913; *The Land* (London 1914) II (Urban), 80–96. A synopsis of judgement in the Lumsden case is in *The Times* 21 July 1914 and a cutting is in PRO IR 83/54.

[55] The Addison Departmental Committee on the High Cost of Building Working Class Dwellings (1921, Cmd. 1447), 5, cited in Yardley, *Land Value Taxation*, 265. E. A. Rawlence, '"A perfectly impartial assessment": some reflections on the land valuations', *The Nineteenth Century* 75 (March 1914), 613. *Land Union, The New Land Taxes*, 36. PRO IR 63/32A, N21.6 notes the consequences of the Lumsden case 1911–13.

[56] Select Committee 1920. Evidence of C. J. Howell-Thomas, 14; the subsequent Plymouth case would have allowed some duties to be collected, but the then Chancellor (McKenna) refrained from proceeding (*ibid.*, 17).

[57] The Commissioners of Inland Revenue v. Smyth (L.R. 1914, 3 K.B., 406). The decision affected Section 25 of the Finance (1909–10) Act 1910. See also above, Chapter 2.

1914 provisional valuations of agricultural land were suspended, although the surveys were still carried out.

Many critics had felt that 'the Inland Revenue Authorities, backed by the law officers of the Crown, have been straining every nerve to retain as many improvements as possible in the Assessable Site Value.'[58]

This, of course, was quite contrary to the original intentions of the Chancellor. Allegations of under-valuation of the original values, in order to enhance any subsequent tax payment, were made from time to time both in the House of Commons and in the Press. These were vigorously denied, but nonetheless instructions (nos. 658 and 713 of 21 November 1911 and 15 March 1912) governing the absolute fairness of valuations were issued by the Commissioners. More fairly perhaps, many experts agreed with Aggs who quoted the truism that 'All valuation is valueless' and went on: 'Owing to the many difficulties which surround the making of a valuation of land . . . it is probable that the valuation made under this Act will be nothing more than a rough approximation and one that in many cases will be very far from being either accurate or correct.'[59]

But the Achilles' heel of the poorly drafted legislation had been the move beyond market values and into theoretical values in order to arrive at the pure rent, moves which had complicated and lengthened the task of the valuers by facing them with complex mathematical problems, invoked the wrath of landlordism and its media, and the disapprobation of the legal establishment. At each step, therefore, owners could be presented with valuations of a novel character, whose amount was frequently impossible to ascertain in advance, and whose complexity made it impossible to understand without paying for professional help. The valuations were made by valuers who themselves were unfamiliar with the methodology, and which bore very indirect relationship to facts and which could be contested at appeal in very sympathetic courts. And as time passed, as hereditaments were divided or amalgamated, the problems of apportioning liability to tax became more severe. Even the Liberal press printed critical letters from supporters of the government. To the extent that taxes need the consent of the populace at large, something demonstrated more recently with the Thatcher government 'Community Charge' (Poll Tax) of the late 1980s, these did not gain that necessary consent.[60]

Summary

The valuation was an enormous undertaking, producing a huge, if temporary, army of bureaucrats who worked to an improbably short timetable in order to

[58] Rawlence, '"A perfectly impartial assessment"', 610.

[59] W. H. Aggs, *The Finance (1909–10) Act 1910* (London 1910), 8.

[60] Offer, *Property and Politics*, 363–4; for an early attack on the complexity of the Act, see Land Union, *The Land Union's Reasons for Repeal of the New Land Taxes and Land Valuation* (Land Union, London 1910), *passim*.

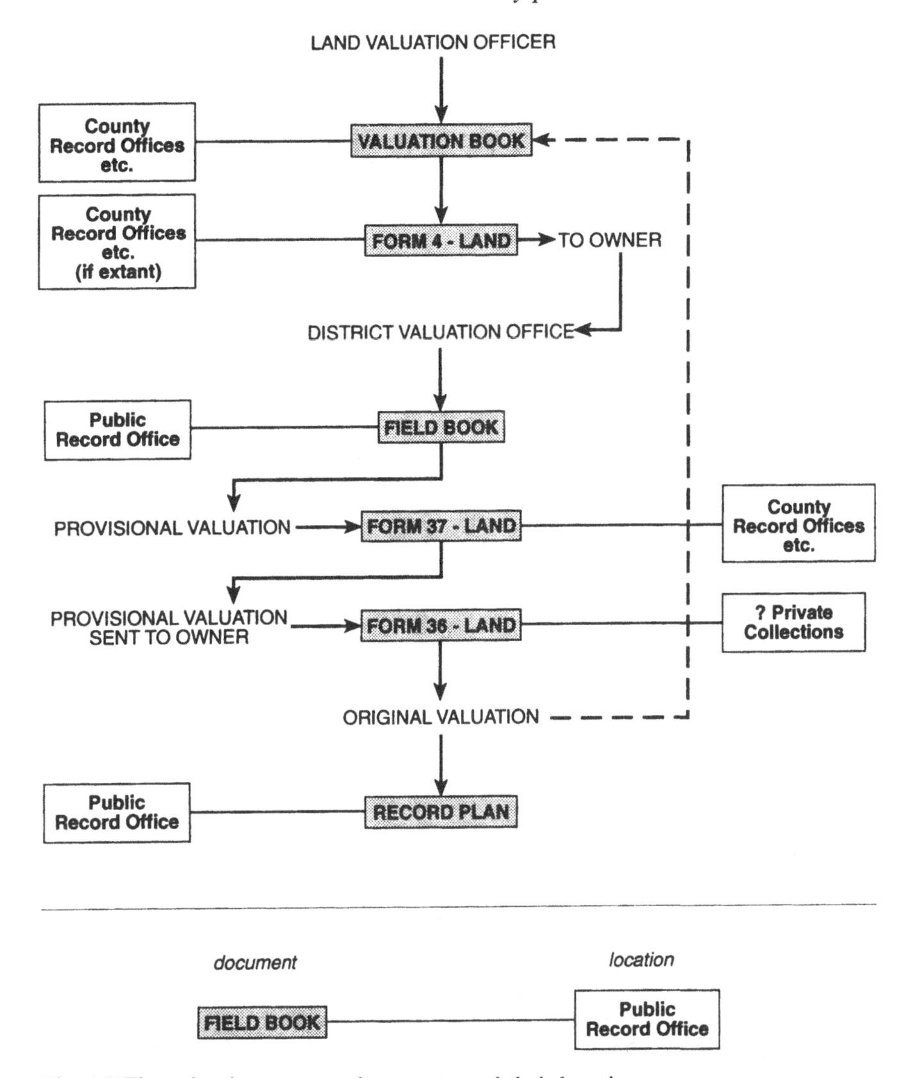

Fig. 4.8 The valuation process, documents and their locations.

attempt to satisfy opponents of the scheme. The process also engendered a large and disparate group of documents, the linkages between them being outlined above. In order to see these linkages more clearly, and to relate them to the overall process, Fig. 4.8 outlines the main stages in the valuation, as well as showing the location of the major documents, a theme to be taken up in the following chapter.

5

The 1910 documents and archival policies

The transfer of the documents to the public domain

Miscellaneous documents in private collections have of course been in the public domain from the time an owner chose to deposit his or her papers in an archive repository. This section will therefore concentrate upon those official documents relating to England and Wales retained by the Inland Revenue until fairly recently.[1]

As early as August 1918 an Inland Revenue destruction schedule provided for the destruction, after a specified period, of routine Increment Value Duty forms and of the surveyors' copies of assessments in respect of worked minerals. In September 1942 a schedule provided for the further destruction of a large amount of similarly routine documentation, including registers of cases presented for the Increment Value Duty stamp, registers of objections to assessments, and registers of provisional valuations made on the occasion of a landowner's death. The decisions to destroy and retain have therefore been made on more than one occasion, and will probably have been implemented with greater or lesser efficiency between offices.[2]

All of the surviving documents being discussed here were originally retained in the Valuation Offices, but were released under Section 3.6 of the Public Records Act 1958. The first to be released were the Working Sheet maps of the Ordnance Survey. These were offered to local repositories in 1968.[3] They were accepted by many as indicated in Appendix 3, although survival of the sheets seems extremely uneven.

The release of these and other documents was supervised by the Public Record Office. The PRO had received the Field Books and the Valuation

[1] Much of this section is based upon PRO circular GEN 33/2/2 dated 19 September 1979, from which the quotations are taken. I am grateful to Mr Alfred Knightbridge, formerly of the PRO, for providing a copy of this document. For the archival situation of the documents relating to Scotland and Ireland see Chapters 11 and 12.

[2] PRO, *Valuation Office: Records Created under the Finance (1909–10) Act* (information leaflet no. 68, 1988). [3] PRO Circular 3/CCA/6 1968.

Books some years previously, and these were housed in their Hayes repository until 1979. The PRO selected the Field Books for permanent preservation and made these available as class IR 58. The Valuation Books and Form 37 were not selected for permanent preservation, and were offered to local record offices as an outright gift 'as an alternative to destruction' under Sec. 3.6. All but two repositories accepted the Valuation Books. The exceptions were the Cities of London and Westminster, and the books for these areas (the latter covering only Paddington) were therefore retained at the PRO in class IR 91.

Forms 37 were still retained in the DVOs and local repositories were invited to collect them if desired. Any which were not collected before 31 December 1979 were to be destroyed by the DVOs, though this destruction was not always undertaken. The PRO gave background information on the material to local archivists in their circular of 19 September 1979. The important passages for our purposes are quoted herewith.

> The Forms 37 Land contain information extracted from the Domesday Books, though in a different format.
>
> The Domesday Books and Forms 37 Land . . . contain no information additional to that in the Field Books, and in some cases less.
>
> Since it may seem an unnecessary duplication to offer both the Domesday Books and the Forms, I should explain that neither series has survived in its entirety and that consequently some record offices may wish to take both records so that items in one series can fill gaps in the other. Some offices may wish to do so even where duplication does occur, perhaps to keep one series at headquarters and the other at branch or district repositories serving the areas covered by individual valuation districts, of which there are normally several within a county. Moreover, in some cases the arrangement of the Forms 37–Land is by street, and this may be more convenient than that in the Domesday Books, particularly for repositories which do not hold the working sheet maps for their areas.[4]

On the basis of this many archivists did not accept the forms or retained only those for which there were gaps in the series of Valuation Books, destroying the remainder. The cautionary words in the last part of the final paragraph above were often insufficient to raise enthusiasm to keep large quantities of the forms. Some retained a sample of forms but others however did accept and retain all that were offered.[5]

It can be seen, then, that the information from the PRO to archivists was explicit and provided clear-cut information upon which to base a decision about acceptance of these documents. The problem for later users of the documents is that some of the information provided was incorrect.

Had a substantial number of archivists, alongside those for the Cities of

[4] PRO Circular GEN 33/2/2 1979. The 2.5 pages of description in this circular subsequently provided the basis for the introduction to the finding aids within many repositories.

[5] The Northamptonshire RO Finding Aid therefore states of Form 37 that 'They are only of importance where the Domesday Book is missing.'

London and Westminster, also refused to accept the Valuation Books, then the possibility remains that the PRO would have enacted its stated alternative – that of destruction. In this event, the Field Books would have lost an enormous amount of their potential usefulness. The destruction or non-acceptance of Forms 37 by many archivists has also created significant problems. Forms of Return did not feature in the offer to local repositories, but DVOs made these available to local archivists in some instances, when Form 37 was collected.

Forms 37 and 39 were retained in the DVOs where some may still remain. They were made available to local repositories in 1979, but not all archivists took up the offer and they were only sporadically transferred. The whereabouts of Forms 47 and 50 and indeed all material relating to minerals is at present unknown.

The Ordnance Survey Record Sheets were retained in DVOs until the mid-1980s, when negotiations between the PRO and the Inland Revenue for their eventual deposition in the PRO were successful. Prior to this, they were available for limited public consultation in the DVOs. The circular authorising this was issued by the Inland Revenue in April 1978, but many local offices seemed unaware of this, and were reluctant to admit the public. It is therefore fortunate that large numbers of these maps have now found their way into the PRO, although the sheer weight of numbers of these maps has defeated attempts to sort and classify them all by 1995.

In short, it can be seen that the transfer of documents to the public domain has been piecemeal and based on a less than adequate understanding of the ways in which the valuation was undertaken. This has resulted in the destruction of much valuable material, and in the splitting up between repositories of much that remains. It is to be hoped that past errors can be partly rectified and that archivists are made aware of the relationships between surviving documents. The ultimate aim, of course, would be that of eventually reuniting the different categories of document, for their more effective use by the public. In recent years the PRO has adopted the very enlightened policy of allowing Field Books to be brought to the Map Room for study alongside the large-scale plans, and this is certainly a step in the right direction.

Problems caused by archival policy

The splitting of the documents connected with the survey has undoubtedly hindered the full appreciation of the worth of the material, and also presented those undertaking full investigations with difficult, arduous and expensive research itineraries. Resolution has been made far more difficult by the policies of the archive service since the documents entered the public domain, and one gains the impression that disastrous choices were only avoided by good fortune. This is not to deny, however, that many archivists have responded extremely quickly and efficiently to the acceptance and classification of the records, particularly given the shrinking resources available to most. In many cases the finding aids are splendid.

One difficulty of the splitting of the records between repositories has, unfortunately, been that there is little consistency in the way that finding aids are presented. For example, the three Cumbrian offices employ different classification systems and styles of listing, although the standardisation of the three series was a hoped-for outcome by Cumbrian archivists resulting from the stimulus to action provided by the 1985–6 survey conducted by the present author. It should be noted, however, that the Kendal office finding aid to the Valuation Books is excellent, giving names of the LVOs and the dates of completion of the volumes (as early as 10 August in the case of Strickland Ketel and Strickland Roger by W. P. Benson of Laithwaite, Burneside, through to 17 December for Newby, Sleagill and other parishes overseen by H. Wilkinson from Hackthorpe). Similarly the survey prompted the Derbyshire RO to list properly the Working Sheet plans which had previously been consigned to an artificial series, and the Oxfordshire CRO to catalogue its Valuation Books.

One senses, however, that the transfer of 'particular instance papers' such as the 1910 series of documents is not without its controversial elements within the archives service. Local record archivists were in some instances resentful that they were being seen as something of a 'waste bin' for 'superior' records not wanted for retention by the PRO, particularly perhaps those which had 'campaigned' for local deposition of the Field Books. Many of the records, such as Forms 37, took up large areas of shelf space and far more knowledge about their significance was really required before sensible decisions could be made about their relative merit and consequent archival policy. Although a liaison officer at the PRO worked with local repositories on matters connected with public records from 1964, one senses that the liaison might have been rather one-way![6]

The first documents to reach the public domain were the Working Sheets in 1968 (see above and Appendix 3). Uneven in quality though they are, they remain the only copies of the Ordnance Survey plans used in the valuation yet deposited in local repositories. Since they are also in many cases revisions of the 1:2500 sheet *c.*1912, they also have an interest in their own right even as 'blanks' without any colouring or addition from the Valuation Office. Large numbers of such maps are therefore to be found, for example, in the Map Department of the British Library.[7] It is clear that the context in which the

[6] Discussion following paper presented by the author to the Southern Region Society of Archivists' Regional Meeting December 1988. The use of the word 'superior' to describe the Field Books retained by the PRO is from the Wiltshire CRO Finding Aid.

[7] I am grateful for information provided by Mr G. Armitage, Map Department, British Library, April 1994.

sheets were compiled, together with their significance, was unknown to most archivists in 1968. Many were merged with non-archival series of maps, and often no cross-linkage was made with the Valuation Books when these reached local repositories. On occasion the sheets went to one repository, while the books later went to another! Thus, for example, in West Sussex, Worthing Reference Library holds Working Sheets for much of West Sussex while the relevant Valuation Books were later deposited in the West Sussex Record Office, unknown to the Worthing librarian! In Hampshire the County Record Office accepted plans from the Bournemouth VO, covering only the area of Bournemouth and the New Forest, while the Southampton City Record Office received those from its own district. In Wales the National Library at Aberystwyth was acting in 1968 as the local record office for the counties of Cardiganshire, Montgomeryshire and Radnorshire, and accepted the Working Sheets from the Welshpool Valuation Office. By the 1990s it still held them [8]

Much more important was the way in which the main transfer of material took place in 1979. As noted above, the Valuation Books were offered to local repositories by the PRO as an alternative to destruction, on the understanding that they contained less information than the Field Books. In the event, the Valuation Books are absolutely essential for understanding the way the Field Books were compiled. Had destruction actually been carried out, then the task of using the Field Books would have been rendered extraordinarily difficult. As it stands today, the separation of the Field Books from the Valuation Books still requires visits to both the PRO and local repositories in order to examine what is fundamentally a single group of documents.

The PRO was even less concerned to emphasise the usefulness of Forms 37 and many repositories failed to collect them, while others destroyed all or most of those that they did collect. Apart from the fact that these forms were often the only document that contained an accurate figure for area and are therefore important in this respect, they are the only document that can be related to the Record Sheets in an unambiguous way. The forms give the provisional valuation of the unit of valuation which, as noted above, following the Revenue Act 1911, might be part of a hereditament or a combination of several, or many, small hereditaments amounting to no more than 100 acres. The unit of valuation – not the hereditament – was indicated on the Record Sheets and the destruction of Form 37 has made the use of the plans more difficult.

The low priority allocated to all documents other than the Field Books by the PRO has often been carried over into the local repositories. The Valuation Books frequently languished for months (and often continue so to languish)

[8] HRO 152M82. See E. M. Brooks, 'Inland Revenue Valuation Records', *Hampshire Field Club Newsletter* (3), 1985 7–8; personal communication, R. Davies, Department of Pictures and Maps, The National Library of Wales, April 1990.

in out-repositories and are often unlisted. Some archivists in urban areas quite reasonably consider them as duplicate rate books and therefore have not listed them. Some archivists in the mid-1980s were unaware of their existence until prompted into searches by the present author's enquiries. In the London Borough of Enfield the Valuation Books arrived in dirty and dilapidated condition and funds were not available for rebinding them. Although it was intended that a name index to owners and occupiers would be compiled for family history research, this was given low priority since there was already a good collection of street directories and electoral registers.[9]

This low-priority status and the problems caused by fragmentation ensure that the material remains unknown to many researchers thereby increasing the sense of insignificance. Only when demand is sufficiently high will the books be prioritised, and until researchers' consciousness of the relevance and worth of the material is raised, there will be no demand. The Valuation Books for the London Borough of Southwark remained in storage for six years without being used. Oxfordshire CRO required a week's notice since the store where the records were kept was only visited once a week.

The previous policy of retention of the Record Plans in the DVOs undoubtedly made public access difficult, although the Inland Revenue did concede the principle of such access. However, DVOs seldom had the space and facilities for worthwhile and prolonged study of the sheets and some offices seem to have had little sense of their significance, although others clearly did and took great care of them. Their full transfer to archive repositories was absolutely necessary both for their preservation, and for the full realisation of the potential of the remaining material. The PRO has liaised with the Inland Revenue on this matter, and the last six or seven years have been spent in transferring all the sheets to the PRO. Nearly all the sheets have now been transferred, and it is to be hoped that such a course of action will facilitate research. The task of listing up to 100,000 sheets has begun speedily, although any research involving particular localities could conceivably be delayed since not all sheets have been transferred and listed. Some 10,000 sheets still await listing by the PRO in 1995. Whilst this is a considerable advance on the previous position, little can now be done to use the sheets in conjunction with Forms 37 (where surviving) or, in their absence, with the Valuation Books, since both have been transferred to local repositories. Many of the maps are now in poor repair and require conservation treatment as their unprotected faces are deteriorating daily.[10]

[9] Personal communication from P. N. Turner, Borough Librarian and Cultural Officer, London Borough of Enfield.

[10] I am grateful for information from Mr William Foot and Miss Geraldine Beech, Map Department, PRO, and Mr Alfred Knightbridge, formerly of the PRO. The maps are classified in the PRO by the area of the regional valuation offices (as at 1988), and form classes IR 121 and IR 124–35, each of which is then sub-numbered by alphabetical order of the DVOs within that region. This yields 135 record classes of maps. The regions used are as follows: Central incl. Beds., Oxford, St Albans (IR 126); E.Anglia, incl. Cambridge, Colchester, Peterborough

Conservation of records was not always a priority however. To cope with changing local government boundaries has always been a problem in archival work, and local offices were faced with myriad difficulties over the valuation material, made all the more complex by the changing boundaries of the Valuation Districts themselves over time. This has not infrequently meant that decisions have had to be made about the location of coverage of transferred areas. Gaps are inevitable as a consequence, but one extreme case has been found of two neighbouring archives offices deciding to cut out the pages in the relevant Valuation Books relating to transferred parishes and exchange them so as to fit with later administrative changes. If this is a common archival practice, it is a deplorable one! Better, surely, to have to travel to find one's document than to have the document emasculated and pages torn out of their contextual volume. The West Yorkshire finding aid, for example, sets out very clearly that the Valuation Books in their care came from eight DVOs (Leeds, Bradford, Pontefract etc.), that reorganisation has entailed some of the material being sent to Cumbria, Humberside, Lancashire, North Yorkshire and South Yorkshire, and that other material for areas now within West Yorkshire has been received into the office, covering four parishes.[11]

Survival and coverage

There are also miscellaneous difficulties, mainly arising from the lack of survival of documents. Owing to the enormous mass of material, the precise spatial extent of survival of material in each category is difficult to establish. The Field Books, selected for retention at the PRO as IR 58, are now well cared for.[12] On the other hand, some local repositories, such as the Berkshire CRO, have planned to microfilm the Valuation Books since the originals were in such poor condition upon their transfer from Hayes.

The general extent of holdings of Valuation Books, Forms 37, Working Sheets and other miscellaneous documents at the county level is indicated in Appendix 3, based on a survey conducted by the present author in 1985–6. Valuation Books are generally well represented around the country (in seventy-six record offices, 87 per cent of the total), although in cases where a

(IR 127); E.Midlands, incl. Boston, Derby, Grimsby, Nottingham (IR 130); Liverpool, incl. Chester, Salop, Stafford, Wigan (IR 132); London (IR 121); Manchester, incl. Bolton, E.Lancs, Preston (IR 133); Northern, incl. Carlisle, Cleveland, Durham, South Lakeland (IR 135); South East, incl. Brighton, Canterbury, Reigate (IR 124); Wales (IR 131); Wessex, incl. Bournemouth, Guildford, NE Hants, IOW, N.Wilts (IR 125); Western, incl. Bath, Bristol, Cornwall, Somerset, Exeter (IR 128); W.Midlands, incl. Birmingham, Hereford and Worcs, Warwick (IR 129); and Yorkshire (IR 134).

[11] West Yorkshire Archive Service C243/1–546.

[12] The IR 58 class is arranged according to the Valuation Office administrative areas as they existed in the 1950s.

historic county has more than one record office, one repository might not hold any Valuation Books. Thus there are none at Cumbria CRO (Barrow), at Essex CRO (Southend), Surrey RO (Guildford) or West Yorkshire CRO (Leeds). There is a virtually full coverage by county in the Welsh Record Offices and in the London Boroughs.[13]

Of those repositories replying to the survey, only forty-one out of eighty-seven (47 per cent) had holdings of Form 37. Counties such as Hampshire and Hertfordshire appear to have a very good coverage, but others such as Durham and Essex decided not to accept these forms, while others such as Dorset, Suffolk or Devon accepted them but subsequently destroyed them. In the absence of any collection by the appropriate repository, some DVOs certainly also destroyed them, as in the case of Loughborough District in Leicestershire, for example. However Leicester itself can be studied through a street-by-street collection of Forms 37, to underline the extreme spatial variability of the source. At Oxfordshire CRO the Forms arrived and were catalogued before the Valuation Books arrived, and the 209 bundles illustrate the difficulty of accessioning such bulky material. Other CROs, such as Derbyshire, East and West Sussex, and Gloucestershire, retained only those Forms 37 to cover parishes for which there were no Valuation Books, destroying the rest. In East Sussex, for example, this entailed the retention of Forms 37 for only Rodmell and Southease parishes. In Gloucestershire only those forms covering the Cheltenham area were preserved to cover the lack of Valuation Books. In Lancashire all Forms 37 and Forms 4 were destroyed with the exception of those for the Blackpool Valuation area for which no Valuation Books survived. Of the London borough libraries, only the Vestry House Museum at Waltham Forest claimed to hold Forms 37, and twenty-five Working Sheet maps as well. Bedfordshire has only a small sample of such forms, and Northumberland likewise has kept samples for such locations as Holy Island or Newbiggin by the Sea, or Berwick, which were then sent to the Berwick branch of the County Record Office. The PRO has a few examples of Form 36, the owners' copies of Form 37.[14]

Of the total responding archives services, forty-five (52 per cent) possessed some coverage of the Working Sheet maps. This coverage may be very limited,

[13] The survey was conducted by the author and Mr Mick Reed, then project Research Fellow on a one-year research project based on the 1910 material. A copy of the questionnaire sent to all archivists in England and Wales is reproduced as Appendix 4. Unless otherwise specified the data contained in the following paragraphs are obtained from the survey or from subsequent correspondence with the repository in question. In some instances, the sending of the questionnaire prompted action at the repository: thus the ten volumes constituting the holdings of Valuation Books for the London Borough of Bexley were listed by Mr Barr-Hamilton, Local Studies Officer, in 1985–6 as a result of receiving the questionnaire.

[14] Information from the questionnaire survey 1985–6; the PRO forms are at PRO RAIL 1057/1714. The Leicester Forms 37 are in LRO DE 2064/1a–1096, plus one box of assorted forms. The Forms 37 in Northumberland are at NRO.1952.

Table 5.1 *Numbers of Working Sheet Maps offered by Valuation Offices to local repositories in 1968*
(Only offerings in excess of 500 sheets are listed)

Valuation Office	No. of sheets	Local repository offered
St Austel	637	Cornwall RO
Carlisle	968	Cumberland Westmorland and Carlisle RO
Dorchester	798	Dorset County Library
Hereford	621	Hereford RO
Maidstone	575	Kent AO
Lancaster	767	Lancashire RO
Boston	741	Lincolnshire AO
Lincoln	692	Lincolnshire AO
King's Lynn	1548	Norfolk RO
Newcastle	770	Northumberland RO
Northumberland North	669	Northumberland RO
Oxford 2	630	Oxfordshire RO
Shrewsbury	914	Salop RO
Bury St Edmunds	592	Bury and W.Suffolk RO
Salisbury	526	Wiltshire RO
Northallerton	606	Yorkshire (NR) RO
Scarborough	1522	Yorkshire (NR) RO
Carmarthen	1044	Carmarthen RO
Wrexham	560	Flintshire RO
Welshpool	1161	National Library of Wales*

Notes:
* In 1968 the National Library of Wales was also acting as the local record office for the counties of Cardiganshire, Radnorshire and Montgomeryshire.
Source: Mr A. Knightbridge (unpublished PRO memorandum).

however, and at least one county (which should remain anonymous) disposed of its Working Sheet maps to a local bookseller! Very large numbers of sheets were offered to local repositories in 1968, totalling 32,930 across the entire country, and with some repositories being offered large numbers (Table 5.1). Numbers varied in fact from the 1,548 offered by the King's Lynn VO down to the single sheet offered by the Redbridge VO to the Greater London Record Office. Some counties, such as Dorset, did not accept any of the 798 maps offered from the Dorchester Valuation Office, where they went instead to Dorset County Library in Dorchester. Others, including Buckinghamshire, incorporated them into general map collections as non-archival series and they often remain unlisted and presumably unnoticed as being anything very different from Ordnance Survey maps. The London boroughs did not appear to have been offered the maps at all, although those from the Croydon VO were transferred to Croydon Library.[15]

At the very least the maps should now be taken out of general map collections and listed separately, as had been done by 1986 in West Sussex and as was then intended by Buckinghamshire RO. Some record offices possess very large numbers of the Working Sheets: the Cambridgeshire RO holds 886 1:2500 sheets, 110 1:1250 sheets and ten 1:500 sheets all with some property boundaries over some part of the sheet, and with the 1,006 sheets covering virtually all of the county, apart from the Newmarket area.

Hertfordshire CRO has a particularly good collection of Working Sheet maps (as well as a complete coverage of Valuation Books except for Cheshunt, and very good coverage of Forms 37) numbering over 600, although there are none for Potters Bar, Barnet or Cheshunt. Some of the Hertfordshire maps show copyholds only, such as those belonging to the manors of Therfield or Anstey. Others have owners' names and coded valuations shown; arable and pasture are often distinguished, valuations of timber and even numbers and species of trees. Other information relating to valuation can also occur in the Hertfordshire collection, such as 'Very good view overlooking miles of country', or notes of 'built after' or 'planted before' the valuation base date of 30 April 1909. Others have been updated – up to 1911 in some cases but much later in others.[16]

Fewer archivists knew of any miscellaneous material connected with the survey but twenty-three (26 per cent) claimed to possess some material of this kind. This is most commonly a copy of Form 4–Land. In the Hampshire RO the DVO has deposited large numbers of Forms 4–Land covering about sixty parishes; Bath City Archives holds Forms 4 (headed 'unofficial') for its substantial holdings of about 600 properties; the East Sussex RO has unofficial copies of Forms 4 for properties belonging to East Sussex County Council, for the Battle Abbey estate, the Glynde estate and elsewhere in the county. Form 4 papers for the estate of the Whitbread family of Southill Park, and the Pyms of Hasells Hall, Sandy (also in Huntingdonshire), are in the Bedfordshire CRO. Similar forms for Jennings Bros., brewers of Cockermouth, are in the Cumbria RO, together with indexed books of returns for estates in the area north of Penrith.[17]

[15] Surrey Record Office 2415 Finance (1909–10) Act 1910: 'Domesday Books' Public Records presented under the Public Records Act 1958, s.3(6), 2.

[16] Hertfordshire CRO IR 1/1–520. The excellent finding aid lists the maps by Inland Revenue reference and also by Ordnance Survey reference. The copyholds of the manor of Therfield are shown on IR 1/20b and 53b; the manor of Anstey is IR 1/55a.

[17] Bath City Council Corporation Finance Committee files, ref. F (77); Forms 4 for East Sussex County Council properties are located at ESRO C/C42/1; Battle Abbey at BAT 2326; Hastings area at SAY 2897; Glynde at Acc. 4262. The Whitbread and Pym papers are respectively at Bedfordshire CRO W3500–7 and PM 2937/11; Cumbria Record Office DB 98/49–50; DB 66/11–15.

Similar papers are held by many Oxford and Cambridge colleges in their capacity as landowners. Solicitors' collections can contain returns, such as that of Metcalfe of Wisbech, Cambridgeshire; Hart Jackson for property in Ulverston and Egton in south-west Cumbria; or Verral, Bowles and Stevens of Worthing, West Sussex. Some government departments also kept copies of Form 4 for reference, such as the Admiralty, Forestry Commission or a railway company such as the Rhymney Railway Company.[18]

Material relating to the difficulties of dealing with the taxes by certain estates is also to be found. The Shere estate of the Bray family in Surrey affords one example where Reginald Bray was forced to re-evaluate the running of the estate in the light of the new burdens thrust upon it by the taxation, and family papers bear testimony to this. Also in Surrey, there are memoranda and letters by the 4th Earl of Onslow dealing with the effects of the legislation. Individuals might also become enmeshed in the process, such as the vicar of Crantock in Cornwall, whose appeal against the Valuation in 1912 is preserved in the Cornwall RO.[19]

Elsewhere copies of Form 4 have turned up unexpectedly: Hove Reference Library possesses them for various East Sussex pubs in its Sussex Industrial Archaeology Society drawer in the Local History Room, filed under 'Breweries'! Northumberland RO has two volumes of Forms 4 for the Matfen and Westwater estates and a file of papers including some Forms 7 and 36.[20]

Some extraneous material which incidentally sheds light on the valuation process can come to light. Thus in the West Yorkshire Finding Aid there are references to forms inside the Valuation Books relating to the remuneration of

[18] Dr Malcolm Underwood, archivist at St John's College, Cambridge, informed the author that there were six boxes of Forms 4 and 36 in the St John's College library (personal communication, November 1986), which would thus form an excellent basis for an examination of the college's Edwardian landownership role; Cambridgeshire RO R84/62; Cumbria RO (Barrow-in-Furness) Bd/HJ Box 311. Ulverston is also extremely well covered by Forms 37, which are an incomplete series, but listed street by street for the town; CRO (Barrow) BT/V/1, Boxes 1 and 2. For solicitors' papers deposited in 1985 and covering many locations in West Sussex and beyond, see WSRO Add. MSS. 32872–6 (five volumes of Forms 4 relating to various parishes in West Sussex but also including Brighton, Guildford, Croydon, Datchet, Langley, Penarth, Hemel Hempstead and Highgate). The Form 4 holdings of the Admiralty are at PRO ADM 116/1279; the Forestry Commission at PRO F 6/16 and the Rhymney Railway at PRO RAIL 1057/1714.

[19] Guildford Muniment Room 85/42/5. And see P. F. Brandon, 'A twentieth-century squire in his landscape', *Southern History* 2 (1980), 191–220. Other Bray papers dating between 1911 and 1914 not cited in the article contain Forms 37 and relevant letters, memoranda etc. See GMR 85/29/9. The Onslow Collection is in GMR 173/19/43–5 and 98. The vicar of Crantock's appeal against the Valuation is in Cornwall RO AD 772/256–83; and see also Crantock DV(1)/136; DV(2)/21,77.

[20] I am grateful to Mr Geoff Mead for drawing my attention to this source. On a similar subject, see also Gloucestershire RO D2428 1/53 – Register of land values of licensed premises, Gloucester Sub-District. This gives names and valuations for licensed premises by parish. The volume was presented by the PRO to the GRO in December 1984. NRO ZBL.62/11.

the LVO at Buckden and Kettlewell with Starbotton; a letter from the LVO to the DVO about the compilation of the Valuation Book at Dent; and memoranda about the valuation of a reading room and alterations to entries made necessary 'since the valuation was copied from the ratebook' dated November 1910 at Gisburn.[21]

It is also clear that there are random gaps at the sub-county level, with the odd Field Book having been lost, but there are other more substantial losses. For example, there are no Field or Valuation Books for the Chichester part of West Sussex, rendering research on the large Goodwood, Cowdray and Petworth estates difficult, and no Field Books or Record Sheet maps for the Isle of Wight, and south-east Hampshire, including Portsmouth and Southampton. These were housed in Portsmouth DVO during the last war, and were destroyed by bombing. Similar gaps appear for Liverpool and parts of Essex. In March 1943 the Valuation Office, from their wartime headquarters in the Imperial Hotel, Llandudno, requested the help of the officials administering the National Farm Survey to seek help in reconstituting their maps of farm boundaries to remedy the loss, a request which was quickly granted. Hampshire RO does, however, hold Valuation Books covering some of the area, but not Portsmouth. There are undoubtedly parallels elsewhere. In Lincolnshire, although the Forms 37 were kept and are arranged by parish, there are many gaps in the series, and in Lancashire there are no surviving Valuation Books for the Blackpool Valuation area.[22]

The extent of coverage for one group of the documents, the Valuation Books, within the historic counties of England and Wales is conveyed in Fig. 5.1. Full coverage in much of Wales, Norfolk and Suffolk, Dorset, Somerset and Wiltshire is complemented by poor coverage in Gloucestershire and only fair coverage in Hampshire and West Sussex (for reasons discussed above) and Surrey. Some border parishes in counties are sometimes to be found within Valuation Books of adjoining counties, and since the counties as at 1910 are frequently so different from those of the present, there will also be other inconsistencies of coverage arising. Thus historic Derbyshire lost Valuation Books to the Greater Manchester Record Office but held books for parishes transferred to South Yorkshire since 1910.

Coverage within one area, that of Cumbria, is shown for the Valuation Books and Forms 37 in Fig.5.2, as an example of the varying spatial extent of survivals of these records. The complexity of county boundary shifts is seen

[21] West Yorkshire RO N269, C275; L280.

[22] Personal communication from Dr Madeleine Beard. The classes PRO IR 125/6 (Portsmouth) and IR 125/10 (Southampton) exist as references only. The Forms 37 in Lincolnshire RO are classified as LAO 6 Tax. Thus the reference for Binbrook is 6 Tax/42/53. See Charles Rawding and Brian Short, 'Binbrook in 1910: the use of the Finance (1909–10) Act Records', *Lincolnshire History and Archaeology* 28 (1993), 58–65. For the correspondence relating to the making good from National Farm Survey maps in 1943, see PRO MAF 38/865, f.23, 28.

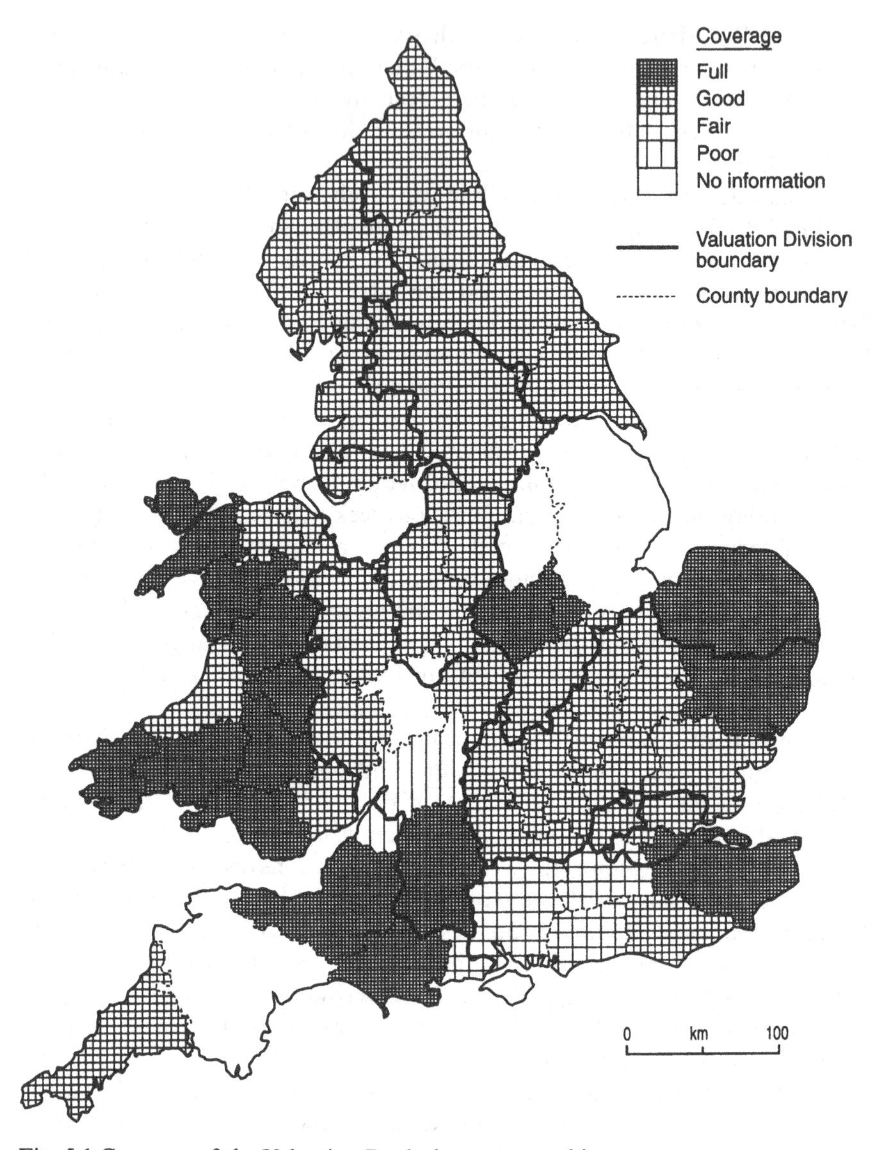

Fig. 5.1 Coverage of the Valuation Books by county archives.

in the fact that the 1910 counties of Cumberland and Westmorland have by the mid-1980s become Cumbria, but the situation is made all the more difficult for these records by the overlaps of the Valuation Districts with county boundaries. Thus four parishes in North Westmorland were in the Lancaster Valuation District. The main deficiencies in the Valuation Books here are in the urban areas such as Whiteside or Workington, with Carlisle and Penrith

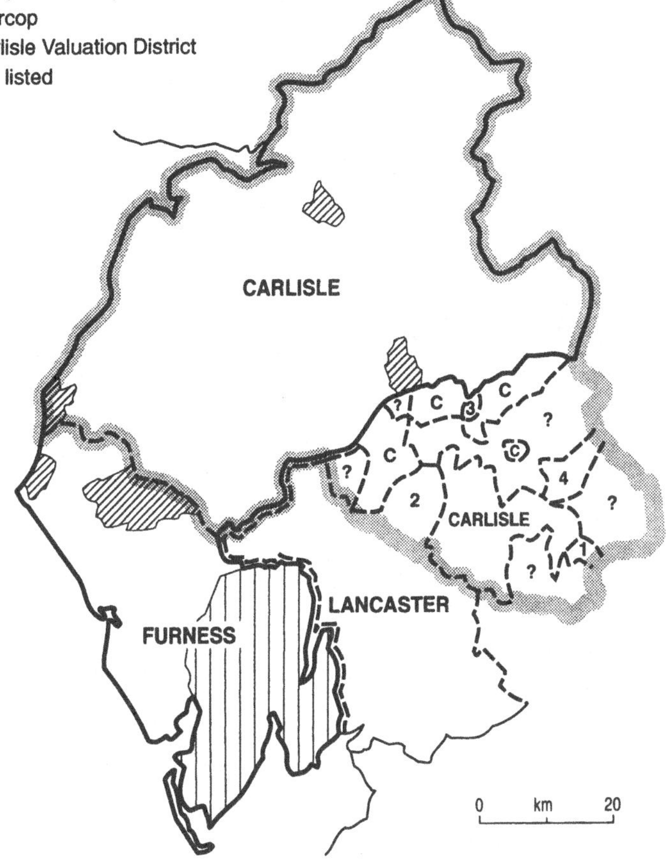

Fig. 5.2 Cumbria: survival of Valuation Books and Forms 37.
Source: Archives Department, Cumbria County Council.

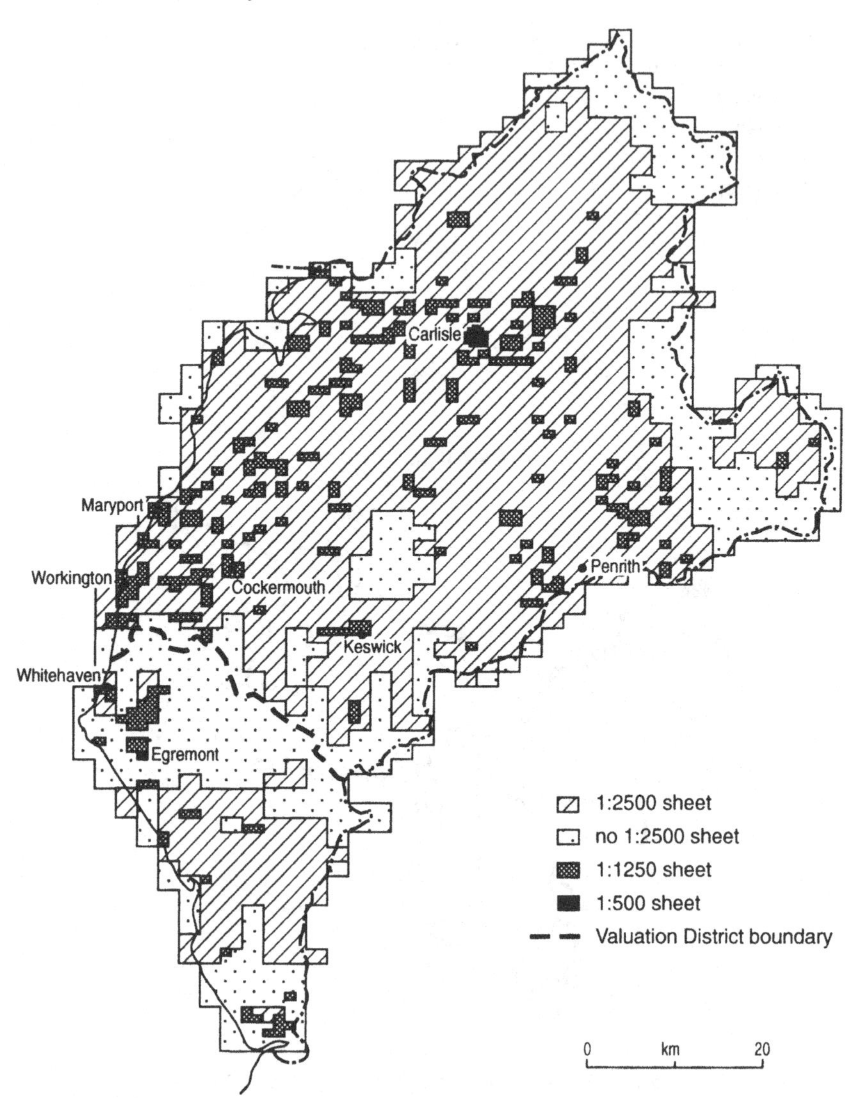

Fig. 5.3 The historic county of Cumberland: survival of Working Sheet Plans.
Source: Archives Department, Cumbria County Council.

also poorly served. About 80 per cent of former Cumberland and 55 per cent of Westmorland are covered by Forms 37 (Fig. 5.2).[23]

Within the historic county of Cumberland (now part of Cumbria), the sur-

[23] I am particularly indebted to the researches of J. G. Higginson, Assistant County Archivist in 1986 for Cumbria, for the information upon which these paragraphs and Figs. 5.2 and 5.3 are based.

vival of Working Sheet plans is shown on Fig. 5.3. The mixture of scales is again evident (see Chapter 4 above) with 1:500 at Carlisle, and 1:1250 sheets scattered throughout the county amongst the surviving 1:2500 sheets. In Shropshire 797 sheets were accepted, out of a possible 1053 sheets (76 per cent) with gaps occurring in rural areas and Bridgnorth. About twenty Valuation Books are missing, including the whole of Chirbury and Purslow Divisions, and most of the Munslow and Overs Divisions, all in the south-west of Shropshire. All Forms 37 were accepted and the ninety-three Income Tax parishes of the county are covered, with 97 per cent of the ecclesiastical parishes covered (all but nine), although the extent of coverage within each parish is variable.

Conclusion

The transfer of the 1910 records, their handling by the archives services, and knowledge of their coverage have been discussed above. To a very great extent, these three issues are interrelated. Advice accompanying the transfer of the documents to the public domain was either completely lacking, as in the case of the Working Sheet maps in 1968, or based on inadequate research as in the case of the Valuation Books and Forms 37 in 1979. As a result, the maps have gone unappreciated in local repositories for many years, for what they really are. Valuation Books can still be consigned to out-repositories with varying degrees of inconvenience for researchers and archivists, and many of the Forms 37 have been destroyed either by the archives services or by the Valuation Offices in lieu of collection by the archivists. The corollary of this is that researchers are often still unaware of the enormous potential of these records, that local studies based on the records are slow to come to fruition, and that as a result of this inaction, we still do not know for certain what gaps in spatial and documentary coverage will be found. It is just one more example of that frustrating moment for researchers when they find, often to their considerable personal cost, that they are confronted by gaps in a run of documents where no gap was previously suspected. Since we are dealing with several classes of document for virtually every hereditament in the country, the amount of material is enormous, but by working at the local level we could at least begin to develop a fair notion of where the major gaps exist.[24]

[24] One attempt to increase broader awareness is contained in Brian Short, *The Lloyd George Finance Act Material* (Short Guides to Records, 2nd Series) (Historical Association and British Association for Local History, London 1994). A recent calculation demonstrates how well such awareness is bearing fruit. In the year 1 March 1995 to 29 February 1996 at the PRO, more Record Sheet Maps were produced for readers (2,787) than the combined totals of all Tithe Maps, National Farm Survey Maps, Colonial Office Maps, Foreign Office Maps, and War Office Maps (2,650). For the period that the PRO is open to the public an average of 1.8 Record Sheet Maps and 1.1 Field Books are produced every hour. (I am indebted to William Foot for this information.)

This is not a happy story. It could have been even worse, for as noted above, the Valuation Books might also have been destroyed in some cases, if more local record offices had refused to accept them, as two London ones so decided. Fortunately, the PRO decided, quite correctly, to retain these particular unwanted Valuation Books. Hopefully the author's survey of local repositories in 1985–86 increased awareness of the value and context of the material. Certainly several archivists were prompted into the compilation of finding aids and it is to be hoped that, ten years on, greater visibility of the records has been achieved in England and Wales.

PART II

Themes and locality studies

6

Projects and problems

It is clear that the information likely to be found in the valuation documents is of great importance and that they present numerous possibilities for the investigation of early twentieth-century geography, society and economy. Some possible uses are included in this chapter, but it should be noted that the suggestions normally assume the full and correct compilation of the documents in accordance with the instructions in force at the time. However, as noted in the previous chapter, this was by no means necessarily the case, and only detailed local studies can now reveal the extent to which local officials complied with the wishes of the Chancellor and Commissioners of Inland Revenue. Some potential or actual problems have already become apparent, and these are also noted below.

Land and farming structures

The fact that the valuers were specifically concerned to identify owners of land for the purposes of assessment for possible taxation ensured the compilation of very full data about the ownership of all hereditaments in the survey. Thus the study of *landownership structures* is among the most obvious uses of the data. The valuation was concerned with the whole of England and Wales, and there is no comparable survey in our history.[1]

Information about owners occurs in the Valuation and Field Books, as well as in the Forms of Return and the Provisional Valuation Forms. The Valuation Books were, of course, primarily copies of the Rate Books, and it might be assumed that the Rate Books will fulfil the same use as the Valuation Books. However, Rate Books in rural areas have frequently not survived, although survival rates in urban areas may be higher. In a study of the landed elite of East Yorkshire, Barbara English has therefore made use of the Valuation Books to yield information on the relative acreages owned by the

[1] See Chapter 1.

ten most wealthy families of the Riding, *c.* 1910, comparing the rankings with data sets which go back to *c.* 1530. Although her study erroneously referred to the Valuation Books as 'rate books' and she does not make it clear that she is dealing with acreages for the Poor Rate which might have been outdated by 1910, compared with more accurate figures available from the Forms 37 or the Field Books, the study is valuable in giving a pointer as to ways in which the material can be used in this comparative way. The acreages *c.* 1910 incidentally show little change from those derived from Bateman's study of 1883, but they do serve to demonstrate that there is a reliable data source for the study of long-term change in landownership between Bateman and the First World War, a point lost on some other recent scholars.[2]

Each hereditament should have its area given and it follows that some precision as to landownership structures ought to be possible. Moreover, the record plans should enable property in the same ownership to be precisely delineated. Since the addresses of owners ought also to be given, it should also be possible to make assessments as to the extent of absentee ownership. The extent of owner-occupation of land and housing can be assessed. Also of great importance is the study of tenurial forms; it is possible, for example, to examine the extent of freehold ownership as well as the extent of lease and copyhold ownership, the existence of Lady Day or Michaelmas tenancies, and any other tenurial forms.[3] Complex mixtures of tenancies and ownerships can be hinted at by examining the contents of the first page of the Laxton, Nottinghamshire, Valuation Book, which shows that although Earl Manvers owns the tenanted farms, the tenant farmer Jonathan Bagshaw is also an owner of small cottages in his own right (Table 6.1).

Similarly, studies can be made of the *occupation of land.* For example, since the boundaries of all farms and other properties can now be precisely located on the large-scale OS sheets, an examination of the dispersal or concentration of individual farming units, and the degree of fragmentation in any one locality, becomes a viable proposition. Studies of farm and field layout at the local, sub-regional and regional level, including comparative work between localities, are thus immeasurably enhanced. The manner in which a rented property was held can also be assessed. Information should be available as to

[2] Barbara English, *The Great Landowners of East Yorkshire 1530–1910* (Hassocks 1990), 32–3 and Table 1.9, and see also note p. 269. The Valuation Books were consulted in Humberside County Record Office (Accession 1291), Hull Record Office, and North Yorkshire County Record Office (which holds the records, in a storehouse in Ripon, for those parishes which were formerly part of East Yorkshire before local government reorganisation in 1974). English notes that some of the Valuation Books relating to Ouse and Derwent wapentake are missing. Campbell's otherwise good study of south-west Scotland would certainly have benefited from using these records in a similar way (see Chapter 11).

[3] See, for example, the use of the records in Olivia Wilson, 'Landownership and land use: continuity and change in the North Pennines' (University of Durham, Department of Geography Graduate Discussion Paper no. 23, 1988).

Table 6.1 *Laxton, Nottinghamshire: the first page of the Valuation Book*

No. of assessment	No. of Poor Rate	Occupier	Owner and residence	Description of property	Size (acres, roods)	
1	1	Morrison and Sons	Earl Manvers, Thoresby	House and building		
				Land	280	0
2	2	Bagshaw, Jno	"	House and building		
				Land	142	1
3	3	Pearson, Hy	"	House and building		
				Land	0	3
4	4	Taylor, Chas	"	House	—	
5	5	Moody, Wm	"	House and building		
				Land	83	3
6	6	Moody, Ar	"	House and garden	0	1
7	7	Butcher	Bagshaw, Jno, Laxton	House and garden	—	
8	8	[blank]				
9	9	Rose, Thos	"	House and garden	—	
10	10	Squires, Thos	"	House and garden	—	

Source: Nottinghamshire RO, Valuation Book for Laxton.

whether a property was held by the week, month, year etc. Rents should be noted, as should details of responsibility for rates, repairs and insurance. Where Forms 4 survive, as in Hampshire for some properties, details of tenure, agreements and rents can be readily noted.

Housing studies

Information on *housing* is a prominent feature of the material. The Field Books in particular present a rich source of detail for both exteriors and interiors of domestic buildings. The numbers and uses of rooms; house rents; sanitation and water supply; repair and general condition etc. are all normally present in the documents. We are told, for example, that Mrs Susan Chalk at Godshill in the parish of Ashley Walk on the edge of the New Forest lived in a cottage which was built of brick, mud and slate, with two living rooms and a pantry on the ground floor and three bedrooms above. There was a wood and corrugated iron stable; piggeries; a wood and thatch wood house, a wood and corrugated lean-to cowpen and similarly constructed cart-shed. But there was no water on the premises and she had to obtain this from the adjacent common. The whole property of about three-quarters of an acre was in a poor state in 1909. The Form 4 had been completed for this property, and from that it can also be ascertained that Mrs Chalk was a tenant for life at £4 per year of Mrs Mary Amstein of Edgement Villa, Worstenholm Road, Sheffield.[4]

In contrast Christopher Servier at Drybrook in the same parish had built between 1909 and 1912 a new three-bedroomed house to replace his old mud and thatch rough-cast cottage, the two buildings now apparently standing adjacently and together with the cart house, piggery, cowpens and stable. Isaac Stainer's old mud cottage had on the other hand burnt down since 1909 and only the retaining walls were left. In such glimpses of living conditions a sense of dynamism can also sometimes be obtained, depending on the timing of the survey in comparison with the benchmark year of 1909.[5]

Comments on interior decoration are also sometimes given, although the relevance of such detail to the valuer's task of defining Site Value (theoretically stripped of buildings and improvements) is surely questionable. Undoubtedly, the value of the documents in this respect will be greatly enhanced when the 1911 Census Enumerators' Books become available, allowing physical structures to be correlated with household information to give a wealth of social and domestic detail unmatched in British historical studies.

Returning to the edge of the New Forest, a further interior description is available for Captain Jeffery Skeffington Smyth's Hale House, set in its own parkland landscape north of Fordingbridge in the parish of Hale (Fig. 6.1). The 10 acre property included the house, a 7 acre park, woods and plantations, fishing and other sporting rights, and enjoyed a 'beautiful view over Avon towards Breamore' (on the west side of the Avon Valley). The rooms are listed in some detail. In the basement were the housekeeper's room, boot and brush room, butler's pantry, servants' hall, coal cellar etc. On the ground floor were the entrance hall/main staircase, smoking room, a lavatory with hot and cold water, coal cupboard, dining room and drawing room. On the first floor were seven bedrooms and two dressing rooms, two nurseries and two bathrooms. On the second floor were another eight bedrooms, a schoolroom, maids' workroom, bathroom and lavatory, box room, linen cupboard, housemaids' cupboard and wardrobe room. A detached left wing had a main entrance hall, two bedrooms and one large room to be divided for bedrooms and bathroom. The right wing had four looseboxes, four stalls and a harness room with four bedrooms over. A brick/slate double garage and similar coach house, toolshed and engine room completed the buildings, all of which were 'in very fair repair' and possessed electric light throughout.[6]

In some cases, comments are included which were certainly not sought by

[4] PRO IR 58/10697 (Field Book); Hampshire CRO 152 M 82/9/1 (Valuation Book for Ashley Walk ITP), and Forms 4 (unlisted). For texts dealing specifically with Edwardian houses and gardens, which would have benefited from consultation of the 1910 material, see R. Gradidge, *Dream Houses: The Edwardian Ideal* (London 1980), H. C. Long, *The Edwardian House: The Middle Class Home in Britain 1880–1914* (Manchester 1993) and D. Ottewill, *The Edwardian Garden* (Yale 1989).

[5] PRO IR 58/10697.

[6] PRO IR 58/10699.

Fig. 6.1 Hale House, Hale, Hampshire.
Source: Hampshire County Record Office Top. Hale 1/4.

the Inland Revenue, but which nevertheless offer invaluable insights into Edwardian perceptions. Mrs Burden, living at 2 Myrtle Cottages, Chessington, Surrey, apparently lived in roomy but uncomfortable surroundings. The Field Book describes her house as:

A very old building of red brick & tiled roof – originally the Priory of a monastery – Its drawback as a residence is its proximity to the farm buildings cowsheds and bull shed being practically next to it. No lighting – all water pumped by hand – drains to cesspool – contains: Hall, Dining Room, Drawing Room, Morning Room, 8 bedrooms, no bathroom, WC, dry cellar, K[itchen], S[cullery], Pantry, Servants sitting room, [?]'s bedroom, Old fashioned walled garden with conservatory. The structure is unsound in places & needs doing up throughout. Area 0–3–16.[7]

In the colliery village of Maerdy, Rhondda Fach, in the parish of Ystradwfodwg, the valuer noted the states of repair of the cottages and shops with great care. Thus we have descriptions such as that for No. 9 Brook Street, Maerdy 'Ground Floor 3 rooms, 1st Floor 3 bedrooms, stone built', while of No. 1 Brook Street we are told that it was a 'Good wkg class ppty'. Shops are also described in detail. We can thus discern that there was at this time a carpenter, chemist, butcher, grocer, sweet shop, boot maker, draper, fruiterer, ironmonger, and a Post Office. Mr L. Walters' single-fronted butcher's shop at 61 Maerdy Road had a two-roomed basement, two rooms and the shop on the ground floor, and four bedrooms on the first floor. J. Rowlands at the Post Office at No. 99 Maerdy Road also had a two-room basement, the shop and a sitting room on the ground floor, and three bedrooms above. D. Lewis had his own chemist's shop at No.1 Cendwen Street – a large shop occupying a corner site and in good repair. At No.4 Wrgant Place, William Edwards from Cardiff had a 99-year lease dating from 1877 on a 'Block of fair working class houses in rubble masonry with brick dressings – double fronted. Not so good as North Terrace houses. Accommodation: ground floor, parlour, kitchen and slope; first floor 2 bedrooms i.e. main building and a small slope lean-to.'[8] Finally, the Maerdy Workmen's Institute, built in 1905 on the site of five demolished houses, had a hall, seating, cinema apparatus, a balcony, library, waiting rooms, an evening class classroom in the basement, a billiard room, and radiator heating with 'no open grates'. A caretaker's house was adjacent.[9]

The Maerdy Workmen's Institute is also illustrative of the fact that the records can also shed light on public or semi-public or official buildings. The Forms 37 for Hertfordshire, for example, include details of rooms in the Hatfield Police Headquarters building, including the Superintendent's house.[10]

[7] PRO IR 58/45215. [8] PRO IR 58/67601 Income Tax Parish of Rhondda.

[9] OS Sheet XVIII NW 1921 Edition shows the buildings facing eastwards onto the recreation ground. The Workmen's club was at the northern end of Richards Street between quarries and tramways. [10] Hertfordshire RO IR 3/33/1–2 Forms 37.

Land use studies

Land use can be studied in urban and rural areas either using the 1910 documents by themselves, or in conjunction with other material. In the urban case, there seem to be few difficulties in this respect, while in the rural areas it is likely always to be possible to determine land use in general terms such as agriculture, woodland, and so on. In addition it is sometimes possible to examine land use on a field-by-field basis, though this cannot be predicted for any one area without consulting the relevant documents. At Ashley Walk, Hampshire, valuers made extensive notes in the Field Books concerning land use, noting use by OS field parcel number. Fields were designated as orchard, pasture, rough land, arable, heath. Occasionally even more detail is available: at Amberwood, Mr Sidney Aston was found to have 'numerous old trees – about 10 bearing fruit trees' in his garden/orchard. In the same county the existing Forms 37–Land frequently have attached schedules giving details of land use etc., the forms themselves surviving for over sixty parishes in the county.[11]

As noted in Chapter 4, land use is also sometimes shown on the valuation maps. Examples of this have come from the 25-inch sheets for the counties of Lancashire, Yorkshire, Surrey and Wiltshire, where arable and pasture are distinguished separately. On the working copy maps for Hertfordshire, arable and pasture, timber valuations and even numbers and species of trees can be found, together with such hints of change as 'built before' or 'planted after' the significant date of 30 April 1909.[12]

Industrial and commercial structures

It is also possible to study the *industrial and commercial structure of towns and cities c.* 1910 as well as shedding light on the spatial organisation of industrial undertakings in urban and rural contexts. The frequently detailed information regarding buildings and equipment enhances our 'archaeological' knowledge of industrial, domestic and other buildings which have since been demolished or heavily altered.

The industrial activities of the town of Horsham *c.* 1910 have been studied by Caudwell, who lists brickmaking, tanning, milling and a timberyard as the main activities at this time. In the 1880s H. and E. Lintott's foundry also expanded greatly and the Field Book entry is one of the more detailed for industrial premises in the town (Fig. 6.2). Within the foundry were two blast

[11] PRO IR 58/10699 (Field Book); Hampshire CRO Forms 37–Land (unlisted). See also E. M. Brooks, 'Inland Revenue Valuation Records', *Hampshire Field Club Archaeological Society Newsletter* 3 (1985), 7–8. Unfortunately this otherwise good summary fails to appreciate the significance of the surviving Forms 37–Land, and no mention is made of the field-by-field land use schedules attached to many of the forms.

[12] Hertfordshire RO IR 1/1 Working Sheet Maps.

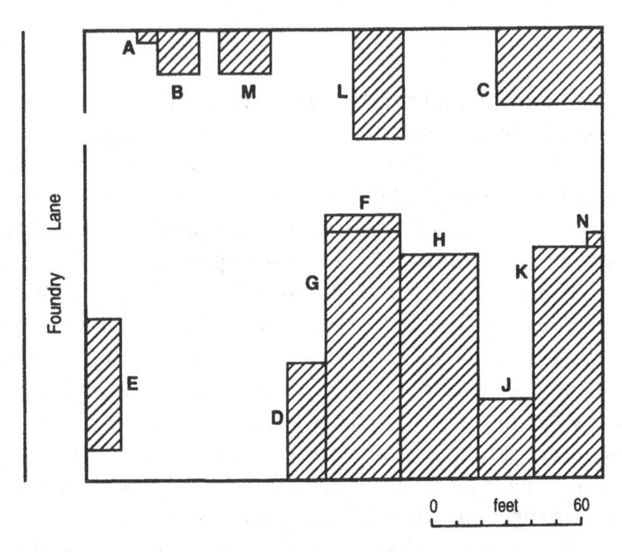

A Shed, in good condition
B Office, in good condition
C Workshop, in fair condition
D Shed, in good condition
E No description in Field Book
F Lean-to, in good condition
G Workshop, in good condition
H Workshop, in good condition
J Store, in fair condition
K Workshop, in good condition
L Shed, in fair condition
M Stables, in fair condition
N Chimney, 85 feet high

Fig. 6.2 Lintott's engineering works, Foundry Lane, Horsham, West Sussex 1910.
Source: Adapted from assessor's Field Book sketch. PRO IR 58/93991–94078.

furnaces and an overhead crane which gave it the capacity to forge castings up to 30 cwt. The products included gratings, manhole covers and railings for council contracts. There was also an elevating pug mill and other brick-making equipment. The mill, powered by steam or horse, was capable of producing 15,000 bricks per day. A wide variety of engineering products was available for sale to the many country houses in the surrounding area, and included heating systems, water pumps and electric light installations which were powered by private generator.

Information is also available on the public utilities. The gas works, opened in 1836, had by 1910 three gasometers which were 60 ft, 40 ft and 20 ft high respectively, three retort houses, a water gas plant, an operator's room and a show room. All were said to be in fair to good condition. The electricity generating system built in 1901 only covered 10 poles but the Field Book carefully detailed its contents: the brick and slate buildings included an engine room, boiler house, battery room, office, workshop and stores for coals and lamps. Within them were three batteries of accumulators which produced 120, 160 and 80 kilowatts. The fuel was burnt in one double- and two single-drawer boilers. Immediately to the south stood the mid-nineteenth-century waterworks, and again the size and dimensions are noted. Within the engine house the two pumps produced 12 and 16 hp to raise 400 cu yd of water per minute from the two boreholes, which were 240 and 570 ft deep.

The largest of the Horsham industrial sites were those belonging to the brickworks, with the 16 acres of Spencers brickworks being the largest. Most were under a yearly tenancy, emphasising the temporary nature of the industry, although the owner-occupied Spencers works had relatively high Rateable and Total Values. William Lilley's works in Littlehaven Lane were seen as unkempt and unprofitable. The 8.75 acre site was described as uneven, and the buildings, apart from two timber and tile sheds, were poor. Nevertheless the works was producing about 200,000 bricks a year, as the assessor noted that a royalty of 1s 6d was being paid per 1,000 bricks, and that this brought in about £15 a year. Finally, the Crossways brickworks, south of the Crawley Road, covered 8.5 acres and produced about 30,000 bricks per annum. The tenant, Mr J. Dinnage, had to place his own buildings on the site (a common practice) but the assessor also noted that the seam would be worked out in about six years' time.[13]

Licensed premises received special treatment from the Valuation Office. In some cases, such as that of the Gloucester subdistrict *c.* 1912–35 or for Warwickshire *c.* 1931, there is a valuation occupying its own separate Valuation Book and listing licensed premises by parish.[14] Copies of Forms 4–Land detail returns for several licensed premises in nine parishes in East Sussex, and the property of Jennings Bros. Ltd, brewers of Cockermouth, mostly in west Cumberland.[15]

Because of their general proximity to developed urban land, or indeed their envelopment within urban areas, market gardens and nurseries were singled out by the political advocates and critics of the legislation for special attention. Certainly in the Field Books they clearly received full coverage, as illustrated by one particularly good example, that of Maybush Nurseries, Ham Lane, near Worthing in West Sussex, which was accompanied by a sketch map as well (Table 6.2).

The identification of farm buildings and their building materials, use and state of repair, is frequent in those valuations carried out prior to mid-1912,

[13] W. G. Caudwell, 'Horsham: the development of a Wealden town in the early twentieth century', University of Sussex, unpubl. MA dissertation, 1986; Brian Short, Mick Reed and William Caudwell, 'The County of Sussex in 1910: sources for a new analysis', *Sussex Archeological Collection* 125 (1987), 199–224.

[14] Warwickshire CRO CR 1978/2/34; Gloucester RO D2428 1/53.

[15] Hove Reference Library: Sussex Industrial Archaeology Society drawer: Duties on Land Values 1910, Form 4–Land. The parishes covered are Hailsham, Hellingly, St Thomas in the Cliffe (Lewes), Wartling, Jevington, Warbleton, Arlington, Pevensey and Alfriston. I am grateful to Mr Geoffrey Mead for drawing my attention to these documents. Cumbria Record Office (Carlisle) DB 98/49–50.

Table 6.2 *Maybush Nurseries, Worthing, West Sussex 1910*

Glasshouse No.	Description
1	Span vinery about 150′×24′, heated with 6 rows of pipes. Ventilators opened by patent gear
2	Span vinery. Ditto
3	3/4 span Cucumber house. 150′×16′. Heated 4 rows 4″ pipes
4	Similar
5	Span vinery heated with 6 rows 4″ pipes. 150′×20′
6	Span cucumber house. Heated with 8 rows of pipes. 150′×14′
7	Similar to 6 but 6 rows of pipes
8	Span vinery. Heated with 6 rows pipes[+] 150′×20'
9	Small house 3/4 span, heated with 5 rows of pipes. 50′×14′
10	3/4 span vinery. The old vines have just been cut down and I understand that new Muscat grape vines are to be planted. 4 rows pipes, 180′×14′
11	Lean-to tomato house. heated by 4 rows pipes
12	(Bean house) lean-to heated with 6 rows of pipes.
	There is a well lighted packing shed. I understand there is a stepped well about 25′ deep, at the bottom of this an artesian well has been sunk 130′ deep. No water could be obtained recently and the men are now sinking it deeper. There is a windpump cistern supplying pipes etc. There are drains from soke holes draining away into ditch beyond southern boundary. There are two orchards, one on each side of packing shed

Notes:
[+]At this point the space allocated for 'Particulars, description, and notes made on inspection' runs out, and the description continues on p. 3 of the Field Book.
Source: PRO IR 58/94184.

and in conjunction with the identified land use patterns should allow for a considerable increase in our knowledge of agricultural practice during this period. The Royal Commission on the Historical Monuments of England was by 1993 working on the 1910 documents as part of a series of case studies to enable links to be made between farm buildings and agricultural practice, as well as to provide information on landownership and estate structures. Thus, the investigation of Village Farm (now Darwood House), Swaton, Lincolnshire used the 1910 material as an entrée both into older documentation through its evidence as to estate ownership, and into the use of the extant buildings through the Field Book notes made by the assessor. The farm *c.* 1910 was part of the Warner estate based at Walsingham Abbey, Norfolk. It contained 143 acres, and brief descriptions of the buildings were included in the Field Books, which enable a modern garage to be identified as a 'bullock hovel' in 1910. A low calf house which was so used in 1910 is still standing, and the house itself which appears to be late nineteenth-century in date was described as 'new' in 1910. A similar investigation of Grange Farm, Little Hale, Lincolnshire, used the 1910 material in the same way.[16]

A case study: Llanfihangel Yng Ngwynfa

One further example is derived from a comparison of information on farmsteads and farmhouses between the Valuation material and a previously published source by Alwyn D. Rees, *Life in a Welsh Countryside* (1950) based on fieldwork undertaken in the late 1930s and early 1940s.[17] This was concerned with the parish of Llanfihangel Yng Ngwynfa, then in Montgomeryshire (later Powys). Llanfihangel formed part of the ITP of Llanfechain which included nearly 1,300 hereditaments overall. All the documentation, whether at the Powys County Library at Llandridod Wells (Valuation Books and Forms 37) or at the PRO (Field Books and Record Plans), is well compiled and detailed.[18]

The Field Books contain field-by-field details of land use and detail on outbuildings, conditions of farms and accommodation, the quality of land and the fairness or otherwise of rents. 'Melwywnfa', a holding of some 26 acres let to Mary Owen at £21 a year (16s 2d an acre), was thought by the valuer to have a rent 'far too high', and his description makes bleak reading. The land near the farmstead was 'poor pasture', though further away he considered it 'much superior'. The premises are most graphically described. Overall he found them 'disgraceful'. The house with a kitchen, a back kitchen, three bedrooms and a dairy was 'old', 'very dilapidated' and 'unfit for human habitation', though apparently Mary Owen actually lived there. The outbuildings were 'very bad' and 'totally unfit for stock'.[19]

[16] Royal Commission on the Historical Monuments of England, Village Farm, Swaton, Lincolnshire (Farmsteads Survey, September 1993) and Grange Farm, Little Hale, Lincolnshire (September 1993). I am very grateful to Dr Paul S. Barnwell for drawing my attention to the RCHME Farmsteads Survey of pre-1914 farm buildings.

[17] A. D. Rees, *Life in a Welsh Countryside* (Cardiff 1950). For a critique of such early studies see Susan Wright, 'Image and analysis: new directions in community studies', in Brian Short (ed.), *The English Rural Community: Image and Analysis* (Cambridge 1992), 195–217. The inspiration for these studies was C. Arensberg and S. Kimball, *Family and Community in Ireland* (Cambridge, Mass. 1940).

[18] Powys Library, Llandrindod Wells M/LVR VB. 23 (uncat.). Valuation Book; Forms 37 (uncat.). The Record Sheet plans were consulted at the DVO in Welshpool prior to their transfer to the PRO, where a full photographic record of those sheets covering Llanfihangel was made. I am grateful to the staff for their kind cooperation. PRO IR 58/90126–7 Field Books.

[19] PRO IR 58/90126, Hereditament no. 298. Rees records this as 'Melinwnfa' (Gwnfa Mill) which became a 'lonely shop' after ceasing to be a mill, no date being assigned for this change in function (p. 104). Melin-y-graig (the mill of the rock) was functioning as a two-storey mill *c.*1910 (Hereditament no. 299, IR 58/90126).

Table 6.3 *Llanfihangel, Montgomeryshire: living accommodation in 1910*

Bedrooms	Non-sleeping rooms (inc. kitchens)					
	1	2	3	4	5	6+
1	6	4				
2	5	24	8			
3		14	23	1		
4		4	30	1		
5		1	6	2	1	1
6+				2	1	1
Totals	11	47	67	6	2	2

Source: PRO IR 58/90126–7, Field Books.

Housing is generally described quite fully. Living accommodation is indicated in Table 6.3. The table shows that the size of houses varied considerably. As Rees points out, this was roughly in accordance with the amount of land attached to each. Rees is very informative on housing, and the Field Books enable comparisons to be made with named properties in *Life in a Welsh Countryside*. Generally, Rees notes, houses had a 'cegin' or living room. Literally this translates as 'kitchen', and the valuers used this literal translation consistently. The 'back kitchen' was also characteristic of most houses, and these two rooms provided the only non-sleeping accommodation for very many homes, although parlours were not infrequently found.[20]

Pen-y-graig, a common type of oblong farmhouse in the parish, is described by Rees as having on the ground floor a living room (cegin), a back kitchen (cegin gefn), parlour, pantry and a lean-to buttery or dairy at the back (Fig. 6.3). The length of the house is exaggerated by the addition of a barn since 1840 which has served to join up two formerly separate buildings. In 1910, all these rooms except the pantry were mentioned. It is possible that Rees' pantry had previously been the dairy referred to in 1910, and that the lean-to structure was a later addition. At the time of the valuation, the occupant here was Evan Evans, farming just under 110 acres for a rent to the Wynn estate of £65 16s. The accommodation was referred to as poor, the outside buildings bad. The land was said to be shallow in part, and 'liable to burn up'. The rest was wet.[21]

The description in the Field Book of the long-house type of building represented by Y Fachwen Ganol (see also Fig. 4.6) is less easy at first to reconcile with Rees. The valuer described a building with parlour, kitchen, two back kitchens and a dairy on the ground floor, with four bedrooms and a granary. Rees describes it as having a living room, a back kitchen, a dairy, and two small

[20] Rees, *Life in a Welsh Countryside*, 32. [21] *Ibid.*, 32–4; PRO IR 58/90127.

rooms now used as a store and a chamber. It seems probable that these two latter small rooms had previously been the granary described by the valuer, and that the back kitchen had been more fully divided in two than is evident on Rees' version of the ground plan (Fig. 6. 4). The 142 acre holding, tenanted by David Humphrey at a rent of £55, had no description of its land included in the Field Book.[22]

Rees gives no clues as to the condition of housing but in 1910 the valuers were less reticent, as Table 6.4 indicates. All but sixteen of these houses were owned by Sir William Watkins Wynn, owner of almost 9,000 acres in the parish or about 80 per cent of the total area, and all but one of those described as 'very poor' or worse were estate property. On the other hand houses owned by the estate were slightly more likely to be described as 'fair' or 'good' than those owned by others (62 per cent compared to 56 per cent). It remains the case, though, that forty-one (39 per cent) of the houses having information about condition were described as 'poor' or worse, and all but seven of these were owned by the Wynn estate, which was responsible for repairs. Even those houses described as 'fair' were not uncommonly also described as 'damp' or 'very damp'.

If housing was frequently in poor condition, then the outbuildings were even more often so described. Details of the condition of outbuildings are available for 110 hereditaments, nearly all owned by Wynn (Table 6.5). Outbuildings on over half of the hereditaments were considered to be in 'poor', 'bad' or 'dilapidated' condition or worse.

In short, the conditions of accommodation for humans as well as live and dead stock in the parish were variable, with large proportions of both being below what the valuers thought was a reasonable standard. In addition, the valuers frequently felt that rent on the estate was excessive. There was little obvious sign of a paternalistic regard for his tenants on the part of the Wynn of Wynnstay at this time, although Rees informs us that by the late 1930s tenants generally felt benign towards 'Old Sir Watkin', preferring to blame his agents for delays or problems, rather than the landlord himself.[23]

Genealogical studies

Genealogists and family historians should be alerted to this important set of documents and maps. Some possible uses which should be of interest to genealogists are included elsewhere in this chapter and those following, but it should be remembered again that the suggestions assume the full and correct compilation of the documents in accordance with the instructions in force at the time. Only detailed local studies can reveal whether this was in fact the

[22] Rees, *Life in a Welsh Countryside*, 34: PRO IR 58/90126 Hereditament no. 258.
[23] Rees, *Life in a Welsh Countryside*, 159.

(a)

(b)

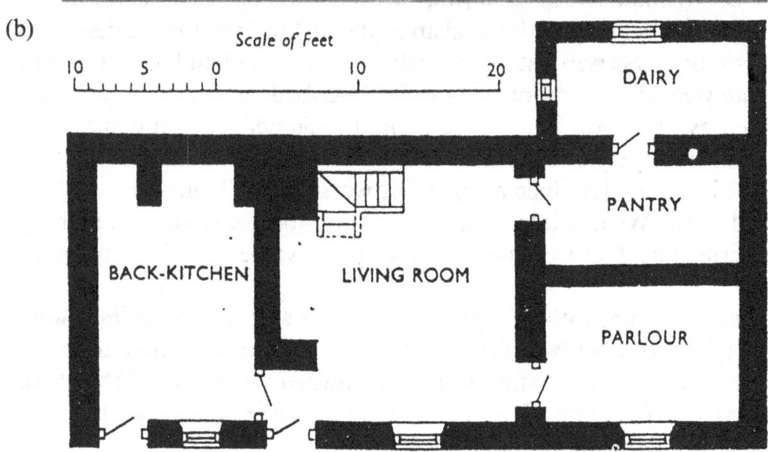
Scale of Feet
10 5 0 10 20
DAIRY
PANTRY
BACK-KITCHEN
LIVING ROOM
PARLOUR

(c)

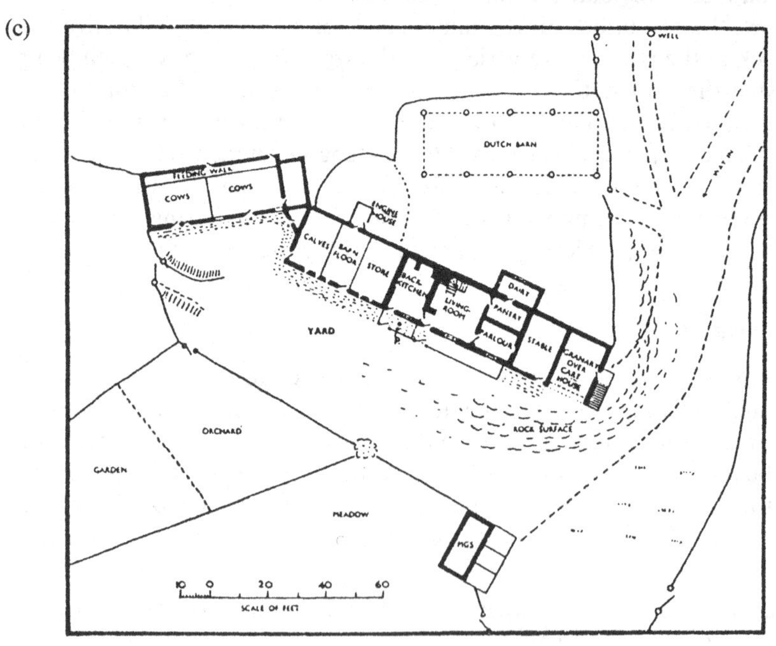
DUTCH BARN
COWS
COWS
YARD
ROCK SURFACE
ORCHARD
GARDEN
MEADOW
PIGS
10 0 20 40 60
SCALE OF FEET

(d)

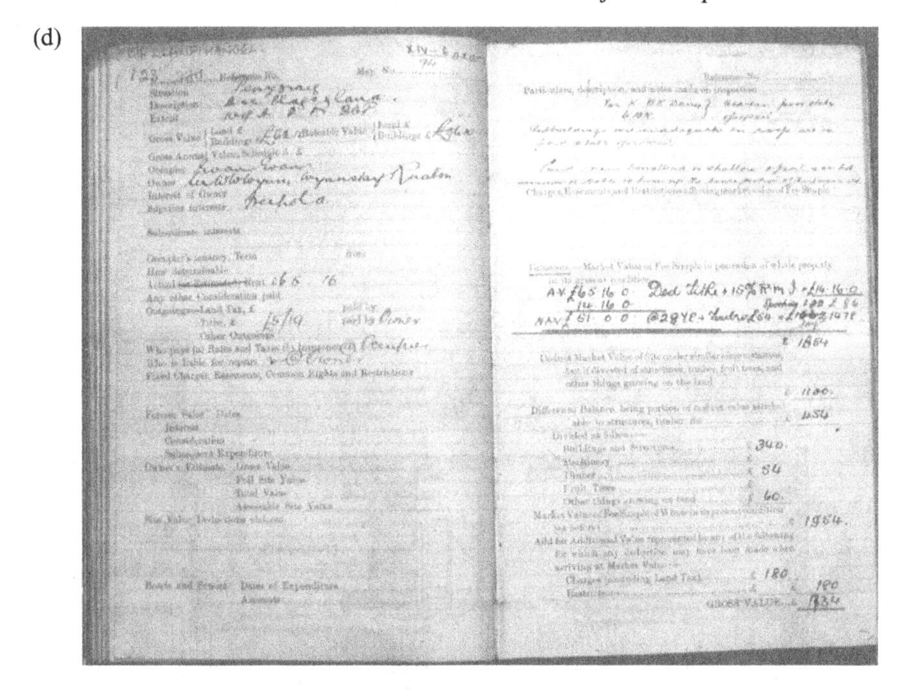

(e)

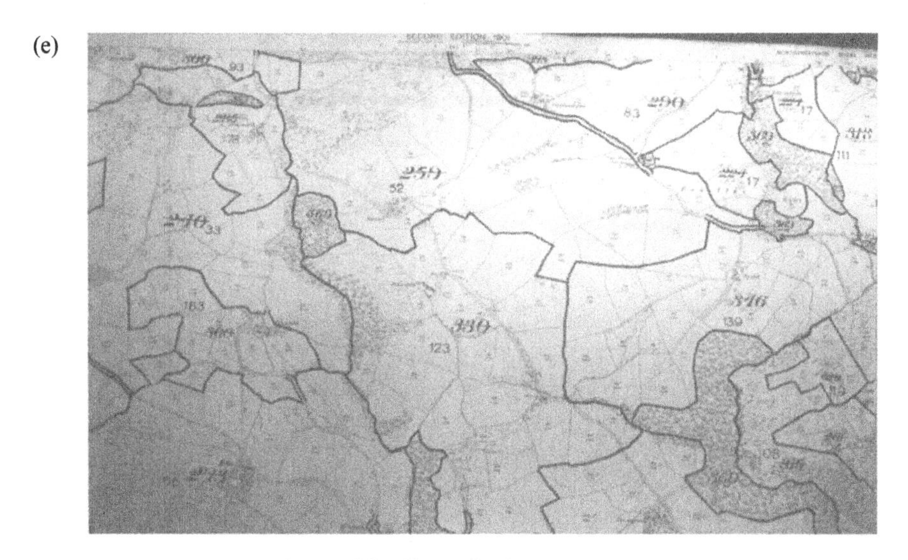

Fig. 6.3 Pen-y-graig farmhouse, Llanfihangel, Montgomeryshire. (a) Photograph from Rees, *Life in a Welsh Countryside*, facing p.32. (b) Ground Plan from Rees, 33. (c) Sketch-plan from Rees, 48. (d) Entry from Valuation Field Book (PRO IR 58/90127). (e) Extract from Record Sheet Map for Llanfihangel (Hereditament no.330).

(a)

(b)

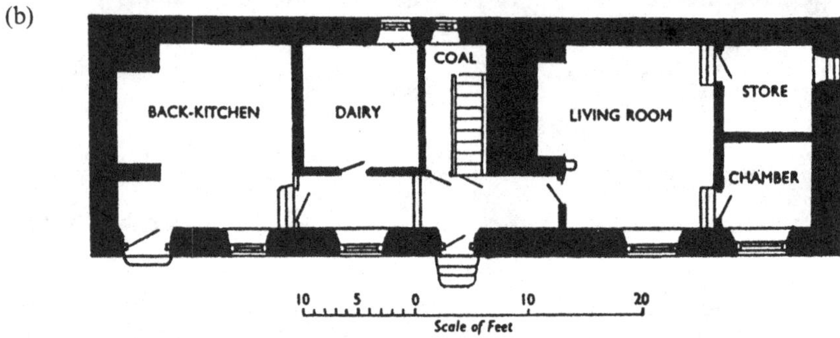

(c)

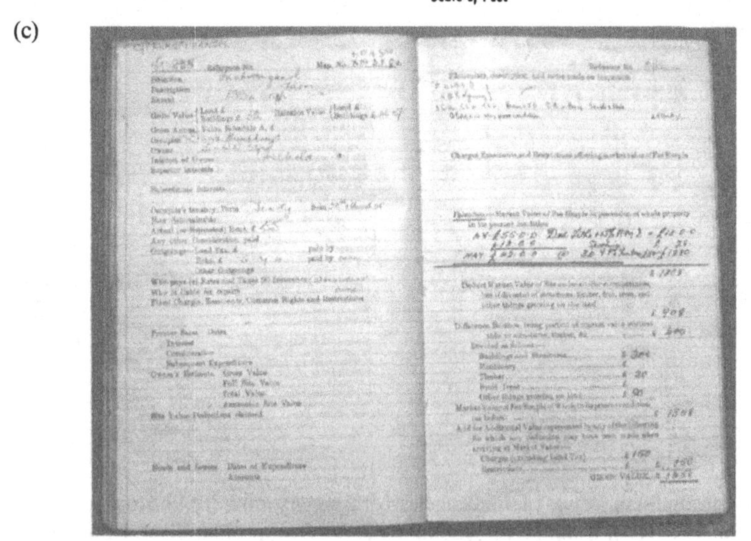

Fig. 6.4 Y Fachwen Ganol farmhouse, Llanfihangel, Montgomeryshire. (a) Photograph from Rees, facing p.33. (b) Ground Plan from Rees, 34. (c) Entry from Valuation Field Book (PRO IR 58/90126).

Table 6.4 *Llanfihangel, Montgomeryshire: condition of housing*

Condition	No.	%
Good	23	22
Fair	42	40
Poor/bad/dilapidated	25	24
Very poor/very bad/or very dilapidated	14	13
Disgraceful	2	2
Total	106	

Source: PRO IR 58/90126–7, Field Books.

Table 6.5 *Llanfihangel, Montgomeryshire: condition of outbuildings*

Condition	No.	%
Good	21	19
Fair	29	26
Poor/bad/dilapidated	26	24
Very poor/very bad/or very dilapidated	26	24
Unsafe and unfit for stock	8	7
Total	110	

Source: PRO IR 58/90126–7, Field Books.

case, but the work of family historians could be very important in this respect.

The fact that the valuers were specifically concerned to identify owners of land for the purposes of taxation ensured the compilation of very full data about *owners' and occupiers' names and addresses and land and property ownership*. The identification of individuals and the study of ownership of former family properties are among the most obvious uses of the data, and indeed the survey represents the most comprehensive set of property records ever compiled in the United Kingdom.[24]

The Record Plans should enable property in the same ownership to be precisely and easily delineated. Since the addresses of owners ought also to be given, it should be possible to make assessments as to the extent of absentee ownership, as well as the derivation of possible family or kinship linkages in towns or parishes hitherto unsuspected. The housing conditions and other

[24] An early example of the perceived significance of the records for family history is to be found in W. J. Taylor, 'Notes on sources: Valuation Records', *Lancashire* 8 (2) (1987), 26–8 (Lancashire Family History and Heraldry Society). Mr Taylor was initially struck by the fourteen hand-coloured Valuation Plans of Darwen examined in a local library.

environmental information about the families under investigation will, of course, also be possible. When used in conjunction with other contemporary sources such as directories, rate books or electoral registers, the potential is indeed great.

When in 2012 the 1911 Census Enumerators' Books become available for public inspection, the family historian will be able to link the two sets of records to obtain an unprecedented view of society on the eve of the First World War.[25] In the meantime, the low priority so far accorded to these records by local repositories could be addressed by family historians whose fast-growing number and influence could well be used to give the records greater prominence.

Use of the documents in legal contexts

In common with other historical documents, the Valuation material can be used in many ways to establish factual points for legal purposes, such as conveyancing or the establishing of landownership and occupation of premises, and the boundaries of hereditaments. In the West Yorkshire Archives Office, for example, the records enable some researchers to use the registered versions of West Riding title deeds 1704–1970, which are indexed by personal name of buyers and sellers. One has to know the name of an owner before 1970, and the 1910 material easily provides that information.[26]

An important use has recently been found by parties interested in rights of way. Since under the provisions of the Finance Act owners were able to reduce original Assessable Site Values by claiming against the Gross Value for any fixed charges, public rights of way or public rights of user, easements etc. across their land, the documents do contain evidence of acknowledged or alleged rights of way. Section 25(3) of the Finance (1909–10) Act 1910 stated that in assessing the value of the land, Total Value: 'means the gross value after deducting the amount by which the gross value would be diminished if the land were sold subject to any fixed charges and to any public rights of way or any public rights of user . . .'[27] Rights of way thus become one of a number of important sources of deduction to be claimed and valued. The Land Union advised owners or their agents to use a phrase such as 'alleged footpath' in cases of disputed public access, and the actual tactics of claiming deductions would doubtless vary from owner to owner. The Valuation Office felt that the

[25] Ongoing work being undertaken by the Cambridge Group for the History of Population and Social Structure into the late nineteenth- and early twentieth-century Census Enumerators' Books has been limited to thirteen places and lacks any nominal information. The OPCS would be very concerned if any attempted linkage through profession, address or other indicator were attempted, and nothing further can therefore be done at this stage to link these two vital sets of data. I am grateful to Peter Laslett for information on the Cambridge Group project. [26] Personal communication, Mr M. Bottomley, West Yorkshire Archive Service.

[27] Finance (1909–10) Act 1910, Section 25(3).

owners' returns were generally very fairly and accurately completed but landowners were caught in a pincer device by Lloyd George. The incidence of death duties and probate would hit harder the greater the value of the estate, but if the estate were valued too low in 1909 the chance of greater incremental values being taxed was that much higher whenever an 'occasion' arose in the future. To minimise Increment Value Duty the two values (1909 and the later 'occasion') had to be as close as possible. As G. Garnons Williams wrote to P. G. Collins at the Ashburnham Estate Office at Barry Port, Carmarthenshire in December 1914: 'It would be quite useless to put a low value on this land for Probate purposes in 1913, and then try to settle for the same land for the Land Values Valuation for 1909 at a greatly enhanced figure.'[28]

There is no evidence to show that exceptionally high or low valuations were sought by landowners and their agents, although there was much debate on this point at the time, and indeed landowners were indignant in some cases that they were not advised by the government as to which way any bias in valuation might eventually work! There is therefore no particular reason to believe that deductions for footpaths etc. were not claimed as of 1909. Certainly it was the case that if they were not claimed in 1909, but could have been, then they could not be claimed for at any later date, so as to minimise Increment Value Duty. In addition, any landowner falsely claiming a way to be public in order to minimise tax liability would be committing a serious criminal offence, but some landowners might take the view that disclosure of a right of way was more of a nuisance than the additional tax liability. Carriageways and bridleways might have significant impacts on income and thus be worth claiming for, but footpaths were less significant in this respect. If no deduction was claimed, the Finance Act material cannot therefore be used as proof that a public right of way did not exist, since no-one could force a landowner to claim tax deductions.[29]

The 25 inch (or whichever scale is locally available) Record Sheet plans will show all private land as being coloured and with boundaries and hereditament numbers. Any public land, or other land outside the scope of the taxation, would be left uncoloured and with no hereditament number. To have shown any land as uncoloured when it was liable for duty would have constituted negligence on the part of the valuers, and can effectively be discounted unless any supporting evidence of nonconformity with the required process of valuation can be demonstrated. Uncoloured land in most cases will be in public owner-

[28] East Sussex Record Office, Ashburnham Mss (uncatalogued), Williams to Collins, p. 2.

[29] 'A person shall not be entitled to claim any deduction for the purpose of ascertaining the site value of any land on any occasion on which Increment Value Duty becomes payable if the deduction is one which could have been, but was not, claimed for the purpose of ascertaining the original site value of the land' (Finance (1909–10) Act 1910, section 12). See also Zara Bowles, 'Rights of Way and the 1910 Finance Act', *Rights of Way Law Review* (September 1990), Section 9.3, 17–18.

ship and vested in the local rating authorities, the District Councils at that time who were the highway authorities, and who were exempt from the taxation. Such uncoloured routeways will frequently therefore be all-purpose public roads. The Record Sheet plans at the PRO constitute the prime source of evidence, since the locally available Working Sheet maps may have been subject to amendment or in other ways be less reliable.

In addition, bridleways, footpaths and other routes through properties can in many instances be equated with claims for reduced values in the Field Books, Forms 4 or Forms 36/37–Land. Where a large property is crossed by several routes, there will, however, seldom be anything in the Valuation Books etc. to pinpoint the exact route(s) for which claims are being made. It is therefore vital to substantiate any modern claims or counterclaims regarding rights of way with supporting supplementary documentary evidence. In parts of the country with smaller farming units and with a surveyor who also recorded OS plot numbers, the information should enable the tracing of public paths across OS plots that tie in to rights of way physically recorded on OS maps. The surest evidence is where a holding contained only one path which was shown on the Record Sheet plan.[30]

The map evidence will still be valuable where a route adjoins or crosses several different hereditaments, even if not all the route is left uncoloured. Different owners may have filled in their forms in different ways, with not all applying for deductions. Any claim can normally be checked in the Field Books, but remembering that it was an offence falsely to claim a public route across land, the existence of uncoloured portions of a routeway on the Record Sheet plan must constitute very good evidence of a public routeway, as also will the absence of a hereditament number on the routeway.

Perhaps the most difficult type of route to assess is that which is uncoloured and wending its way totally between a number of different properties which front onto it The question to be addressed in such cases is whether such a route is a genuinely public one and therefore correctly left uncoloured, or whether the adjoining property owners either: (a) were ignorant of its status and thus failed to claim it as being within their boundaries (which legally extended to the centre part of the routeway), or (b) did not wish to acknowledge its status which was known to them as a public road, thus forfeiting any rights to deductions from their Site values, or (c) were simply not interested in the status of the route, since it crossed nobody's land. Whose input, in other words, led to

[30] Bowles, 'Rights of way', 19; and see also Bill Riley, 'Highways and the 1909–10 Finance Act maps', *Byway and Bridleway: Journal of the Byways and Bridleways Trust* 1 (1992), 4. Riley has calculated that out of nearly 1,000 Working Sheet maps in the Wiltshire CRO, twenty-three have annotations indicating a mistake on the sheet, e.g. 'Road should not be coloured' (Sheet no. 15/10), 'Public road not to be coloured' (Sheet 52/5). In correcting these, the LVO would ensure that the correct information was transferred to the Record Sheet Plans. (I am indebted to Bill Riley for this information.)

the map showing an uncoloured route? Clearly the original onus was on the owner(s), whose Forms 4–Land, when completed, would indicate whether or not a public right of way existed *and was being claimed for*. In this case, recourse to the Field Book entries might confirm that someone or no-one was claiming for the routeway. There was no public consultation on the 1910 maps, so the community was unable to challenge rights of way omitted. On the other hand, there was consultation between valuer and landowner (or, in the case of large estates, their agents), so that objections could be made to the omission of a right of way. The Valuer would also have to be satisfied on his surveying visit that the hereditament boundaries were properly claimed and that no private land was being left unclaimed. Certainly valuers actively had to discover for themselves most of the facts necessary to make a valuation, including the existence of public rights of way, matters on which they did not only rely on evidence given to them by owners, one of the reasons given for the overall slow speed of the valuation process by Edgar Harper in his evidence to the Select Committee on Land Values 1920.[31] There is no general answer to this difficulty. Much depends on the local context, and the importance of considering the 1910 material alongside supporting documents is again emphasised.

Nevertheless, the records are now considered by many authorities to be a reliable source of evidence on rights of way and other matters. The *Instructions to Valuers* issued in 1910 made it clear that the valuers must strive for accuracy, uniformity of treatment between hereditaments, and 'expedition'. They were warned that: 'The Valuation Book and the Returns must be carefully preserved, and must under no pretence whatever be taken out of the office. As a permanent record they will be invaluable, and in some cases may form the basis of future litigation.'[32] A similar entreaty was made regarding the Record Sheet plans, upon which care must be lavished to ensure that 'all property within the limits of the Valuer's district has been recorded',[33] as also for the Field Books which 'must in all cases be kept neatly and carefully preserved, as it may be required for reference many years hence, and in some cases may have to be produced as evidence'.[34]

Therefore, Inspectors for the Department of the Environment have been willing to take the documents in proof of public rights, and during the late 1980s and early 1990s the number of such cases has multiplied. Examples from Essex, Suffolk, Wiltshire and Devon can be cited. At Salford (Bedfordshire) in August 1992, the Inspector noted that the lane in question was uncoloured on the 1910 map and that this was 'consistent with it being regarded as having public status', although the Inspector cautiously went on

[31] Evidence of Mr Edgar Harper, Select Committee on Land Values, para. 15, 43–4.
[32] *Instructions to Valuers* Part I, 23–5 (para. 111, 118). [33] *Ibid.*, 29 (para. 149).
[34] *Ibid.*, 31 (para. 159).

to say that this needed corroboration by other evidence.[35] An early county court judgement to incorporate the Finance Act material was that of the Poole County Court (Case No. 89 02274) of January 1993. This involved the establishment of land as roadside waste and therefore part of the public highway at Alderholt, Dorset. This was duly established, with the 1910 map being used to demonstrate that such areas of waste were not shaded and were not therefore regarded as private land.[36]

Further analysis

More attention will be devoted to specific analyses in the following chapters. For example, many of these topics can be studied over time. Comparisons with the earlier Tithe Surveys should allow assessments to be made of changes in ownership and occupation structures, land use patterns and farm fragmentation or consolidation over the period from the beginning of Victoria's reign to the First World War. Future comparisons might also use material derived from the local Rate Revaluation of the 1930s, and perhaps more importantly, with the National Farm Survey (1941–3), available for England and Wales from 1992 and for Scotland from 1994. Comparisons between *c.* 1910 and the present can also inform the studies of those seeking to examine the changing nature of twentieth-century town and country.

Temporal comparisons can be integrated with spatial ones. The wide range of empirical information contained in the various documents is amenable to a greater or lesser extent to spatial analysis, both quantitative and qualitative. In the case of the former, the provision of a range of values for each property as well as rateable value will be of use to, for example, geographers conducting analyses of social areas in both town and country. However, the descriptive 'back-up' for each property in the Field Books can give added qualitative substance to the quantitative framework within which these studies have normally been situated.

The examination of some of these potential uses of the material will be made in greater detail below, in relation to the themes which have been chosen to illustrate the vast array of possibilities opened up by the documents. However, one caveat must be entered. Initial case study work indicates that in many parts of the country the ideas presented above may not be feasible, whether owing to the contemporary administration of the valuation, or to modern archival policies.

Research based on the valuation material is liable to be seriously mis-

[35] Bowles, 'Rights of way', 19, fn 10 gives references to public enquiries. Inspector's decision under Wildlife and Countryside Act 1981 – Section 53 and Schedule 15 County Council of Bedfordshire (Definitive Map and Statement for Bedfordshire) (Hulcote and Salford: Byway open to all traffic Nos. 10 and 11) (no. 2) Modification Order 1990. Dated 26 August 1992.

[36] Poole County Court Case no. 89 02274, para. 27 (7 January 1993).

leading if the many problems with the documentation are not anticipated. Problems will always arise owing to the disjuncture between the original purpose for which the documentation was intended, and the needs and aspirations of modern researchers. This is true of virtually all historical material, but there are nevertheless problems that are specific to the 1910 data. Some of these problems originate in the way the valuation was conceived and framed in legislation, and in the instructions issued to valuation staff. Other problems derive from the idiosyncratic interpretation of instructions by the different District Valuation Offices and by individuals in those offices. And archival policies since the first documents reached the public domain in 1968 have created a third category of problems, which were described in Chapter 5.

Problems caused by the legislation and its administration.

Like all legislation, the Finance (1909–10) Act 1910 is extremely difficult to interpret, especially for the non-lawyer. The numerous exceptions in the Act, and the several judgements based upon it, created a minefield after 1909 for would-be interpreters of the law. However, there are a few unproblematic issues of fundamental importance to modern researchers and the discussion of these and their implications now follows.

Initially, and of great importance for the study of landownership, is the definition of 'owner'. This is perhaps the most fundamental piece of information yielded by the documents, and is of concern to a wide range of contemporary researchers. Section 41 of the Act defines the 'owner' as:

> the person entitled in possession to the rents and profits of the land in virtue of any estate of freehold, except that where land is let on lease for a term of which more than fifty years are unexpired, the lessee under the lease or if there are two or more such leases the lessee under the last created under-lease shall be deemed to be the owner instead of the person entitled to the rents and profits as aforesaid.[37]

In other words long leaseholders or even long under-lessees may be defined as the owner for the purposes of the Act. Where this is the case, it ought to be stated in the Field Books or Forms of Return since owners were supposed to indicate the existence and identity of superior and subordinate interests. Therefore identification of ultimate ownership ought to be possible.

Secondly, there is the question of what exactly comprised the unit of land being valued. In Section 26.1 of the Act there was provision for the valuation of all land to be made. The basic unit of valuation was to be the hereditament or each piece of land which is in separate occupation. However, this straightforward definition was complicated from the point of view of

[37] Finance (1909–10) Act 1910, S.41.

researchers, in that owners could require, under Section 26.1, that the hereditament be divided into as many parts as he/she wished and that each part be separately valued. Moreover, this section was amended by the Revenue Act 1911 Section 5 so that the owner could request the Valuation department to value together 'any pieces of land which are contiguous, and which do not in the aggregate exceed one hundred acres in extent . . . although those pieces of land are under separate occupation'.[38] As long as the Department was satisfied that this was equitable, they were obliged to agree to such a request. In practice, this means that the 'hereditaments' delineated on the record plans are actually 'units of valuation', and that they could comprise part of a hereditament, or several contiguous hereditaments. It may therefore become difficult to identify a farm or industrial concern *per se*, since it may well be that the 'owner' has decided to have it valued together with another piece of land. It is normally relatively easy to pick out such occurrences when examining the records in detail, but it should serve as a warning against accepting too easily the units of valuation on maps or in Valuation Books as being meaningful in any economic or functional sense. Theoretically the subdivision or combining of hereditaments ought to be indicated in the Valuation and Field Books. However, it often is not, since valuation staff frequently omitted to note this information. Such information is sometimes found partly in the Field Books, partly in the Valuation Books, but sometimes not at all.

The administration of the valuation has also created less obvious problems for researchers. As has been stated, the Income Tax Parish (ITP) was chosen as the smallest administrative unit.[39] ITPs could comprise a single civil parish, two or more normally contiguous civil parishes, or part of a civil parish. The example of Ashley Walk, Hampshire was cited in Chapter 3, and in Gloucestershire, for example, the parishes of Winchcombe, Toddington, Stanway, Sudeley, Alderton and Prescott were grouped together as an ITP named Alderton after the first name alphabetically in the list of parishes (Fig. 6.5).[40]

The county of Leicestershire, in the same way, comprised 165 ITPs incorporating 324 civil parishes *c.* 1910 (Fig. 6.6 and Table 6.6). Little geographical rationale for the ITPs was found in Leicestershire. There was no obvious linkage between the composition of an ITP and its geographical extent and neither did the ITPs have roughly similar numbers of hereditaments. Excluding Leicester Borough with some 60,000 hereditaments, the number of hereditaments per ITP ranged from fourteen in Carlton Curlieu ITP (a single civil parish) to 7,269 in Loughborough ITP, which included the civil parishes of Loughborough, Garendon, Thorpe Acre, Woodthorpe and Nonpanton.

[38] Revenue Act 1911, S.5. [39] See Chapter 3.
[40] Gloucestershire RO D2428: a plan of the county's ITPs is available under D2428.

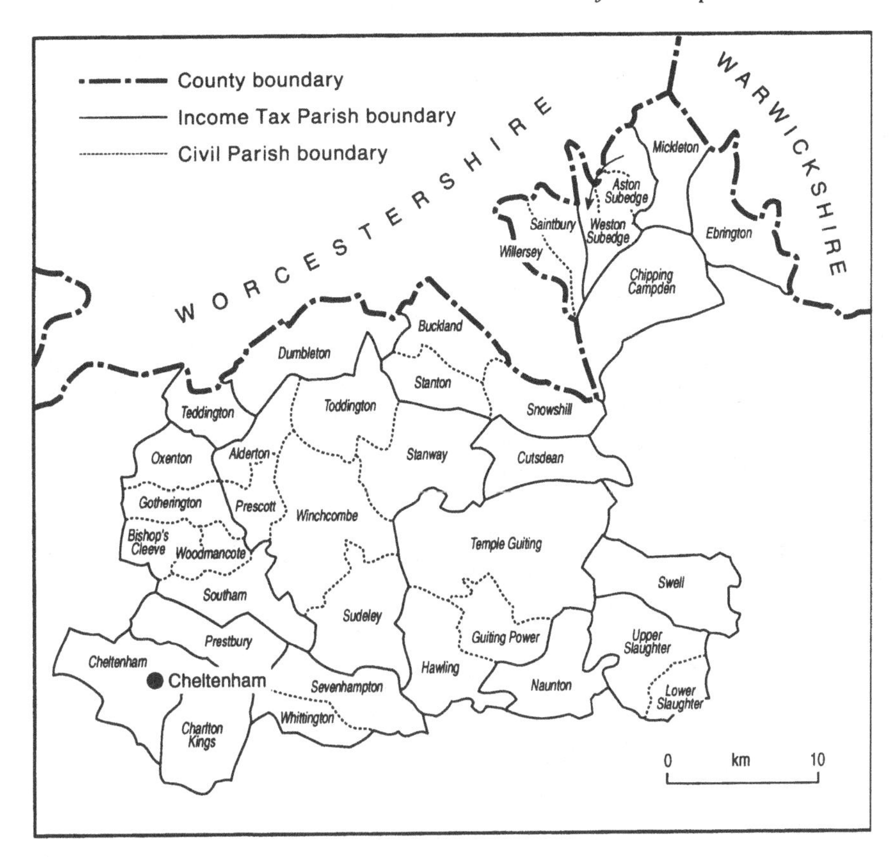

Fig. 6.5 Income Tax Parishes in north Gloucestershire: Alderton ITP and its neighbours.
Source: Gloucestershire Record Office D2428.

Presumably there was a list or lists of the 'poor law' (civil) parishes that constituted each ITP in 1910, but none has yet been located. In their absence, the only documentation that enables identification of the ITP is the Valuation Book. Into this the Valuation Officer transcribed the 1909–10 Rate Books for the constituent parishes. Each hereditament was then numbered consecutively, and the first parish in the group gave its name to the entire ITP, as described above.

The ITP was thus synonymous and spatially identical with a civil parish only in those ITPs that themselves comprised only a single civil parish. Each unit of valuation within the ITP was identified by the name of the ITP and a number. Therefore a unit of valuation in civil parish X may actually be identified by the name of parish Y and a number.

To illustrate something of the difficulties arising from this point, one may

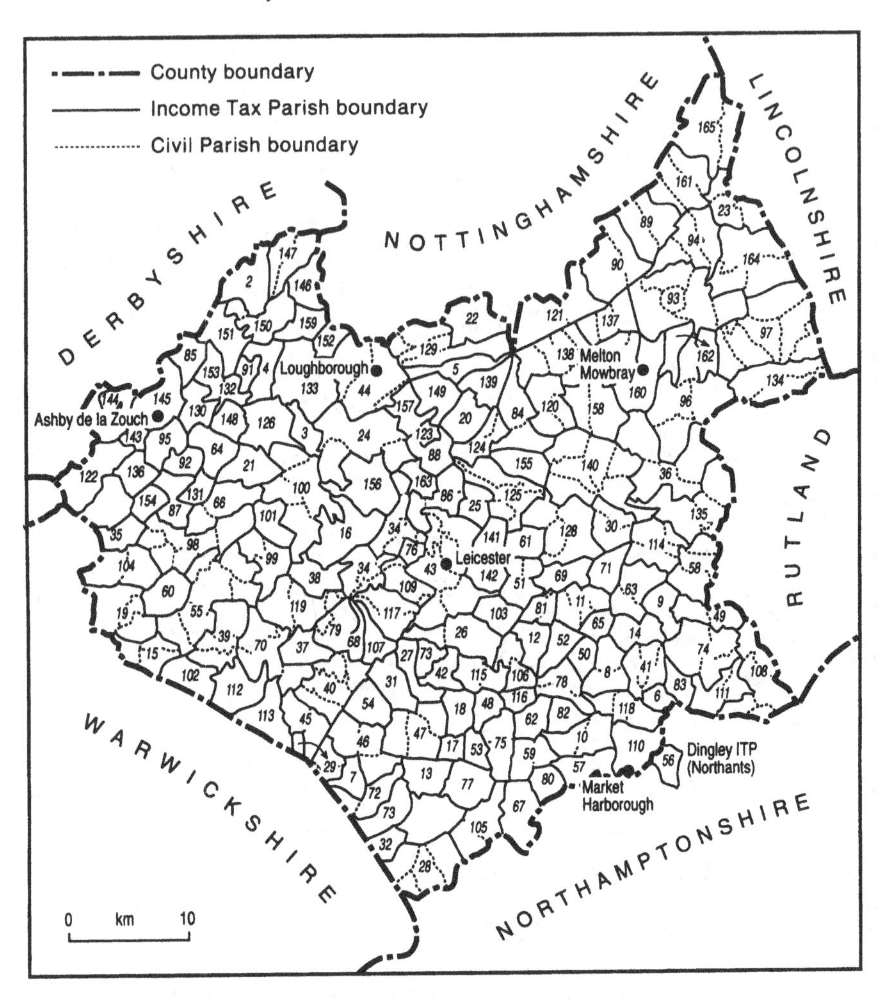

Fig. 6.6 The Income Tax Parishes of Leicestershire in 1910.
Source: Leicestershire CRO: Valuation Books.

Income Tax Parish	Civil Parish	Hd. nos.	Remarks
1. Misterton	Misterton	1–152	
2. Castle Donington	Castle Donington	1–857	
3. Charley	Charley	1–53	
4. Belton	Belton	2–268	
5. Walton On Wolds	Walton On Wolds	1–100	
6. Welham	Welham	1–38	
7. Ullesthorpe	Ullesthorpe	1–175	
8. Langton Tur	Langton Tur	1–98	
	Shangton	99–128	

Income Tax Parish	Civil Parish	Hd. nos.	Remarks
9. Tugby And Keythorpe	Tugby And Keythorpe	1–109	
10. Foxton	Foxton	1–123	
	Gumley	124–194	
11. Frisby By Galby	Frisby By Galby	1–18	
	Galby	19–47	
	Kings Norton	48–72	
12. Glen Magna	Glen Magna	1–298	
13. Gilmorton	Gilmorton	1–245	
14. Goadby	Goadby	1–45	
	Nosely	46–70	
15. Ratcliffe Culey	Ratcliffe Culey	1–50	
	Sheepy Magna	51–145	
	Sheepy Parva	146–164	
16. Ratby	Ratby	1–570	
	Groby	571–816	
17. Peatling Parva	Peatling Parva	1–67	
18. Peatling Magna	Peatling Magna	1–88	
19. Orton On The Hill	Orton On The Hill	1–79	
	Twycross	80–165	
20. Sileby	Sileby	1–1109	
21. Coalville	Coalville	1–2175	2 Books
22. Wymeswold	Wymeswold	1–367	
23. Belvoir	Belvoir	1–130	
	Harston	131–175	
	Knipton	176–177	
24. Swithland	Swithland	1–63	
	Ulverscroft	64–96	
	Woodhouse	97–588	
25. Thurmaston	Thurmaston	1–723	
26. Wigston Magna	Wigston Magna	1–2700	
	Glen Parva		
	East Wigston		
27. Whetstone	Whetstone	1–432	
28. Catthorpe	Catthorpe	1–47	
	Shawell	48–117	
	Swinford	118–233	
	Westrill And Starmore	234–247	
29. Claybrook Magna	Claybrook Magna	1–143	
	Claybrook Parva	144–189	
	Wigston Parva	190–222	
30. Cold Newton	Cold Newton	1–41	
	Lowesby	42–71	
31. Cosby	Cosby	1–499	
32. Cotesbach	Cotesbach	1–50	
33. Lutterworth	Lutterworth	1–670	With 32
34. Anstey	Anstey	1–797	

Income Tax Parish	Civil Parish	Hd. nos.	Remarks
	Anstey Pastures	798–814	
	Gilroes	815–827	
	Cropston	828–926	
	Glenfield	927–1367	
	Leicester Frith	1368–1392	
	Glenfield Frith	1393–1407	
35. Appleby Magna	Appleby Magna		
36. Brentingby & Wyfordby	Brentingby & Wyfordby	1–39	
	Burton Lazars	40–145	
	Dalby Parva	146–190	
	Stapleford	191–230	
37. Earl Shilton	Earl Shilton	1–1350	
38. Desford	Desford	1–341	
39. Dadlington	Dadlington	1–65	
	Stoke Goldings	66–249	
	Sutton Cheney	250–340	
40. Croft	Croft	1–196	
	Potter Marston	197–208	
	Stoney Stanton	209–688	
	Sapcote	689–1015	
41. Cranoe	Cranoe	1–30	
	Gloosten	31–70	
	Stanton Wyville	71–109	
42. Countesthorpe	Countesthorpe	1–451	
43. Leicester	Leicester	38 Books	
44. Loughborough	Loughborough	1–7045	
	Garendon	7046–7088	
	Thorpe Acre or Dishley	7089–7151	
	Woodthorpe	7152–7177	
	Nonpanton	7178–7269	
45. Aston Flamville	Aston Flamville	1–29	
	Burbage	30–857	
	Sharnford	858–1031	
46. Ashby Parva	Ashby Parva	1–80	
	Frolesworth	81–182	
	Leire	183–354	
47. Ashby Magna	Ashby Magna	1–83	
	Dunton Bassett	84–263	
	Willoughby Waterless	264–365	
48. Arnesby	Arnesby	1–36	
49. Alexton	Alexton	1–36	
	East Norton	37–108	
50. Carlton Curlieu	Carlton Curlieu	1–14	
51. Bushby	Bushby	1–45	
	Stoughton	46–100	
	Thurnby	101–177	

Income Tax Parish	Civil Parish	Hd. nos.	Remarks
52. Burton Overy	Burton Overy	1–148	
53. Bruntingthorpe	Bruntingthorpe	1–96	
54. Broughton Astley	Broughton Astley	1–484	
55. Osbaston	Osbaston	1–78	
	Upton	79–114	
	Carlton	115–180	
	Shenton	181–239	
	Cadeby	240–297	
	Market Bosworth	298–570	
56. Little Bowden	Little Bowden	1–768	
	Dingley	769–867	(Northants)
57. Lubenham	Lubenham	1–209	
58. Launde	Launde	1–11	
	Loddington	12–54	
	Withcote	55–69	
59. Laughton	Laughton	1–63	
	Mowsley	64–160	
60. Sibson	Sibson	1–86	
61. Scraptoft	Scraptoft	1–65	
62. Saddington	Saddington	1–80	
63. Roleston	Roleston	1–19	
	Skeffington	20–90	
64. Ravenstone & Snibstone	Ravenstone & Snibstone	1–463	
65. Illston	Illston	1–65	
66. Ibstock	Ibstock	1–1183	
	Odston	1184–1283	
67. Husband Bosworth	Husband Bosworth	1–324	
68. Huncote	Huncote	1–231	
69. Houghton On Hill	Houghton On Hill	1–133	
70. Barwell	Barwell	1–816	
	Stapleton	817–886	
71. Billesdon	Billesdon	1–237	
72. Bittesby	Bittesby	1–14	
	Bitteswell	15–16	
73. Blaby	Blaby	1–716	
74. Blaston	Blaston	1–36	
	Hallaton	37–242	
	Horninghold	243–321	
75. Knabtoft	Knabtoft	1–12	
	Shearsby	13–99	
76. Kirby Frith	Kirby Frith	1–4	
	Kirby Muxloe	5–327	
	Leicester Forest East	328–373	
77. Kimcote & Walton	Kimcote & Walton	1–208	
78. Kibworth Beauchamp	Kibworth Beauchamp	1–208	
	Kibworth Harcourt	542–737	

Income Tax Parish	Civil Parish	Hd. nos.	Remarks
79. Thurlaston	Thurlaston	1–162	
80. Theddingworth	Theddingworth	1–78	
81. Stretton Magna	Stretton Magna	1–17	
	Stretton Parva	18–50	
82. Smeeton Westerby	Smeeton Westerby	1–140	
83. Slawston	Slawston	1–72	
84. Rearsby	Rearsby	1–129	
	Thrussington	130–298	
85. Staunton Harold	Staunton Harold	1–59	
86. Beaumont Leys	Beaumont Leys	1–53	
	Birstall	54–382	
	Wanlip	383–418	
87. Swepstone	Swepstone	1–179	
88. Rothley	Rothley	1–995	
89. Harby	Harby	1–210	
	Strathern	211–429	
90. Long Clawson	Long Clawson	1–250	
	Hose	251–400	
91. Osgathorpe	Osgathorpe	1–137	
92. Normanton Le Heath	Normanton Le Heath	1–38	
93. Goadby Marwood	Goadby Marwood	1–41	
	Scalford	43–236	
	Stonesby	237–311	
	Waltham	312–470	
	Wickham Cum Chadwell	471–524	
94. Branstone	Branstone	1–53	
	Eaton	54–156	
	Eastwell	157–191	
95. Packington	Packington	1–146	
96. Burrough	Burrough	1–67	
	Dalby Magna	68–170	
	Pickwell & Leesthorpe	171–243	
	Somerby	244–399	
97. Buckminster	Buckminster	1–85	
	Coston	86–118	
	Garthorpe	119–141	
	Saxby	142–159	
	Sewstern	160–231	
	Sproxton	232–334	
98. Barton In The Beans	Barton In The Beans	1–68	
	Gopsall	69–78	
	Nailstone	79–183	
	Shackerstone	184–263	
	Bilstone	264–295	
	Congerstone	296–362	
99. Barlestone	Barlestone	1–264	

Income Tax Parish	Civil Parish	Hd. nos.	Remarks
	Newbold Verdon	281–548	
100. Bardon Park	Bardon Park	1–122	
	Markfield	123–635	
	Stanton Under Bardon	636–824	
	Thornton	825–1002	
101. Bagworth	Bagworth	1–254	
102. Atherstone	Atherstone	1–16	
	Fenny Dratton	17–48	
	Witherley	49–200	
103. Oadby	Oadby	1–870	
104. Norton By Twycross	Norton By Twycross	1–90	
105. North Kilworth	North Kilworth	1–156	
	South Kilworth	157–384	
106. Newton Harcourt	Newton Harcourt	1–62	
	Wistow	63–84	
107. Narborough	Narborough	1–468	
108. Bringhurst	Bringhurst	1–29	
	Drayton	30–88	
	Easton Magna	89–307	
	Stockerston & Holyoaks	308–339	
109. Braunstone	Braunstone	1–55	
	Braunstone Firth	56–61	
	New Parks	62–94	
110. Market Harborough	Market Harborough	1–562	
	Bowden Magna	1949–2236	
111. Holt & Bradley	Holt & Bradley	1–13	
	Medbourne	14–189	
112. Higham On The Hill	Higham On The Hill	1–166	
113. Hinckley	Hinckley	1–3811	
114. Halstead	Halstead	1–66	
	Tilton On The Hill	67–119	
	Whatborough	120–131	
115. Foston	Foston	1–16	
	Kilby	17–112	
116. Fleckney	Fleckney	1–547	
117. Enderby	Enderby	1–879	
	Lubbesthorpe	880–961	
118. East Langton	East Langton	1–98	
	West Langton	99–131	
	Thorpe Langton	132–183	
119. Peckleton	Peckleton	1–97	
	Elmsthorpe	98–114	
	Kirkby Mallory	115–179	
	Leicester Forest West	180–190	
120. Brooksby	Brooksby	1–10	
	Hoby	11–94	

Income Tax Parish	Civil Parish	Hd. nos.	Remarks
	Rotherby	95–137	
121. Nether Broughton	Nether Broughton	1–162	
	Dalby Super Wolds	163–278	
122. Chilcote	Chilcote	1–32	
	Stretton En Le Field	33–52	
	Oakthorpe & Doniston	53–742	
123. Mountsorell	Mountsorell	1–827	
124. Cossington	Cossington	1–126	
	Ratcliff On Wreak	127–173	
125. Barkbey	Barkbey	1–250	
	Barkbey Thorpe	251–267	
	Syston	268–1401	
126. Whitwick	Whitwick	1–1162	
127. Hugglescote	Hugglescote	1–1335	
128. Beeby	Beeby	1–26	
	Hungarton	27–170	Includes Baggrave Ingarsby Quenby Keyham
129. Burton On Wolds	Burton On Wolds	1–91	
	Prestwold	92–109	
	Cotes	110–124	
	Hoton	125–205	
130. Coleorton	Coleorton	1–204	
131. Heather	Heather	1–215	
132. Thringstone	Thringstone	1–427	
133. Shepshed	Shepshed	1–1746	
134. Edmondthorpe	Edmondthorpe	1–53	
	Wymondham	54–365	
135. Cold Overton	Cold Overton	1–41	
	Knossington	42–140	
	Owston & Newbold	141–198	
136. Measham	Measham	1–722	
137. Ab Kettleby	Ab Kettleby	1–97	
	Holwell	98–183	
	Wartnerby	184–246	
138. Ashfordby	Ashfordby	1–367	
	Grimstone	368–432	
	Ragdale	433–457	
	Saxelbye	458–482	
	Shoby	483–494	
	Sysonby	495–555	
	Welby	557–577	
139. Seagrave	Seagrave	1–140	
140. Ashby Folville	Ashby Folville	1–67	

Income Tax Parish	Civil Parish	Hd. nos.	Remarks
	Barsby	68–161	
	Gaddesby	162–247	
	South Croxton	248–344	
	Thorpe Satchville	345–454	
	Twyford	455–561	
	Marefield	562–567	
141. Humberstone	Humberstone	1–152	
142. Evington	Evington	1–234	
143. Willesley	Willesley	1–21	
144. Blackfordby	Blackfordby	1–180	
145. Ashby De La Zouche	Ashby De La Zouche	1–1672	
	Ashby Wolds	1673–2481	
146. Kegworth	Kegworth	1–287	
147. ? (missing Book ?)	Hemington	864–1001	
148. Swannington	Swannington	1–645	
149. Barrow On Soar	Barrow On Soar	1–879	
150. Diseworth	Diseworth	1–163	
	Ipsley Walton	164–178	
	Langley Priory	179–189	
151. Breedon On The Hill	Breedon On The Hill	1–280	
152. Hathern	Hathern	1–448	
153. Worthington	Worthington	1–392	
154. Snarestone	Snarestone	1–96	
155. Queniborough	Queniborough	1–306	
156. Newton Linford	Newton Linford	1–166	
157. Quorndon	Quorndon	1–812	
158. Frisby By Wreake	Frisby By Wreake	1–177	
	Kirby Bellars	178–279	
159. Long Whatton	Long Whatton	1–233	
160. Melton Mowbray	Melton Mowbray	1–2490	Includes Eye Kettleby
161. Barkestone	Barkestone	1–108	
	Plungar	109–191	
	Redmile	192–344	
162. Freeby	Freeby	1–47	
	Thorpe Arnold	48–92	
163. Thuraston	Thuraston	1–138	
164. Bescaby	Bescaby	1–16	
	Croxton Kerrial	17–150	
	Saltby	151–228	
165. Bottesford	Bottesford	1–479	
	Muston	480–644	

Table 6.6 *Numbers of civil parishes per Income Tax Parish in Leicestershire c.1910*

No. of civil parishes	No. of ITPs	ITPs as %
<1	nil	nil
1	87	52.1*
2	33	20.0
3	30	18.2
4	8	4.8
5	2	1.2
6	3	1.8
7	3	1.8
Total	165	99.9

Notes:
*Includes Borough of Leicester.
Source: Leicestershire CRO, Valuation Books for Leicestershire and Rutland (uncat.).

consider an investigation into the use of the valuation material in attempting to shed light on the chainmaking industry of Cradley Heath in the Black Country.[41] The first point that had to be recognised was that Cradley Heath was part of the large ITP of Rowley Regis. The ITP contained almost 8,900 hereditaments, and in order to locate Cradley Heath within these, a visit to Dudley Library was made and the Valuation Books examined. Only with the aid of a street plan was it possible to isolate Cradley Heath within its ITP. The Valuation Books were compiled on a street-by-street basis making identification easy at first sight, but it soon became clear that the book was also arranged by categories of hereditament (Table 6.7). This arrangement complicated the investigation, and Cradley Heath was found somewhat scattered through the books, although identification was still possible.

The hereditaments were classified under the headings outlined in Table 6.7, where the numbers in each that relate to Cradley Heath are indicated. The Valuation Books directly indicate nineteen hereditaments devoted to chainmaking. *Kelly's Directory* for Staffordshire (1916) indicates thirty-three firms involved in the industry. This may indicate an increase in the size of the industry in the six years between the two sets of data – quite probable given the onset of war in 1914 – but it may also plausibly suggest that the Valuation Books do not always accurately indicate the purpose to which a hereditament

[41] For a slightly fuller account of the case study, see Brian Short and Mick Reed, *Landownership and Society in Edwardian England and Wales* (University of Sussex 1987), 44–5.

Table 6.7a *Classification of hereditaments in Rowley Regis ITP, Warwickshire*

Rail and other utilities	8028–43
Manufactories	8044–200
Collieries and quarries	8021–244
Land	8245–460
Schools and public buildings	8461–80
Government property	8481–2
Churches and chapels	8762–93
Unrated property	These are untidily entered and positive identification is seldom possible.

Table 6.7b *Hereditament numbers identified as covering Cradley Heath, Warwickshire*

General	6293–8027
Rail and other utilities	8033 and possibly others not clearly defined
Manufactories	8071–128
Collieries and quarries	8024–5, 8221, 8223
Land	8373–94
Schools and public buildings	8477–80
Government property	8481
Churches and chapels	8787–93
Unrated property	These are untidily entered and positive identification is seldom possible.

Source: Dudley Public Library: Valuation Books for Rowley Regis.

is put, or that firms in *Kelly's* were not necessarily operating in Cradley Heath, or were entered more than once.

However the Valuation Books give no direct indication of the existence of a substantial outwork industry. What was evident, though, was the existence of 'workshops' attached to a great many small hereditaments – often houses – and it is here that we could expect to locate the outwork industry. Unfortunately the study could go no further using the valuation material since the Field Books proved to be missing at the PRO.[42] This tends to reinforce the view that the separation of Field and Valuation Books by the PRO was extremely unfortunate. Given the arrangement of data by ITPs, it is

[42] For a vivid account of the conditions of work in this industry, see Peter Keating (ed.), *Into Unknown England, 1866–1913* (London 1976), 174–84; for a photograph of the chainmakers, see Marie Rowlands, *The West Midlands from A.D. 1000* (London 1987), 310.

impossible to ascertain whether the Field Books for a particular location exist, until an examination of the Valuation Books has been made. This may require a long trip to a local repository, both to identify the ITP within which a particular locality is situated, and to isolate it within that ITP, before consulting the PRO class lists.

In July 1986, the Search Room staff at the PRO were given a tutorial on the valuation material, and told to inform readers of the need to consult the Valuation Books for most purposes, before attempting to use the Field Books. This necessary advice possibly highlights the original error of considering the Valuation Books to be of only peripheral importance.

This system of organisation causes two types of problem unless one is familiar with the system. When consulting the Field Books in the PRO, one needs to know the ITP in which a civil parish is situated. Unless the civil parish gave its name to the ITP, the class lists will give no indication that the Field Books for the parish survive and searchers may assume that their quest is a fruitless one. Furthermore, even if the civil parish sought is of the same name as the ITP and the relevant Field Books are found, they may well contain material actually situated in other civil parishes, and there will be no indication of this in the PRO finding aids, nor in the Field Book itself. This problem is quite easily overcome in terms of documentary linkage in that the Valuation Books provide the vital entry into the Field Books. It will be recalled that the Land Valuation Officers (LVOs) compiled the Valuation Books by entering the information from the Rate Books of each civil parish in the ITP. The next step was to number the hereditaments consecutively throughout the ITP beginning at number 1. Thus in the Cradley Heath example, since the ITP of Rowley Regis contained almost 8,900 hereditaments, Cradley Heath was located only after a visit to Dudley library to examine the Valuation Books.

Only in one instance has an exception to this rule so far been found, presumably owing to a clerical error. This occurred in the case of an investigation of Cromford, Derbyshire. According to the PRO finding aid and to the Field Books themselves, Cromford was the name of an ITP comprising some 3,500 hereditaments. On this basis, Cromford should have been the first parish covered in the ITP and the hereditaments in the civil parish should have been numbered from no. 1. In order to isolate the hereditaments within the ITP that were actually in the civil parish of Cromford, it was necessary to consult the Valuation Books in the Derbyshire CRO. These showed that Cromford civil parish actually comprised nos. 2957 to 3283 within the ITP. They also gave the ITP as 'Matlock', yet Matlock hereditaments were numbered from 1126 to 2956. In fact the first civil parish in the Valuation Books was Matlock Bath which should therefore normally have been the name of the ITP.[43] Probably

[43] Derbyshire CRO D595R/4/1/71–3. For a fuller treatment of the anomaly in its localised setting see Chapter 9.

the error was made in 1910. Perhaps the Land Valuation Officer copied the Rate Books of the component parishes of the ITP in the wrong order, or else the clerical staff in the DVO assigned an incorrect name to the ITP. In any event the Valuation Books give the separate civil parishes in the ITP and also the number of each hereditament within the ITP. Valuation Books were supposed to be confined to a single ITP, though in practice this was not always complied with. However, in these exceptional cases, the numbering of the hereditaments gives the necessary clue to the identity of the ITP.

Another potential difficulty arises where a unit of valuation extended into two or more parishes or Valuation Districts. In such a case, the whole unit was normally, though not necessarily, valued and recorded under that parish or district in which the greater portion of the property was situated. The instructions for making returns on Form 4, which were sent out on Form 2, make it clear that:

If any piece of land under one occupation extends into two or more parishes, separate returns may, if the owner thinks fit, be made for the parts lying within each parish, or, one return, relating to the whole of the land, may be made in the parish in which the greater part of the land is situate. In the latter case notes should be made on the forms for the other parish or parishes affected, 'Included in return for . . . parish' (stating the name of the parish in which the inclusive return has been made).[44]

The Instructions to Valuers issued in 1910 is more clear cut, and gives no choice:

Where a unit of valuation extends into two or more parishes of Valuation Districts, the whole unit should be valued and recorded under that parish or district in which the greater portion of the particular property is situate. In all such cases an advice must be sent to each District Valuer affected where any property partly in his district has been included in the valuations of an adjoining district.[45]

The latter appears to leave the owner of any hereditament so affected with no choice but to include the unit in the parish or district where the majority of the holding was situated.

Nevertheless, from a geographical viewpoint, the problem is obvious. In many cases there is another 'invisible' unit that might be called the 'hereditament parish' that is neither identical with a civil parish nor with an ITP. Hereditaments in any given parish may: (a) have a larger area than that actually within the parish boundaries because of the inclusion of land from an adjoining parish or parishes, or (b) be excluded from the parish entirely because the portion in an adjoining parish is of a greater extent so that the whole is included in the adjoining parish. Thus land may be 'imported' into or

[44] Inland Revenue Valuation Department, *General Instructions to Land Valuation Officers* (HMSO, London 1910), 2.

[45] 'Instructions relating principally to particulars which it is compulsory to furnish', Section I, Instructions for making returns on Form 4 (Form 2).

'exported' from a parish for the purposes of valuation. However, for rating purposes, the actual parish in which a piece of land was situated was the important point, and each part of a unit of valuation should be listed and numbered in the Valuation Book in the first instance, and where combined with another portion in an adjoining parish, this should be indicated in the Valuation and Field Books. In practice this is not always the case. Moreover, an accurate figure for area is only likely to be found for the combined units, so that the area of each of the component parts will be unknown. Full resolution of the problems caused by this instruction are in principle straightforward, though the task is likely to be an arduous one and requires access to the Record plans as well as the Valuation and Field Books. An example can be cited from the Ashburnham estate in Sussex, where the estate, covering 19,000 acres according to the Field Books, was centred on the parish of Ashburnham itself. But the acreage stated to be within the parish actually amounted to 131 per cent of the total area of the parish![46]

Clearly this figure is absurd, but it is explicable in terms of the 'importing' of land into the parish on the Forms of Return. Access to the Record Sheet plans resolves the problem in the case of Ashburnham, at least in general terms, and suggests the need for easy access to the plans for most studies. In this case, Ashburnham Park overlapped into Penhurst parish, while substantial areas of woodland overlapped into Dallington, yet were included in Ashburnham. There were other less dramatic cases of discrepancies between hereditament boundaries and that of the parish, as Fig. 6.7 shows. It follows from this that the figures for those parishes adjoining Ashburnham, especially Penhurst and Dallington, are understatements, and suggests that this exercise can only be reliably attempted with access to the plans. It will be obvious that even a close overall correlation between 'revealed' and actual area between two parishes, may conceal substantial differences between civil/ITP boundaries and 'hereditament' parish boundaries. With little coincidence between property boundaries and parish boundaries, one must expect the Valuation material to yield acreages which are only approximations to 'reality'. Similar discrepancies have so far been found for the Ashley Walk (Hampshire) area and Lockinge (Berkshire).

Problems derived from 'unofficial' practices of valuation staff

The most universal of these problems concerns unrated property and some other entries of uncertain status. It will be remembered that the LVOs had to list all unrated properties in a parish once the entries in the Rate Books had been transcribed. These entries are almost always very sparse in terms of

[46] PRO IR 58/29198–29769. See Chapter 8 for a fuller account of the Ashburnham Estate *c.* 1910.

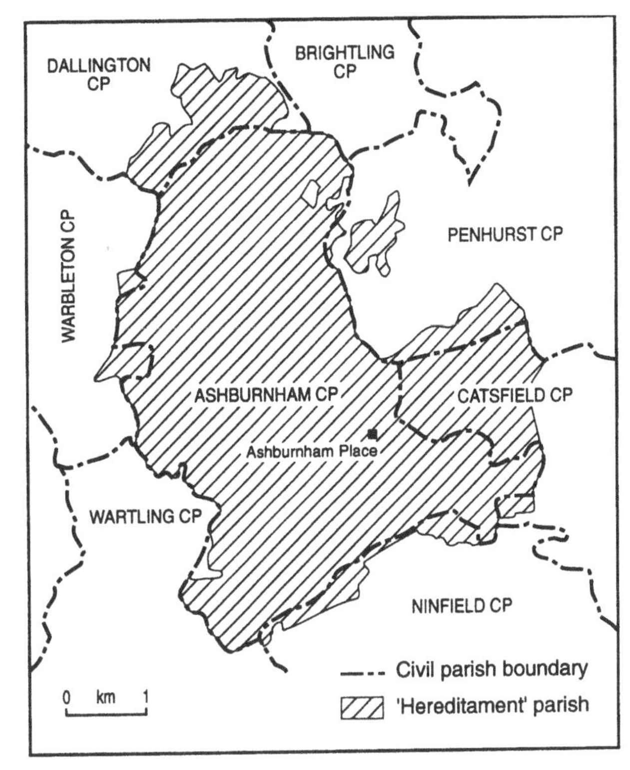

Fig. 6.7 Ashburnham, East Sussex, 1910: civil and 'hereditament' parishes.
Source: East Sussex RO: Valuation Books; PRO IR 58/29198–769.

detail. Moreover it is frequently difficult to assign individual properties to a particular civil parish except where obvious buildings such as churches or chapels are involved. Sometimes the LVO entered these properties immediately following the rated entries for the relevant civil parish. Frequently though, the LVO made these entries at the end of the Valuation Book following the last civil parish in the ITP. In this case it is commonly very unclear to which civil parish an individual hereditament belonged.

Another class of entry is also frequently indistinguishable at first sight from the unrated hereditament. If units of valuation were subdivided after valuation, the original number was supposed to be retained with the addition of a subordinate number. If a hereditament in Leeds was numbered 1485, its correct identification was Leeds 1485. If it was subsequently subdivided (or pieces were 'lopped off' in the words of the Instructions to Valuers) the subdivisions might be referred to as Leeds 1485/1, Leeds 1485/2 and so on. Any further subdivision should then have been referred to as Leeds 1485/1/1 and so on.[47] In the event valuation staff sometimes assigned a completely new number to the subdivisions but it is not always clear that this is so. Valuation Books continued to be used as working documents for very many years after the valuation ended, and there is seldom any indication as to the date of these additions.

Another common problem is that of the inadequate compilation of the Valuation Books. Frequently there is little or no information in these other than that transcribed from the Rate Books (plus unrated property) and the essential hereditament number. The most immediate problem arising is that accurate information as to the area of a hereditament is not easily available. The estimated areas taken from the Rate Books were often extremely inaccurate and are therefore only to be used with great care. Owners were asked to provide information on the area of each hereditament in their ownership, and this was normally copied into the Field Book. However this was by no means always accurate. The valuer calculated the precise area of a hereditament by aggregating the parcel areas printed on the Ordnance Survey sheets. Unfortunately, there was no space provided for this figure in the Field Book, and valuers therefore entered it anywhere – or nowhere – in the Field Book. It was intended that it be entered in the Valuation Book but was frequently omitted. The Provisional Valuation Form (Forms 36/37) always contains the accurate figure, but these have frequently been destroyed. In these cases reference to the Record Plans is necessary to calculate the missing areas, unless relying on the Rate Book or owners' figures.

There are also sometimes major omissions in the documents. In parishes around the New Forest, for example, such as Ashley Walk, only some 13 per cent of the civil parish was accounted for in the documents. The reasons for this remain unclear, but are probably linked to the fact that Crown lands such as much of the New Forest were exempted from the payment of Increment Value Duty.[48] Under such circumstances, Crown land might reasonably be expected to have taken a back seat with priority being given to the valuation of land which could yield tax in the near future, and coverage is therefore more patchy. The various problems which brought the valuation to a near halt in 1914 may have meant that much Crown land had not been valued by this time. The valuers were given clear instructions that they must first attempt a complete valuation of all land in their districts which would be subject to

[47] Inland Revenue Valuation Department, *Instructions to Valuers*, Part I (HMSO, London 1910), 25.

[48] Finance (1909–10) Act 1910, Section 10.1: 'Any increment value duty in respect of the fee simple of, or any interest in, any land held by, or in trust for, His Majesty or any department of Government, which would have been collected on any occasion had it been held by a private person, shall for the purposes of the provisions of this Act . . . be deemed to have been paid.'

Undeveloped Land Duty, before beginning on 'covered land or purely agricultural land'.[49]

Although not found in the Valuation Office literature, it becomes clear that there was a preferred order of dealing with the land – to start with land liable for Undeveloped Land Duty, to go on to other agricultural and built-up land, and then to deal with land which was not immediately liable to any of the taxes, such as Crown land or land belonging to rating authorities and statutory companies. Because land belonging to the latter might at some future time be subject to the taxes, they would certainly have to be valued, but given the urgency with which the valuers were required to work, such land might reasonably be deferred. And thus, for the reasons advanced above, it might never be valued at all.

So, if one turns to a quite different area to investigate industrial structures, one must beware of ownership by statutory companies which were similarly exempted from duty under Section 38.1 of the Act. Thus:

> Neither increment value duty, reversion duty, nor undeveloped land duty shall be charged in respect of any land whilst it is held by a statutory company for the purpose of their undertaking.[50]

> The Commissioners shall not require a statutory company to make any returns with respect to any such land . . . as to valuation other than as the actual cost to the company of the land, and that cost shall, for the purposes of this Part of this Act, be substituted for the original site value of the land.[51]

> For the purposes of this section the expression 'statutory company' means any railway company, canal company, dock company, water company, or other company who are for the time being authorised under any special Act to construct, work, or carry on any railway, canal, dock, water, or other public undertaking . . . [52]

In attempting to investigate the industrial structure of the Hartlepools, for example, the Valuation Books were first examined at Cleveland County Archive Office. Unfortunately, the Books contain only the Rate Book information in almost every case, with any additional information on values etc. present for only a handful of entries. No areas were given.[53] On turning to the Field Books it can be discerned that many of the industrial premises were owned by the North-Eastern Railway Co. (NER), and other statutory companies. There were thirty-three hereditaments, including sawmills, shipyards,

[49] *Instructions to Valuers*, Part I, 32.

[50] Finance (1909–10) Act 1910, Section 38.1. [51] *Ibid.*, Section 38.2. [52] *Ibid.*, Section 38.4.

[53] Cleveland County Archives Department (CAD), Middlesbrough. Valuation Books (there are thirty-three volumes covering the Hartlepools area in total, although these include parishes within County Durham such as Shotton, Easington and Seaham). In addition there are approximately 60 linear feet of shelving taken up by Forms 37–Land, which have not been inspected but which may shed more light on the industrial structure of the area.

docks and quays, engine sheds, gas and water works, which were said to be owned by the statutory companies. Another three, including sawmills, engine yards and docks, were said to be owned by other people, but a pencilled note 'NER' is all there is to be found elsewhere in the Field Book. Four other hereditaments were said to be owned by other individuals, but in fact the 'owners' claimed on Form 4 to be merely short-notice tenants of the NER. In the case of T. W. Watson (hereditament nos. Hartlepool 3582–3) it was stated that 'he has no knowledge of his interest in the property' and furthermore: 'The deponent is a tenant subject to 3 mths notice from any date. He has asked the NER Co for the necessary information to complete the document. They replied you must refer the Government to our Head Office at York.'[54] The Valuation Books give T. W. Watson & Co.'s Boiler-covering works as being in Hart Road, with their fish warehouse in an adjoining building.[55]

Overall the Valuation Books conveyed the information that there were sixty-nine large industrial premises in Hartlepool and West Hartlepool. But since forty (58 per cent) of the hereditaments here were owned by the NER or other statutory companies, no information is available for any of them in the Field Books. Ownership by NER implied no necessary functional connection with the railway, and could therefore include tenants such as Thomas Robinson and Sons with their Egg Preserving Works at Middleton.[56] Of the remaining twenty-nine hereditaments, details of five were said to be in other files, such as the information for Ralph Barker's Howcroft Carriage Works in Oxford Road, for which it was said 'see details filed' or Harrison and Singleton's Baltic Sawmills in Baltic Street 'all set out in file with Form 4';[57] two were composed of derelict land or had buildings which had been demolished on them, and all but two of the remainder had virtually no information, apart from values. The remaining two – a paper works and an expanded metal factory – had details of the use to which the various buildings on the complex were put, but little else.

It would therefore appear that work from the 1910 Valuation material on large industrial premises is likely to be problematic even without the added complication of the presence of statutory companies as landowners. The sheer size and complexity of the premises would certainly have defeated attempts to encapsulate the information within the confines of the Field Books anyway, and most of the information would have been contained in files. The use of separate files was commonplace not only for industrial complexes but also for large country houses where the relevant detail was too copious to enter in the

[54] PRO IR 58/38121. [55] CAD Valuation Books, vol.3.

[56] CAD Valuation Books. Hereditament no. 15586. All hereditaments between 15479 and 15708 were stated to have been wrongly entered in vol. 9 of the West Hartlepool volumes, and should be in one of the Hartlepool Valuation Books.

[57] PRO IR 58/38274; 38302; and see CAD Valuation Books for Hartlepools, vols. 1 and 2.

limited space provided in the Field Book. These files are now unavailable and have probably been destroyed.

In Lockinge (Berkshire), for example, the practice of entering descriptive material into a file instead of the Field Book was carried out in every case so that there is no descriptive information available for this parish. The Field Books relating to the properties of the Lockinge estate are thus very disappointing in that not a single description of any property was included. References are made here to 'the file', to the 'notebook', or to the 'schedule' attached to Form 37. The two former have presumably not survived, while the latter may or may not be still attached to the forms, and until access to these is possible, we cannot know. Thus a re-examination of the Lockinge estate, Berkshire, originally studied by Michael Havinden and published as *Estate Villages* (1966), could add frustratingly little of significance, since only the Valuation Books yield information at present.[58] The Forms 37 are in the Berkshire Record Office but are unlisted, and the Field Books contain not one description of a property, referring instead to 'the notebook' or 'the file' or 'the schedule'. From the Valuation and Field Books one can obtain information on hereditament size, tenancy terms and rents. One can discern that Lady Wantage owned the vast majority of the land in the three parishes (two ITPs), but the gap in Havinden's original 1966 publication for the years between the 1890s and 1918 can only partially be filled by the valuation material. This should therefore serve as a warning not necessarily to expect good coverage, and to check with both county repository and PRO before commencing a research project.

This case study certainly points to one finding, however. The practice of attaching schedules holding important information to Forms 37 seems quite common, and the widespread destruction of these is therefore likely to have been detrimental to research.

Conclusion

This chapter has pointed to some potential uses of the rich variety of documents, and has specifically focussed on issues of landownership and landholding structures, housing studies, commercial and industrial premises, genealogical studies and the use of the material in legal contexts. The following chapters will expand on more uses in greater detail. The chapter has also, however, pointed to the dangers of using the material without a working knowledge of its internal assumptions and inconsistencies. There is nothing

[58] M. A. Havinden, *Estate Villages: A Study of the Berkshire Villages of Ardington and Lockinge* (University of Reading 1966); Berkshire CRO P/DVO 6 Acc. 2487/55 Valuation Book; PRO IR 58/68884–5, 68986, 69369 Field Books.

unique to the valuation material about such matters. There can be no historical documents which do not have such problems, but it is vitally important in using the material in innovative ways that the researcher is fully aware that, in these as in all records, there is no neutrality. These were prepared for reasons of political policy, and in the heat of political and social change.

7

Urban social area analysis 1909–1914

The intention in this chapter is to suggest various themes within the broad topic of the geography of Edwardian towns. All the themes might be broadened and deepened by further research, and what follows should not be seen as an attempt to present polished empirical findings on particular urban areas. The study of contrasting internal social geographies within towns is one possible theme, and will be the subject of the following discussion, drawing on four very different towns from north and south, both large and small, and from industrial and service centres. Brighton, Cromer, Bradford and Camberwell in South London form the case study areas.

The streets of Brighton

By 1910 Brighton had over a century of steady urban growth behind it. Gone were the days of royal splendour, when the new urban resort basked in the reflected glory of the Prince Regent, and the town now looked to salaried middle-class trippers and holiday makers for its main sources of income and to the provision of services and light industries serving a growing professionalisation within the London orbit. Annuitants, clerks and white-collar workers joined businessmen and their families to give an interesting mixture of raffish excitement and middle-class solidity in a town now large enough to accommodate a great variety of social and economic interests. Growth now pushed the town physically northwards and westwards into the medieval parishes of Hove, Preston, West Blatchington and Aldrington. In 1873 the borough boundaries were extended to cover much of Preston, and lacking a paternalistic developer, substandard housing grew across former open-field strips around workshops and the railway engineering works. Slums were being demolished as early as the 1860s, but by 1911 Brighton had one of the highest population densities for any county borough in England and Wales

To illustrate the social variation between Brighton's streets, a sample of 302 hereditaments in seven streets in Brighton used data taken from Field Books

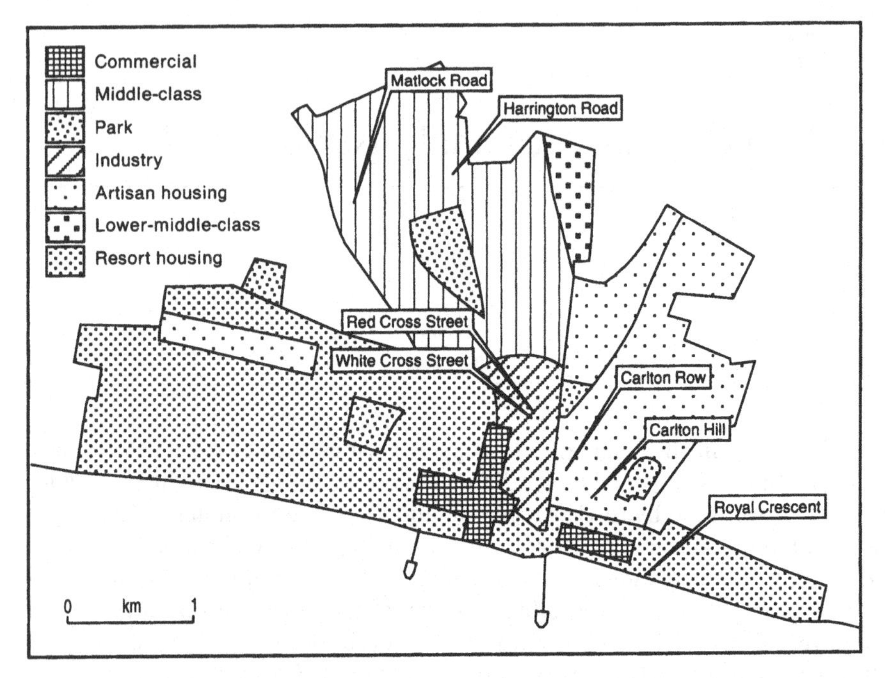

Fig. 7.1 Social areas in Brighton and Hove, East Sussex *c.*1910.
Source: based on Farrant, Fossey and Peasgood, *The Growth of Brighton and Hove 1840–1939*, 26.

and Valuation Books, the latter to identify relevant hereditament numbers for each of the seven streets, and to note owner-occupation with some ease.[1] These streets were fairly representative of the major residential areas in the town, though they cannot of course be assumed to be typical of the town in every respect. For example, no major commercial areas are studied here (Fig. 7.1 and Table 7.1).

Carlton Hill and Carlton Row were amongst the poorest and most infamous areas of Brighton, in the eastern working-class suburbs built in the early nineteenth century – 'like a dark and noisome corner in a beautiful garden' (Fig. 7.2). Thus we find No.7 Carlton Hill to be a common lodging house, described in the Field Book as 'Hse damp, distempered walls poor', with two rooms on each of four floors but with a ground floor which had in addition a scullery, yard and WC. No.15 was 'untenantable'. Old and poor housing was common here, mixed with old shops, taverns such as the Devonshire Arms, the Rising Sun, or the John Bull, 'old in poor neighbourhood', or the Alma

[1] East Sussex Record Office: Valuation Books IRV 1/18, 23, 29, 110; PRO Field Books IR 58/12718–19, 12622, 12701–2, 12947, 12453, 12626.

Table 7.1 *Brighton, East Sussex: streets included in the case study*

Street	No. of hereditaments
Carlton Hill	83
Carlton Row	51
Red Cross Street	47
White Cross Street	42
Matlock Road	29
Harrington Road	36
Royal Crescent	14
Total	302

Fig. 7.2 Housing prior to demolition in the Carlton Hill area of Brighton, East Sussex, in the 1930s.
Source: Brighton Urban Studies Centre.

Tavern beer house which stood close to the wooden Mission Hall with mixtures of cobbles, brick, stucco and slate materials. Carlton Row similarly had many small tenements in poor condition, with No.14 having a yard at the back with two WCs being shared between three houses.[2] This housing, in streets steeply rising from the central Steine area, was subsequently demolished under slum clearance orders in the 1930s.[3]

White Cross and Red Cross Streets were also in a working-class area, near the Brighton Station and London Road and to the north of the Steine, and in an area of mixed small industrial units and artisan housing in the narrow

[2] PRO IR 58/12701–2.

[3] The overcrowded common lodging houses (doss houses) provided shelter for those such as tramps, vagrants, itinerant workers who found it difficult to keep a permanent home. After 1881 bye-laws restricted numbers of lodgers and segregated the sexes, unless married couples were involved. See K. Fossey, 'Slums and tenements 1840–1900', in S. Farrant with K. Fossey and A. Peasgood (eds.), *The Growth of Brighton and Hove 1840–1939* (Centre for Continuing Education, University of Sussex 1981).

London Road valley on cheaper land away from the fashionable seafront. Brick, flint, stucco and slate building materials were used to create workshops and small houses which dominated the Field Book descriptions for the two streets. The condition of the accommodation was perceived to be very much better.

Harrington Road and Matlock Road were in the Preston Park area, on either side of the London Road and London railway line. The former was on the Harrington estate, being developed for housing prior to the First World War, and houses were large, detached and comfortable. The latter was also new in 1910, although being developed as middle-class terraced housing. The Harrington Road properties could be large; 'Burmah', purchased by Mr Edward Geere in 1905, was a 'very good' detached house with two attics and a box room on the second floor, three bedrooms, a drawing room, bath (with hot and cold water) on the first floor, two large rooms and a small room and kitchen, scullery etc. on the ground floor. Outside there was a WC, coal cellar, potting shed, small greenhouse and a store.[4] Servants' rooms, stables, half-landings and other accoutrements of solid middle-class life were all to be found here. Matlock Road lacked the refinements but provided comfortable, standardised brick-and-slate-built three-bedroomed terrace properties with bathroom, WC, two living rooms, kitchen, scullery and rear garden.

Finally, Royal Crescent, completed in 1807, had splendid, large four-storey bay-windowed houses faced with black mathematical tiles. The standard design, varying a little from house to house, was that of two rooms on each of the third and second floors with bathrooms and WCs, two rooms and a slip room on the first floor, two rooms, pantry and WC on the ground floor, and sitting room, kitchen and scullery in the basement. Stables were attached off the adjoining mews. No.10 had been modernised by the insertion of a large oak-panelled hall and staircase.

Documentation is not always as full as one would like. The East Sussex Record Office destroyed all copies of Form 37 for Brighton that came into their possession, and no copies of Form 4 are known to exist for the streets examined. Thus the study is reliant upon the Valuation and Field Books. It was expected that the latter would be very detailed since the Superintending Valuer of the Division commented in 1912 that the work of the Brighton Valuation Office presented 'some difficult problems' owing to the extreme attention to detail of the staff, resulting in very slow progress being made.[5] And indeed, the Field Books are very detailed. The Valuation Books also do appear to have been fairly carefully compiled. The valuers' figures for extent were entered consistently, though for the purposes of this case study they were ignored. Also entered were the figures for the various values, though the steps taken to arrive at each value were not indicated.

[4] PRO IR 58/12453. [5] PRO IR 74/148, 91.

Table 7.2 *Types of ownership in selected Brighton streets*

	Freehold	Leasehold	Not stated
Carlton Hill	80	1	2
Carlton Row	50		1
Red Cross Street	46		1
White Cross Street	42		
Matlock Road	26		3
Harrington Road	36		
Royal Crescent	14		
Total	294	1	7

Source: PRO Field Books IR 58/12718–19, 12622, 12701–2, 12947, 12453, 12626.

However, for this study the Valuation Books were not used in any detail. Their chief use was for the identification of the relevant hereditament numbers for each street. In addition a note was made of the numbers of owner-occupiers in each area. Neither were the Ordnance Survey sheets used, although a handful of Working Sheets are at the East Sussex Record Office.

In short, the Field Books were used on the assumption that they would provide all of the required information, but no assessment of their accuracy was therefore possible. Alterations to the detail extracted from Form 4 are frequent, though this seems mainly to have been due to the time-lapse between the completion of the form by owners, and the valuation itself. Alterations are mainly to the names of owner and/or occupier, and also of amounts of rent. Since the status of the alterations is not always clear, the original figures for rent have been retained for the purposes of Table 7.7 below.

Property in Brighton was almost all owned on a freehold basis (Table 7.2) and ownership within the sample was widely dispersed, the 302 hereditaments being owned by 190 different people. In every case except Carlton Row, a massive majority of owners owned but a single property in the streets studied, though there is no way of knowing without further study whether they owned property in other parts of the town. Only in the poorest areas, Carlton Hill and Carlton Row, were there significant numbers of people with more than two hereditaments in their ownership, and in the former there was one person owning fourteen properties. By contrast, in the wealthiest area, Royal Crescent, each of the large houses had a separate owner (Table 7.3).

Owner-occupation was uncommon. Even in the wealthy Harrington Road, Preston Park, and in the elite Royal Crescent, owner-occupiers were in a minority. There is some discrepancy between the Valuation and Field Books in this respect, although it is only in Red Cross and White Cross Streets that

Table 7.3 *Ownership patterns in selected Brighton streets*

	No. of hereditaments owned							
	1	2	3	4	5	6	7	14
Carlton Hill	28	10	6	3	1			
Carlton Row	7	6	3	1	1			1
Red Cross Street	32	5			1			
White Cross Street	20	8	2					
Matlock Road	11	3			1		1	
Harrington Road	20	3	1				1	
Royal Crescent	14							
Total	132	35	12	4	4		2	1

Source: PRO Field Books IR 58/12718–19, 12622, 12701–2, 12947, 12453, 12626.

Table 7.4 *Owner-occupiers in selected Brighton streets*

	Field Books	Valuation Books
Carlton Hill	7	7
Carlton Row	nil	nil
Red Cross Street	5	nil
White Cross Street	3	nil
Matlock Road	5	5
Harrington Road	9	8
Royal Crescent	5	5
Total	34	25

Source: PRO Field Books IR 58/12718–19, 12622, 12701–2, 12947, 12453, 12626; East Sussex Record Office: Valuation Books IRV 1/18, 23, 29, 110.

it is at all significant, and even there it does little to alter the overall picture (Table 7.4).

Most owners were local people. Over 70 per cent lived in Brighton and Hove, and there were several more from Sussex. Outside of the county, only London was the residence for as many as ten owners, indicative perhaps of the flow of capital from London to this Edwardian seaside resort (Table 7.5).

Tenanted property was let on several bases. As one would expect, security was lowest in the working-class areas, where the great majority of properties were let on a weekly basis. At the other end of the scale was Royal Crescent, where all were let annually or for longer periods. In fact six of the seven with details of this were let on three-year tenancies or longer. Matlock Road, while

Table 7.5 *Residence of property owners: selected Brighton streets*

	A	B	C	D	E	F	G	Total
Brighton/Hove	35	12	30	17	15	20	5	134
Sussex (other)	5	2	2	6		1		16
Berkshire		1						1
Devon			2					2
Dorset				1				1
Glamorgan	1							1
Hampshire				1		1		2
Kent	2	1				1		4
Leicestershire			1	1				2
London	3	1	1	1		1	3	10
Somerset							1	1
Surrey			1	1			1	3
Wiltshire		1						1
Yorkshire					1			1
Not stated	2	1	1	2		1	4	11
Totals	48	19	38	30	16	25	14	190

Notes:
A Carlton Hill; B Carlton Row; C Red Cross Street; D White Cross Street; E Matlock Road; F Harrington Road; G Royal Crescent
Source: PRO Field Books IR 58/12718–19, 12622, 12701–2, 12947, 12453, 12626; East Sussex Record Office: Valuation Books IRV 1/18, 23, 29, 110.

having several weekly tenancies, was mainly a road of quarterly or annual tenancies. Only Harrington Road had a substantial number of properties without such information, though it seems likely that longer lettings were usual here (Table 7.6)

Rents varied enormously. Carlton Row had the lowest with about half the properties being let at less than £15 per annum, or less than 5s 9d per week. For Carlton Hill, Red Cross and White Cross Streets, rents were mainly between £15 and £25 a year, while the remaining streets had no properties let as low as the latter figure. Harrington Road homes were let mainly at between 20s and 30s a week, while Royal Crescent houses could be obtained mainly at rents of more than £2 per week, although considering the difference in the quality of accommodation, this would seem very reasonable (Table 7.7).

Tenants of the poorer housing had some respite. Overwhelmingly, landlords took responsibility for payment of rates and insurance, as well as assuming responsibility for repairs, although it seems unlikely that the latter were high on their list of priorities. Occupiers usually had to pay the rates in Matlock Road, though in Harrington Road and Royal Crescent this was less marked owing to the increased incidence of owner-occupation. In the wealthier streets,

Table 7.6 *Tenancy terms (of rented property) in selected Brighton streets*

	A	B	C	D	E	F	G	H
Carlton Hill	54	4	4	8		9	2	2
Carlton Row	42			3		1		5
Red Cross Street	41							1
White Cross Street	25		1	2		3		1
Matlock Road	7	1	6	5	1*	2	1	2
Harrington Road			1	2	5+		1	19
Royal Crescent				1	1+			2
					2$			
					3^			
Totals	169	5	12	21	12	15	4	32

Notes:
* 2-yearly; +3-yearly; $7-yearly; ^21-yearly
A Weekly; B Monthly; C Quarterly; D Yearly; E More than one year; F Leasehold (i.e. leases of unstated length); G Empty; H Not stated
Source: PRO Field Books IR 58/12718–19, 12622, 12701–2, 12947, 12453, 12626.

Table 7.7 *Annual rents in selected Brighton streets*

Rent (£)	A	B	C	D	E	F	G
<10	1	1					
10 to 14 19s 11d	4	23	3				
15 to 19 19s 11d	17	18	10	7			
20 to 24 19s 11d	22	4	17	14			
25 to 29 19s 11d	17	1	8	8	14		
30 to 34 19s 11d	6		8	6	8		
35 to 39 19s 11d	5			1			
40 to 49 19s 11d	4				3	4	
50 to 74 19s 11d	1			1*	2	15	
75 to 99 19s 11d						5	4
100+				1*		1	10
Not stated	6	4	1		2		
Totals	83	51	47	38	29	25	14

Notes:
* includes two hereditaments let as one.
A Carlton Hill; B Carlton Row; C Red Cross Street; D White Cross Street; E Matlock Road; F Harrington Road; G Royal Crescent
Source: PRO Field Books IR 58/12718–19, 12622, 12701–2, 12947, 12453, 12626.

Table 7.8 *Responsibility for rates, repairs and insurance in selected Brighton streets*

	Rates	Repairs			Insurance
		All	External	Internal	
Carlton Hill					
A	61	70	2		71
B	14	3		2	1
C	8	8			11
Carlton Row					
A	45	47	2		49
B	5	2		2	2
C	1				
Red Cross Street					
A	40	45			45
B	6	1			1
C	1				
White Cross Street					
A	30	30			31
B	10	8			7
B1	2	3			2
C		1			1
Matlock Road					
A	6	22			18
B	19	2			2
C	4	4			9
Harrington Road					
A	15	23	2		24
B	16	4		2	4
C	5	7			8
Royal Crescent					
A	5	6	4		9
B	9	4		4	5
C					

Notes:
A Owner; B Occupier; B1 Joint; C Not stated
Source: PRO Field Books IR 58/12718–19, 12622, 12701–2, 12947, 12453, 12626.

owners occasionally made tenants responsible for repairs, particularly those internal within the house (Table 7.8).

The valuer of Red Cross Street entered figures into the Field Book that seem to relate to length of residence of each occupier (Table 7.9). Although certainty on this point is not possible, and such information was not

Table 7.9 *Length of residence: Red Cross Street, Brighton*

Years	No.
<1	7
1 to 1.9	4
2 to 4.9	9
5 to 9.9	5
10 to 24.9	3
>25	7
Subtotal	35
Not stated	6
Empty	6
Total	47

Source: PRO Field Books IR 58/12718–19, 12622, 12701–2, 12947, 12453, 12626.

Table 7.10 *Types of hereditament in selected Brighton streets*

	A	B	C	D	E	F
Carlton Hill	32	30		9	2	9
Carlton Row	43	1	1		1	1
Red Cross Street	40	4	2			1
White Cross Street	25	1	3			13
Matlock Road	23	4	2			
Harrington Road	36					
Royal Crescent	14					
Total	213	40	8	9	3	24

Notes:
A House only; B House/shop; C House/workshop/store; D Beerhouse/public house; E Lodging house/tenement block; F Other
Source: PRO Field Books IR 58/12718–19, 12622, 12701–2, 12947, 12453, 12626.

actually required by the Valuation Department, the context makes it highly probable that this is the case. The figures suggest that some 20 per cent of tenants had lived in their houses for over twenty-five years, while a similar proportion had been in the same house for between five and twenty-five years.

The largest single group of hereditaments in each street was used as living accommodation only (Table 7.10). Indeed in Harrington Road and Royal Crescent, all were in this category, as were most in Matlock Road. In White

Table 7.11 *Number of rooms in hereditaments in selected Brighton streets*

	2	3	4	5	6	7	8	9	10	11	12	n.s.
Carlton Hill	2	16	16	18	9	4	4	1			2	2
Carlton Row		8	36	2								
Red Cross Street		1	26	14	1			1				3
White Cross Street		1	11	15		2						
Matlock Road				2	26		1					
Harrington Road					2	10	12	6	1	1		
Royal Crescent								2	6	2	4	

Notes:
n.s. not stated
Source: PRO Field Books IR 58/12718–19, 12622, 12701–2, 12947, 12453, 12626.

Cross Street there was a substantial number of workshops and stores without living accommodation, but generally hereditaments included living space. Only in Carlton Hill were there large numbers of shops, with nine beerhouses and public houses comprising over 10 per cent of the total hereditaments.

Living accommodation was sparse in the poorer streets. Including kitchens, three or four rooms were the norm in Carlton Row and Carlton Hill, while in Red Cross and White Cross Streets, houses were a little larger, with most premises having four or five rooms. Almost all houses in Matlock Road had six rooms, while in Harrington Road and Royal Crescent, there was even more space, though servants would undoubtedly have occupied rooms in these houses (Table 7.11).

Not only did the poorer part of the population have limited accommodation, but the condition of that accommodation was frequently very poor. Assessment as to condition is likely to be subjective, so that comparison between the assessments of different valuers may be difficult. Nevertheless Table 7.12 suggests stark distinctions between the different areas, with Carlton Hill and Carlton Row having very substantial numbers of premises considered to be 'poor', 'bad' or worse. Red Cross Street and White Cross Street would appear to have been better, and conditions improved as one moved into the more prosperous areas.

Many worthwhile statements about urban property and property relations can be made from the valuation documentation. Assessments of the extent of owner-occupation are easily made in the case of Brighton, where it would seem uncommon in all social groups. Economic activity can be gauged to some extent: the existence of shops, workshops and so on is usually clearly indicated, while the precise character of the enterprise is sometimes described. Used in conjunction with directories and other sources, the material will yield

Table 7.12 *Condition of hereditaments in selected Brighton streets*

	A	B	C	D	E	F
Carlton Hill	6	26	14	11		26
Carlton Row	2	29	9	1		12
Red Cross Street		2	39			6
White Cross Street		5	30	2		7
Matlock Road		14	15			
Harrington Road		6	12	7		7
Royal Crescent			2			12

Notes:
A Very poor/very bad; B Poor/bad; C Fair; D Good; E Very good; F Not stated
Source: PRO Field Books IR 58/12718–19, 12622, 12701–2, 12947, 12453, 12626.

even greater insights.[6] Assessments of living conditions are possible and there are likely to be other sources available for cross-checking in many areas. Reports from the Medical Officers of Health are likely to be useful. Ownership patterns, rents, tenures and accommodation, then, seem to be among the most readily available data from this material.

Cromer

Other methods of approaching the analysis of urban social areas *c.* 1910 might be adopted. Cromer, a small north Norfolk fishing town (population 4,073 in 1911) developing as a coastal resort during the nineteenth century, can be used to illustrate this. The writer Clement Scott had coined the phrase 'Poppyland' for the area around Cromer, an epithet which helped to convey 'romance on a coast ripe for exploitation' in the 1880s and 1890s. The long, level beach, surrounding hills and woods and pleasant, quiet surroundings led to a boom and considerable expenditure on sea walls, esplanade, pier, hotels and other resort amenities. The town was clearly in process of change in its dominant functions from fishing to tourism. New streets were developed, filled with boarding houses and private hotels, and red brick, bay windows, gables and dormers were much in evidence. Most development was to the west and south-east where a golf course was established at Overstrand, itself developing as a high-class suburb. The small town had two stations, so important had access become by 1910.[7]

[6] See, for example, W. Caudwell, 'Horsham: the development of a Wealden town in the early twentieth century', University of Sussex, unpubl. MA dissertation, 1986.

[7] *Kelly's Directory for Norfolk* (1912), 102–3; D. Dymond, *The Norfolk Landscape* (London 1985), 177–8, 238; M. Girouard, 'The birth of a sea-side resort: Cromer, Norfolk', *Country Life* 150 (1971), 424–6, 502–5; E. J. Burrow (ed.), *Cromer and Sheringham as Holiday and Health Resorts* (Cheltenham nd [*c.* 1920]); P. Wade Martins, *An Historical Atlas of Norfolk* (Norfolk Museums Service, Norwich 1993), 166.

Fig. 7.3 Cromer, Norfolk, *c.*1910: the Hotel de Paris after its rebuilding in the 1890s. *Source:* Norfolk Museums Service MUS/W22.

This study relied on the Valuation Books in Norfolk County Record Office, the Field Books in the PRO, and the 1912 *Kelly's Directory*. Cromer's growth at this time as a developing seaside resort was indicated by some 300 'building plots' being included amongst the hereditaments in the valuation documents. Information on these was sparse in the extreme, and they were therefore excluded from the analysis. Neither did the few unrated properties have detailed information entered against them, and these were also excluded from the following study. This lack of information, or its ambiguity, on unrated property is a frequent feature of the material, and inclusion of this category of hereditament in detailed analyses would normally require more detailed study.[8]

The study is concentrated therefore upon the 1,295 rated hereditaments in the civil parish of Cromer (the ITP included the civil parishes of Cromer and Overstrand). A 20 per cent sample was taken, including every fifth hereditament in the Valuation Book, commencing with hereditament No.5. Thus the sample included 259 hereditaments in total.

The documentation on these was generally well-compiled in both Valuation and Field Books, and no fresh problems with the data were noticed, although an idiosyncratic method of compiling the Rate Books, from which the Valuation Book was transcribed, was observed. In Cromer, it seems to have

[8] Norfolk County Record Office P/DLV 1/114 Valuation Book (Cromer ITP); PRO IR 58/62393–407 Field Books.

been the practice to place the more highly rated properties at the beginning of the book; and the poorer properties at the end. Thus some streets are found intermittently throughout the valuation documents on the basis of the rateable value of the specific hereditament. However, there seems no reason that this practice should seriously distort the sample.

Ownership was extremely widespread. In the mid-nineteenth century the largest interests were those of a 'formidable family clan of rich, philanthropic and intensely worthy Gurneys, Hoares, Barclays, Buxtons and Birkbecks', living at Cromer and Overstrand. In the 1880s and 1890s ownership became more diffuse with the sale of building land by Lord Suffield at Overstrand and from the Bond-Cabbell estate in 1890–1 – to the fury of the wealthy cliques.[9] The vast majority of owners in the sample owned just one hereditament, although a few owned two or three. Almost the only substantial owners of property in the town were the Trustees of Benjamin Bond-Cabbell II (d.1892), an enthusiastic developer of the town who had died at the early age of 34, with twenty-four hereditaments included in the sample (9.3 per cent). These included the expansive and gothic Cromer Hall, with some 90 acres of land attached, the residence of E. Bond-Cabbell in 1910. The family were lords of the manor. The remainder of the Trustees' properties consisted of houses, shops and cottages. The family, in the shape of Mrs Bond-Cabbell, also owned four houses in West Street. Unfortunately the Field Books contain few details of Cromer Hall, except that the 80 acres included 9 acres of garden, 36 acres of park and 35 acres of woodland. There were two lodges, a bailiff's house, coachman's cottage and gardener's house, but for more detail we are enjoined to 'see the file' – alas now lost.[10] The only other person to own a substantial number of hereditaments was Mr John Smith of London who owned thirteen cottages in the sample (5.1 per cent), and who had himself developed a good deal of the town.

Owner-occupation was more common than in Brighton, with 28.6 per cent of the sample compared with 8.3 per cent in Brighton being owner-occupied, although the differential basis of the two samples should be noted. Ownership in Cromer was almost entirely (88.8 per cent) freehold.[11] Excluding owner-occupiers, the addresses of the owners of only sixty-nine hereditaments are included in the documentation. Cromer and nearby towns and villages were the home of the majority. Norwich, some 20 miles distant, was home for nineteen owners, while John Smith, already mentioned, came from London, as did three other small owners. The owner of one hereditament lived in Canada. Thus property ownership in Cromer was probably mainly small-scale, and

[9] Girouard, 'Cromer', 425–6. [10] PRO IR 58/62396.

[11] Of the 295 properties sampled, 262 were freehold, three leasehold, two copyhold, three were a combination of copy and freehold, and twenty-five had no tenure stated. The owners of the combined copyhold/freehold premises were unable to differentiate 'on the ground' which parts of their properties were freehold, and which copyhold (PRO IR 58/62393–407).

Table 7.13 *Period of tenancies in the Cromer, Norfolk, sample*

Weekly	2
Monthly	33
Quarterly	27
Half-yearly	3
Yearly	58
Lease	19
Total	142

Source: PRO IR 58/62393–407.

locally based. Finance capital for development had come from London, Norwich and the town itself.

Despite the relatively high levels of owner-occupation, it was still the case that over 70 per cent of sampled hereditaments (185 in number) were tenanted. For these, details of letting periods are available for 142 properties. Monthly, quarterly and yearly tenancies were the most usual as Table 7.13 shows.

This again contrasts strongly with Brighton, where weekly letting was by far the most common except in the more prosperous districts. Monthly lets in Brighton were very unusual, and quarterly tenancies were common only in Matlock Road, where they comprised one-quarter of the properties. Whether this reflects a more substantial rentier economy in Cromer, with higher-quality properties compared with Brighton, or whether the two case studies reflect other regional tenancy conventions, could only be resolved with more research.

Annual rents are known for 186 hereditaments in the sample. Table 7.14 indicates the ranges of rents in the town, but does not distinguish between types of hereditament. It also includes figures for estimated rent in owner-occupied hereditaments.

Well over a quarter of the hereditaments were let at rents lower than £15 a year. In Brighton, only thirty hereditaments (around 10 per cent) were let at rents as low as this, and all but seven were in Carlton Row, where almost half the hereditaments were let at these lower rents. It may be that rents were generally lower in Cromer than in Brighton, or it may be – less plausibly – that properties were of even poorer quality than in Brighton. Without knowledge of other factors, the level of 'real' rents cannot be determined.

On the other hand, over 7 per cent of hereditaments in Cromer were let at rents above £100 a year. Apart from those in Royal Crescent, only two hereditaments in the Brighton sample were let at such high levels of rent. Of course, it should be remembered once again that the different sample techniques between the Cromer and Brighton studies make direct comparisons hazardous.

Table 7.14 *Annual rents in the Cromer, Norfolk, sample*

Rent (£)	No.	%
<10	25	13.4
10 to 14 19s 11d	29	15.6
15 to 19 19s 11d	35	18.8
20 to 24 19s 11d	12	6.5
25 to 29 19s 11d	9	4.8
30 to 34 19s 11d	14	7.5
35 to 39 19s 11d	9	4.8
40 to 49 19s 11d	18	9.7
50 to 74 19s 11d	17	9.1
75 to 99 19s 11d	4	2.2
>100	14	7.5
Total	186	

Source: PRO IR 58/62393–407.

Table 7.15 *Living accommodation (including kitchen) in the Cromer, Norfolk, sample*

No. of rooms	No. of hereditaments
2	8
3	3
4	22
5	19
6	43
7	8
8	11
>8	74
Total	188

Source: Field Books, PRO IR 58/62393–407.

Living accommodation in many houses was ample. Table 7.15 shows the numbers of living rooms in inhabited hereditaments, with the necessary details. Unfortunately, the use of each room is frequently omitted, so only the numbers of rooms can be shown.

The most noticeable feature of Table 7.15 is that almost 40 per cent of the houses had over eight rooms. Only 13 per cent of the Brighton sample had as many rooms as this, though again the different sampling techniques doubtless invalidate many comparisons. Houses of this size were found only in Royal

Crescent and Harrington Road in Brighton. It should be said that some of the Cromer examples were apartment houses and a handful were hotels, neither of which made more than a token appearance in the Brighton case study. Nevertheless, it is a fact that there were very substantial numbers of large houses in Cromer. One example is No.17 Cliff Avenue, a brick and tile-hung detached house with tiled roof set within a 'nice garden and good lawns'. On the ground floor there was a long hall with fireplaces, a fair-sized drawing room, morning room, a 'nice' dining room. There was also a larder, WC, butler's pantry, servants' hall, kitchen, scullery, larder etc. Outside was a coal and boot house. On the first floor were four bedrooms, two dressing rooms, linen cupboard etc., and on the second floor were four more bedrooms. Other properties in Cliff Avenue were equally large.[12]

At the other end of the scale, there were rather more smaller and basic cottages in Cromer than in even the poorest areas of Brighton. Thus among small shops and living premises was No.62 Church Street, described as 'very small' with a 'tiny yard', and described as 'two up, two down' and one WC between two houses.[13] But overall, accommodation seems to have been slightly larger than in Brighton, where the majority of houses had five or fewer rooms. In Cromer, a large majority (72 per cent) had more than five rooms.

Of the 259 hereditaments in the sample, the occupiers of ninety-two (35.5 per cent) were located in *Kelly's Directory* for 1912. Of these, only six were 'private' residents, while eighty-six were entered in the commercial section, such as the Ship Electric Laundry, proprietor Eyre W. Edwards of Church Street, which had an old brick, flint and tile cottage adjoining.[14] This suggests that one-third of Cromer occupiers were involved in some kind of business on their own account. Thirty-one of these occupiers were hoteliers, boarding-house keepers, or otherwise ran apartment houses. Of the sampled hereditaments, 12 per cent were providing accommodation for visitors to the town.[15] Cromer certainly seems to have been well established as a seaside holiday town before 1914. The remaining commercial hereditaments were devoted to the usual business activities of the day: occupants included builders, tailors and shopkeepers of various kinds.

The Field Books occasionally also give insights into social matters. For example, the 'comfortable and well appointed' Hotel de Paris in the High Street (Fig. 7.3) had a very full entry. Rebuilt in brick and slate *c.*1894 at a cost of £8,000 by the local Jarvis family, the old portion was now used for staff quarters. The ground floor contained the usual lounges, dining and drawing rooms, hotel bar etc. On the first floor were sixteen bedrooms and a few private

[12] PRO IR 58/62395. [13] *Ibid.* [14] *Ibid.*

[15] Boarding houses were sufficiently numerous by this time that one house, 'La Maisonette' on a corner site in Cabbell Road, was described as '*not* a boarding house' to avoid any misunderstanding in a street which otherwise had quite a large number of such establishments (PRO IR 58/62393).

rooms, together with four rooms set aside for visitors' chauffeurs, and for their maids, in the old building. On the second floor were twenty bedrooms, including five chauffeurs' rooms and five staff rooms, and on the third floor another twenty bedrooms. The basement contained the kitchens and storerooms, servants' hall and cellars.[16]

The sampling technique used here has enabled a reasonably representative assessment of some aspects of life to be obtained for Cromer as a whole. The documentation is good on ownership, on rentals and tenancy periods, and gives insights into the accommodation of many homes, and occasionally into the valuation process itself. The tenant of 'Bradford Villa' in Cabbell Road was obviously uncooperative, and in the Field Book we find 'Tenant refused to say what rent was'.[17] But without access to other material, notably the 1911 Census schedules, assessments of, and comparison with, other towns on matters such as overcrowding cannot properly be made. The data seem much less problematic to use in an urban parish than a rural one and this may well be a general finding. The small area of most urban hereditaments ensures that few overlap parish boundaries, and it is seldom found that vital details are contained in an entry in the documents of a neighbouring parish, which is so frequent in rural areas, or that an entry may also refer to large areas from another parish. A comparative study of pre-1914 seaside resorts might well be mounted, using the valuation records as a basis, if not as the sole source of information.

Housing in the Bradford area of West Yorkshire

It is, however, also possible to undertake comparative work in a variety of environments, and to illustrate this one may turn to the Bradford area and a study first undertaken by M. J. Mortimore in 1969.[18]

This study investigated the influence of landownership on the speed and spatial pattern of housing development. One element of Mortimore's study was based on two blocks of streets on the outskirts of Bradford (Fig. 7.4). One was in the village of Heckmondwike, and the other in Cleckheaton, both mining villages to the south of Bradford, and further up Spen Valley. Both were within the rapidly growing and large parish of Birstall, whose population overall increased 4.6 times during the nineteenth century, as a constituent part of the rapid population growth associated with textiles and iron mining in the Bradford area at this time.[19]

Both blocks had been substantially built up by 1850 and Mortimore was able to show how former property boundaries influenced Bradford's street

[16] PRO IR 58/62395; Girouard, 'Cromer', 504–5. [17] PRO IR 58/62393.

[18] M. J. Mortimore, 'Landownership and urban growth in Bradford and its environs in the West Riding conurbation, 1850–1950', *Transactions of the Institute of British Geographers* 46 (1969), 105–19. [19] D. Hey, *Yorkshire from AD 1000* (London 1986), 247.

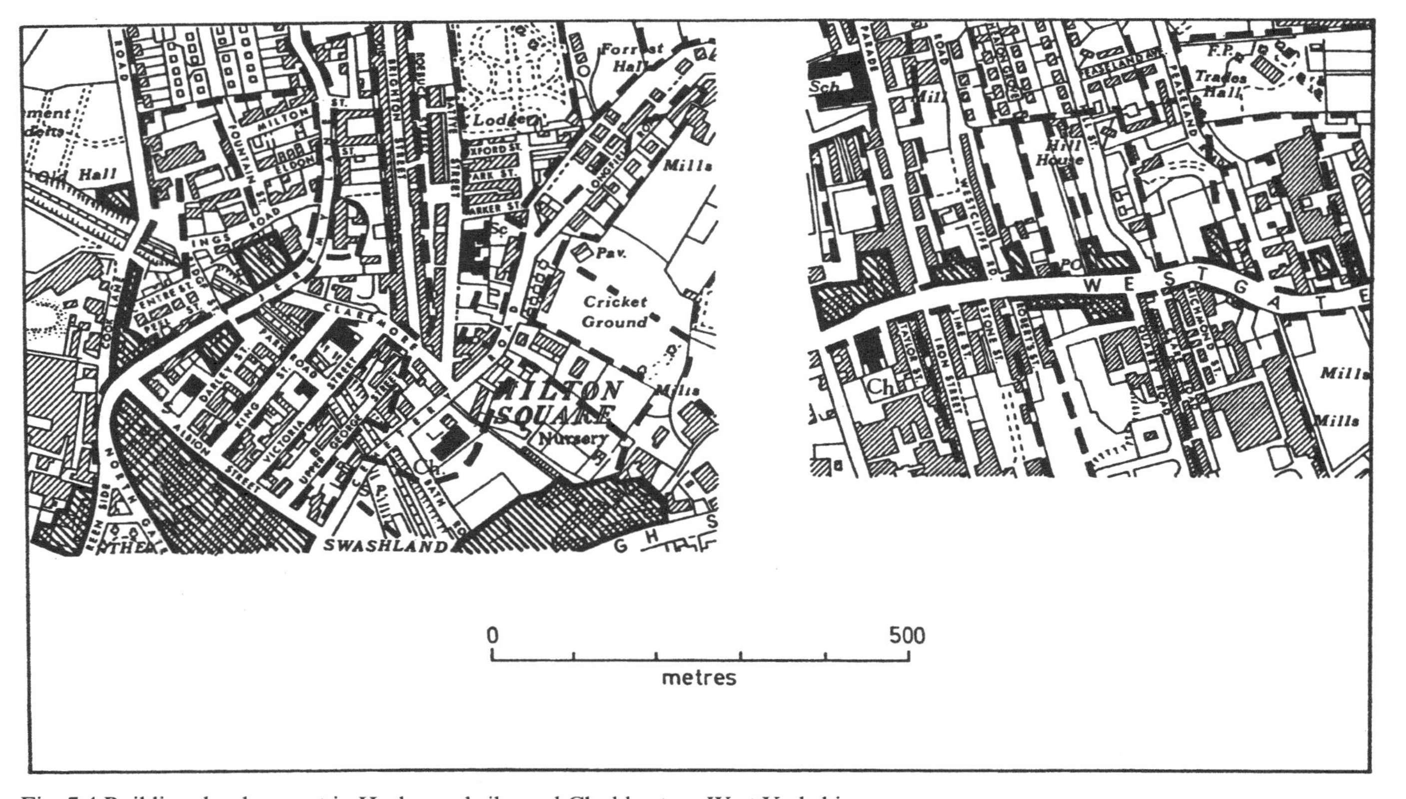

Fig. 7.4 Building development in Heckmondwike and Cleckheaton, West Yorkshire.
Source: After Mortimore, 'Landownership and urban growth', 108.

layout and thus urban morphology. Thus 'existing property units became an invisible skeleton for the growing body of the town'.[20] In Heckmondwike streets followed the older ('antecedent') field patterns of the eighteenth century, which themselves betrayed an older structure relating to intermixed occupation of open-field strips. This gave long, sub-parallel streets which had been built within the property boundaries of eighteenth-century stone walls which themselves derived from the 'activities of the medieval land surveyors'. By contrast, Cleckheaton's fields had been enclosed by Parliamentary Act in 1795, and the streets followed the rectangular pattern of post-enclosure holdings.

The spatial pattern of large estates and smallholdings was fundamental in determining the speed of development of building around Bradford, and in the manner in which landowners reacted to the great demographic pressures to sell and develop land for housing. Many larger landowners chose not to develop, and large open areas were thus preserved by 1910 to the south of Bradford, separating Heckmondwike and Cleckheaton from Bradford itself. The large landholdings of the Low Moor Company in North Bierley and the Bowling Iron Company remained open, with valuable underground mineral rights. Otherwise the smallholder was dominant, and such holdings provided a ready supply of building land with no legal restrictions on their sale, as more and more smallholders gave up their land for cash, and their domestic weaving employment with its requirement of land for tentergarths, agricultural land and water for a proletarian lifestyle.[21]

The smaller outlay necessary by small builders on the freehold land meant that there was a brisk trade in such hereditaments, and building therefore progressed more quickly in areas of predominantly smaller holdings. The free supply of small building plots with few restrictions as to its use was a fundamental feature of Bradford's growth as an urban-industrial complex by 1910. Maximisation of profits on the small plots could frequently lead to overcrowding and back-to-back development in Bradford, where there were said to be over 26,000 back-to-back dwellings in 1884, over 20,000 in 1910 and 33,000 remaining in 1920.[22]

For the purposes of illustration, use was made here of the Valuation Books

[20] Mortimore 'Landownership', 106.

[21] Tentergarths were enclosures in which frames for stretching finished cloth could be erected.

[22] Mortimore, 'Landownership', 116. The correlation between freehold tenure and high-density housing has been observed elsewhere, but has been challenged by M. J. Daunton, *House and Home: Working Class Housing, 1850–1914* (London 1983), 66–8. Daunton cites arrangements from Gateshead whereby full payment for plots was deferred until the construction or sale of the housing, and material from several provincial towns in 1911 to demonstrate a lack of relationship between tenure and overcrowding. For the 1884 and 1910 figures of back-to-backs see A. Jowitt, 'Late Victorian and Edwardian Bradford', in J. A. Jowitt and R. K. S. Taylor (eds.), *Bradford 1890–1914: The Cradle of the Independent Labour Party* (Bradford Centre Occasional Papers 2, 1980), 14.

Table 7.16 *Hereditament ownership in Cleckheaton and Heckmondwike, West Yorkshire*

No. of hereditaments	No. of owners	
	Cleckheaton	Heckmondwike
1	11	28
2	11	17
3	4	7
4		10
5		6
6	6	8
7	3	5
8	1	2
9	2	1
10 and over	8	8
Total	46	92

Source: West Yorkshire Archives Service, Wakefield. C243/106–10; C243/121–2 Valuation Books; PRO IR 58/26814–18, 26977, 26979–83, 26991–3 Field Books.

to ascertain the hereditament numbers and the Field Books to obtain maximum information on the hereditaments themselves. The documents were generally well compiled and the only problem was with the layout of the Valuation Books which was such that hereditaments in a particular street were scattered and intermingled with those in other streets. It was thus necessary to survey a considerable number of Valuation Books in order to isolate the hereditaments in the streets selected for study.[23]

The Cleckheaton block, in the smaller of the two settlements, included 247 hereditaments, although no information at all was available for one of these. In Heckmondwike, the larger of the settlements and an Urban District, twenty hereditaments had no information, leaving 366 with at least partial information. Property ownership was dispersed in both areas as shown in Table 7.16.

Ownership was more concentrated in the Cleckheaton survey area than in Heckmondwike. In the latter 27.6 per cent of all hereditaments for which ownership data are available were owned by the eight owners with ten or more hereditaments. The eight largest owners in Cleckheaton on the other hand owned no fewer than 47.8 per cent of the hereditaments. Ownership was almost always freehold, and there was only a single leasehold owner in the two

[23] West Yorkshire Archives Service, Wakefield. C243/106–10; C243/121–2 Valuation Books; PRO IR 58/26814–18, 26977, 26979–83, 26991–3 Field Books.

Table 7.17 *Tenancy periods in Cleckheaton and Heckmondwike, West Yorkshire*

	No. of hereditaments			
Period	Cleckheaton		Heckmondwike	
	No.	%	No.	%
Weekly	54	28.3	43	14.9
Fortnightly	30	15.7		
Monthly/4-weekly	72	37.7	186	64.6
Quarterly	16	8.4	29	10.1
6-monthly	8	4.2		
Yearly	7	3.7	26	9.0
Over one year	4	2.1	4	1.4
Total	191	100.1	288	100.0

Source: PRO IR 58/26814–18, 26977, 26979–83, 26991–3 Field Books.

areas. According to Mortimore, this situation pertained a century or more earlier.

Owner-occupation was uncommon. In Cleckheaton, only nine of the forty-six owners (19.6 per cent) occupied their own properties, amounting to thirteen hereditaments (5.3 per cent). In Heckmondwike, sixteen owners occupied their own property (17.4 per cent), amounting to seventeen hereditaments (4.6 per cent). Thus the great majority of hereditaments were tenanted. Most hereditaments for which information is available were let on a monthly or shorter basis, as shown in Table 7.17. Weekly periods for letting were proportionately twice as frequent in Cleckheaton as in Heckmondwike. In the latter, monthly lettings were by far the most common.

House rents are usually given in the documents, but often several hereditaments were merged together for valuation purposes, and a total rent given. An example was Hereditament no. 2408 in Cleckheaton. This included Hereditament nos. 2409 to 2419 inclusive (Nos. 1–10 Platts Square, and Nos. 72 and 74 Westgate – comprising ten houses and two 'cellar dwellings'). The total rent was £86 2s 6d, but it cannot be assumed that each property was let at the same rent, i.e. one-twelfth of the total. Accordingly in Table 7.18 rents are given only for those individual houses for which there are specific figures for the amount of rent charged.[24] Given the problems already referred to, firm statements about rents are perhaps premature. Nevertheless it would seem that house rents in Heckmondwike tended to be considerably higher than in Cleckheaton. Nearly a quarter of the houses in the former area were let at £15

[24] PRO IR 58/26814.

Table 7.18 *House rents in Cleckheaton and Heckmondwike, West Yorkshire*

Annual rent (£)	No. of houses	
	Cleckheaton	Heckmondwike
<5	1	1
5 to 5 19s 11d		6
6 to 6 19s 11d	33	5
7 to 7 19s 11d	32	21
8 to 8 19s 11d	14	7
9 to 9 19s 11d	3	8
10 to 12 9s 11d	8	13
12 10s 0d to 14 19s 11d	8	8
15 and over	1	20
Totals	100	89

Source: PRO IR 58/26814–18, 26977, 26979–83, 26991–3 Field Books.

a year or more, compared to only a single house in Cleckheaton. In Cleckheaton two-thirds of houses were let at less than £8 a year, whereas in Heckmondwike the proportion was only a little over one-third.

The amount of living accommodation is available for many houses, and is detailed in Table 7.19. It can be seen that houses tended to be larger in Heckmondwike and this may well account – partially at least – for the higher rents payable there. Half of the houses in Cleckheaton had only two rooms, while another quarter had only three. Fewer than three-fifths of houses in Heckmondwike were as small.

Toilet facilities were frequently shared in both districts, but apparently more so in Cleckheaton. Usually two houses shared a facility, but occasionally up to half-a-dozen houses shared a single privy. More often the privies for a block of six houses would be situated together at one end of the block. The six houses belonging to Herbert Clark, Nos. 6–16 (even numbers only) James Street, Cleckheaton, each had a bedroom, a living room, a cellar, and a jointly used privy at the end of the block. The three houses belonging to the Springwell Brewery Company used privies in the yard of the neighbouring Upper George Hotel. In some cases, WCs had been added 'since 1909', as in Mr Hirst's property at No.8 Quarry Road, Cleckheaton. Next door at No.6, probably not quite as it appears in the Field Book, there was said to be a 'WC in street since 1909'.[25]

[25] PRO IR 58/26815. For more detail of the Upper George Hotel we are unfortunately advised to 'See sheet with folder' – now presumably lost (PRO IR 58/26982).

Table 7.19 *Rooms in houses in Cleckheaton and Heckmondwike, West Yorkshire*

No. of rooms	Houses			
	Cleckheaton		Heckmondwike	
	No.	%	No.	%
1			7	2.3
2	96	49.2	61	20.3
3	50	25.6	109	36.3
4	31	15.9	57	19.0
5	15	7.7	39	13.0
6	2	1.0	13	4.3
7			5	1.7
8 and over	1	0.5	9	3.0
Totals	195	99.9	300	99.9

Source: PRO IR 58/26814–18, 26977, 26979–83, 26991–3 Field Books.

Although often small, housing was generally said to be in 'fair', 'moderate' or 'good' condition in both localities. Only in Heckmondwike, where, as we have seen, houses tended to be larger and more expensive, were there pockets of housing described as being in 'poor' or 'bad' condition. Such subjective comments by the valuers should therefore always be seen in the context of the particular area being surveyed. A property deemed 'bad' in Heckmondwike might not have been thought so bad elsewhere. Hereditament nos. 1915 to 1924 in Upper George Street comprised ten back-to-back houses described as 'old and poor'. Each comprised a cellar, a living room, a bedroom and an attic. Also in Upper George Street a block of six houses (Hereditament nos. 1957–62) were in 'poor repair & (with a) poor class of tenant'. However, such adverse comments were very infrequent. The development on old strip fields in Heckmondwike had clearly meant that many properties were owned which fronted onto two roads. Thus the property of H. M. Preston which comprised six houses, all in very good repair, included four in King Street (cellar, living room, two bedrooms and WC) and two in Victoria Street which ran parallel, to the south-east. Mr J. W. Ackroyd's five-house property was placed in exactly the same relationship to the two streets[26] (Fig. 7.4). Not all were owned by individuals: the newly built freehold property in Albion Street, facing onto the macadamised street, had a cellar and basement, kitchen, living room, two bedrooms and an outside WC, and was owned by the Heckmondwike and District Cooperative Society. In Cleckheaton, the local

[26] PRO IR 58/26980.

Cooperative Society also owned, for example, a 'fairly new' shop and stores in Westgate.[27]

Tippler closets, for the safeguarding of pigeons, were frequently noted in both townships: in the case of A. E. Rhodes' five houses off Victoria Street, Heckmondwike, there were two closets to the five houses, and they were noted, for example, in two of the six properties with 'poor class of tenants' noted previously in Upper George Street.[28]

Thus there were substantial differences in ownership patterns, house sizes and rents between the two areas at this time, a time which has been described for Bradford as 'the last majestic flowering of Nineteenth century provincial life'.[29] The Valuation material clearly sheds more light on the districts and would have been a most useful addition to the 1969 study had the documents then been available. Cleckheaton was characterised by somewhat more concentrated house ownership, smaller houses and lower rents, than its neighbouring township. To what extent this has its origins in the field patterns that underlay the street layouts described by Mortimore, is of course impossible to assess, at least on the basis of this study, though it is tempting to hypothesise that the more concentrated ownership in Cleckheaton may have had its roots in the dispossession of small owners at the time of the 1795 enclosure.

Camberwell: a South London slum

Finally, it is also possible to examine particular social areas in some detail. For the sake of illustration, the Sultan Street area of Camberwell is presented, partly because of the link-up with the work of H. J. Dyos as set out in his *Victorian Suburb*.[30] In particular he was concerned with the box of streets between Crown Street, Wyndham Road, Pitman Street and Bethwin Road (later Avenue Road), to the west of the Camberwell Road and somewhat isolated by the Herne Hill and City Branch of the London, Chatham and Dover Railway (Fig. 7.5).[31]

Dyos described the core of this area in graphic terms, and concluded that this was probably the worst building estate to be found in Victorian Camberwell. Many of the speculatively built houses were inspected in 1889 by Charles Booth's investigators, by which time they had become abodes of

[27] PRO IR 58/26814. [28] PRO IR 58/26981–2. [29] Jowitt, 'Bradford', 5.

[30] H. J. Dyos, *Victorian Suburb: A Study of the Growth of Camberwell* (Leicester 1961). The relevant section is that headed 'The making of a suburban slum' (pp. 109–13).

[31] For a highly class-ridden perception of slum conditions in London, and the origins of slums as being through the isolation caused by canals and railways in London, see the address by Charles Booth's co-worker G. H. Duckworth, 'The making, prevention and unmaking of a slum', *Journal of the Royal Institute of British Architects* 33 (1936), 327–37. Among tips given to architects, Duckworth suggested that they should 'avoid sharp turns in the stairway so as to allow the coffin to be carried down decently. A coffin slung from a window will disaffect a neighbourhood' (p. 333).

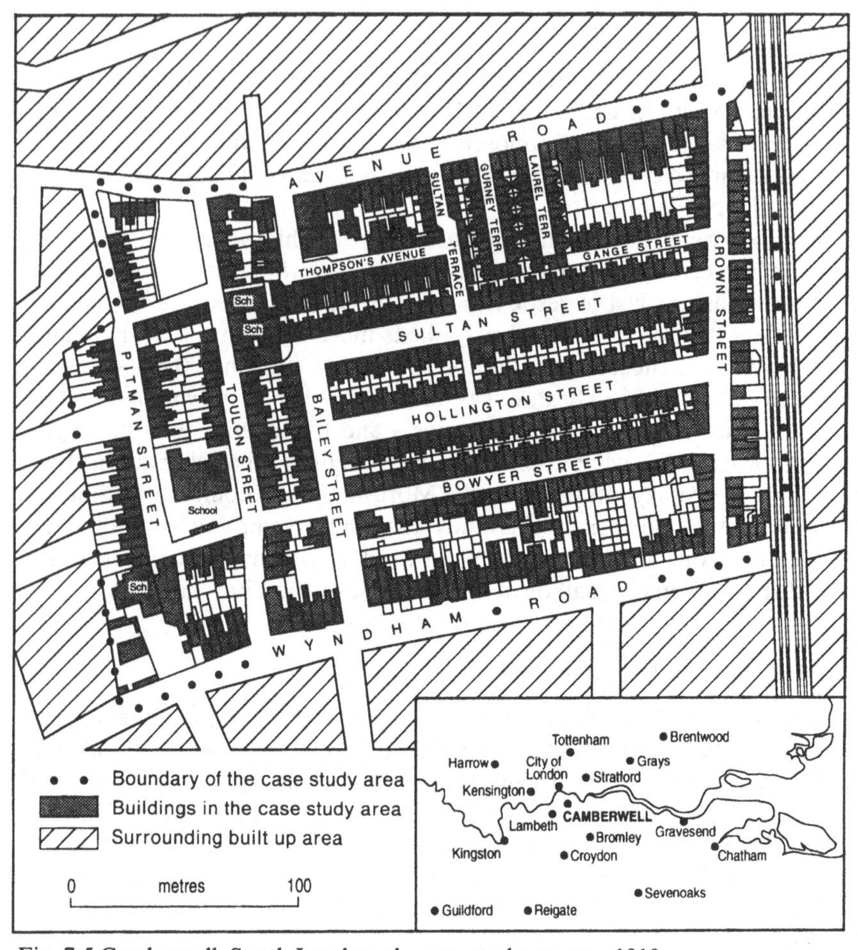

Fig. 7.5 Camberwell, South London: the case study streets *c.*1910.

squalid poverty. By 1904 Booth's *Life and Labour of the People of London* was being reprinted for the English League for the Taxation of Land Values, alongside the work of Henry George, James Dundas White and the Red Van Reports of 1894–7.[32]

The district lay in the ITP of Camden, and the sources used were the Valuation Books and Field Books. The former, held by Southwark Local Studies Library, were used again to establish the relevant hereditament numbers for the various streets, so that the entries in the Field Books could be

[32] Charles Booth, *Rates and the Housing Question in London: An Argument for the Rating of Site Values* (English League for the Taxation of Site Values, London 1904).

examined. A full index in these Valuation Books made it a straightforward task to extract the hereditament numbers, despite the fact that entries for each street were frequently scattered throughout the books, and intermingled with entries from other streets. The Valuation Books were generally well compiled, with the values entered in each case, although the extent of each hereditament was seldom entered. Occupants were named as usual, often with the occupants of the different floors being indicated. It is interesting to note that the Field Books do not contain this information.[33]

One apparent error in the Valuation Books posed a slight problem. The Land Valuation Officer appeared to have sometimes entered the Poor Rate number from the Rate Book into the column for the hereditament number by mistake. The latter were accordingly entered by the LVO into the column for the Poor Rate numbers. This practice was inconsistent and could have caused substantial difficulties. However the problem was minimised by the fact that both the Poor Rate and the hereditament numbers had been noted down, and in conjunction with the address, the identification of each hereditament in the Field Book was relatively straightforward.

The Field Books provided very detailed information on most hereditaments, though there are problems. Like the Valuation Books, the Field Books provide no data on the extent of each hereditament, and none on the occupier(s). Multiple occupation is frequently indicated, but the extent of it in any hereditament is seldom clear. Neither is the absence of any reference to multiple occupation proof that it did not occur in a given property. The Valuation Books are more consistent in this respect, almost invariably naming the occupier. The difficulties caused by the separation of the Valuation and Field Books is further illustrated in this case study, although admittedly there are few instances so far discovered of the Valuation Books containing more information than the Field Books. The valuer frequently entered a figure for rent that differed from that supposedly extracted from Form 4, but the status of these alterations is unclear, although it may be due to the valuer obtaining from the occupier an amount that was paid at the time of the survey. In any event the figures for rent are problematical, and will be discussed below.

The 20 per cent sample of hereditaments illustrates much of the life of the neighbourhood, as well as the potential of the material. The area had changed little in terms of its physical layout since 1871. A new street, Bailey Street, had been driven through a few properties from Hollington Street to Wyndham Road, and the linoleum factory in Bethwin Road had vanished. This last road had been renamed Avenue Road. With a few exceptions therefore, the valuation was concerned with the same buildings as those described by Dyos.

[33] Southwark Local Studies Library. Valuation Books for Camden ITP. PRO IR 58/78101–9 Field Books.

Table 7.20 *Number of hereditaments examined: Sultan Street area, Camberwell, South London*

Street	No. of hereditaments
Avenue Road	6
Bailey Street	2
Bowyer Street	12
Browns Terrace	2
Crown Street	17
Gange Street	2
Gurney Terrace	5
Hollington Street	13
Laurel Terrace	4
Pitman Street	9
Sultan Street	16
Thompson's Avenue	6
Toulon Street	8
Wyndham Road	15
Total	117

Source: PRO IR 58/78101–9 Field Books.

Table 7.21 *Property ownership in the Sultan Street area sample, Camberwell, South London*

Hereditaments owned	No. of owners
1	31
2	13
3	1
4	2
5	1
6	1
40	1
Total	50

Source: PRO IR 58/78101–9 Field Books.

The 20 per cent sample resulted in the number and distribution of hereditaments being examined in the streets as shown in Fig.7.5 and Table 7.20.

Property ownership as reported in the Field Books was fragmented (Table 7.21). Apart from Camberwell Borough Council, which owned one-third of the hereditaments in the sample, the great majority of owners had only one or two properties.

Table 7.22 *Types of 'ownership' in the Sultan Street area sample, Camberwell, South London*

Freehold	62
Leasehold (including underleasing)	50
Not stated	7
Total	119

Source: PRO IR 58/78101–9.

According to Section 41 of the Finance (1909–10) Act 1910, the 'owner' of a hereditament was the freeholder, or lessee or under-lessee, whose lease or sub-lease had more than fifty years to run. Leasing and sub-leasing was common in this district, and Dyos described the complex system of leasing and sub-leasing that arose out of the original terms for leasing the estate in 1781, and how the situation had got increasingly convoluted during the nineteenth century. Indeed, he considered this to be the major cause of the appalling slum conditions that existed there in the latter part of the nineteenth century.[34]

The situation was so complex that it is far from certain that the information in the Field Books relating to ownership, and superior and subordinate interests, is always reliable. For instance, Hereditament no. 50 (Nos. 1–10 Beaconsfield Mansions, Avenue Road) was 'owned' by L. A. Somers of Coldharbour Lane, Brixton, while W. B. Jenner of Lower Mitcham had an unspecified superior interest. A note in the Field Book points out that Jenner had a 99-year lease from Ladyday 1888, while Somers had only a weekly interest in the different tenements in the hereditament, for which he paid a variety of rents. From this it would appear that Jenner was actually the 'owner' as defined in the Act. We do not know from whom he leased the premises. In Hereditament no. 45 (No. 5 Avenue Road) M. Strugnel was said to be the 'owner', but he was also said to be an 'annual tenant'. This would seem to be a contradiction in terms. Table 7.22 shows the extent of leaseholding (including under-leasing) in the area as a whole.[35]

The complexity is well illustrated by Camberwell Borough Council's interests in the area. As already shown, it was by far the largest single 'owner' in the sample with 40 hereditaments. In fact it had an interest in 49 hereditaments as shown in Table 7.23.

The Council had extensive interests in the heart of the 'slum', including fifteen of the sixteen Sultan Street hereditaments in the sample. The freehold (or sometimes the head lease) of all of these was held by one J. D. Grummant. Six were leased from him by the Council, and let directly to the inhabitants of

[34] Dyos, *Victorian Suburb*, 110–12. [35] PRO IR 58/78101.

Table 7.23 *Types of interest of Camberwell Borough Council in the sample hereditaments*

Interest	No. of hereditaments
Freehold	22
Leasehold 'owner'	14
Under-lessee 'owner'	3
Lessor (superior interest)	Nil
Under-lessor (superior interest)	10
Total	49

Source: PRO IR 58/78101–9.

the buildings. Another was leased by him to Preston and Francis, who sub-leased to the Council, who in their turn let the premises to the actual occupiers. The remaining eight, though, were leased by Grummant to the Council, who sub-leased them to other individuals who were themselves the 'owners' according to the Field Books, and therefore presumably, according to the Finance Act itself. An example of the latter case was No. 18 Sultan Street. Grummant leased the property to the Council, who sub-leased it to Mrs Taylor of Montgomeryshire, the actual 'owner'. She presumably gathered the rents from the actual occupiers, undoubtedly through an agent. It is certainly not surprising that the properties deteriorated to the slum conditions reported by Dyos.[36]

The condition of the accommodation varied enormously. The outer streets, such as Avenue Road and Wyndham Road, tended to be very much better than the core district. Laurel Terrace, off Avenue Road, was made up of 'small, poor class' cottages but No.2 Gurney Terrace was a 'poor class cottage' with one room on the ground floor, a scullery and WC and one bedroom on the first floor. The brickwork was sound but 'hse dirty. very poor tenant of low class', and most of the rest of Gurney Terrace was described in the same way.[37] No. 25 Thompson's Avenue was, according to the valuer, an old property let to 'poor class tenants'.[38] Hollington Street was said to be a 'v. poor class street let in small tenements. G[rea]t many cases in single R[oo]ms. H[ou]ses flush to pavement with 3 floors and flat front.'[39] The tenants were said to be 'v. poor and dirty', and many floors and rooms were vacant. Crown Street was described simply as a 'slum street [with] filthy tenants'. At No.59 it was noted that the house was very old and that the 'doors open into front room'.[40] There

[36] PRO IR 58/78108. [37] PRO IR 58/78101. [38] PRO IR 58/78102.
[39] PRO IR 58/78106–7.
[40] PRO IR 58/78108. The remark is made against the entry for No.11 Crown Street. No. 59 is in PRO IR 58/78109.

is of course the clear problem here that the perceptions of the condition of the inhabitants of a house or street were coloured by the observer's own social background and expectations, and although they were as far as possible local people, they would clearly have been most unlikely to have been well acquainted with 'low life'.

To some extent, the Field Books here prevented full analysis of problems of overcrowding, since they give little detail on the occupation of any property. This again represents a prime example of the need for access to the 1911 Census for full exploitation of the valuation data. Nevertheless the Field Books do frequently hint at multiple occupation, but one cannot assume that those houses were the only ones in which multiple occupation occurred. At No.12 Wyndham Road the three-storeyed house was let in three separate units of occupation, but with one WC for the whole house.[41] In fact only some twenty-six hereditaments in the sample were explicitly said to be let in tenements. These included every property in Hollington Street, and several in Sultan Street. Also included were the purpose-built flats of Beaconsfield Mansions, said to be identical to other blocks in Avenue Road, including Edinburgh, Devonshire, and Cadogan Mansions, all of which were untypical of the tenement buildings in the core area. Beaconsfield Mansions comprised ten tenements (two per floor). The lower six each had two bedrooms, a sitting room, a kitchen, a scullery and a WC. There was a dust chute from each, and each was self-contained. The upper four tenements were the same but with only a single bedroom.[42]

In contrast, houses in the core area, typified by Hollington Street, were often let in single rooms. Typical houses were of three floors with two rooms per floor, and, in addition, a scullery on the ground floor and an outside WC. There was also water and a sink on the top floor. No.11 Sultan Terrace had a copper (for washing clothes) in the yard.[43] Although one cannot know, without access to the 1911 Census enumerators' schedules, how many people lived in each house, it was presumably a considerable number. Often a family in all probability inhabited each room, sharing the washing and toilet facilities. No. 21 Hollington Street had been visited by Charles Booth's investigators, twenty-one years earlier. The six rooms were then let to five different groups of occupants, totalling nineteen people. The two rooms on the top floor each housed five people.[44] There seems no reason to suppose that the situation was markedly different in 1910, despite the various attempts made in the interim to clear the inner London slum areas.

Rents for these premises seem high, but accurate assessment is often difficult. The stated rents are usually for the whole building, and without information on the number of households in each, the actual rent per household

[41] PRO IR 58/78109. [42] PRO IR 58/78101. [43] PRO IR 58/78102.
[44] Dyos, *Victorian Suburb*, 103.

cannot be ascertained from this documentation. Moreover, it is not always clear whether the stated rent is that which was received from the actual occupants, or whether it was that paid by a lessee or under-lessee to the person with the superior interest. Apparently identical premises in Hollington Street, for example, had annual rents ranging from £18 to £37 14s.

These differences may be due to the fact that some rooms were empty in the 'cheaper' buildings. In fact, the 'cheapest' – No.48 – with a rent of £18 was said to be 'totally empty'. The valuer stated that rents varied according to the style of tenancy, but made a 'fair' estimate of rents as 5s per week for the entire ground floor; 4s 6d for the whole first floor; and 4s 0d for the top floor. This amounted to £35 2s per year, and suggests that a tenant of a single room should have paid between 2s and 2s 6d per week.[45]

Dyos, in a graphic passage, described the:

> intermixture of cowsheds and piggeries . . . the glue and linoleum factories, a brewery, and the . . . haddock-smoking and tallow-melting yards . . . The sickly smell of costermongers' refuse combined . . . to make an atmosphere which seemed . . . to be a concoction of haddocks and oranges, of mortar and soot, of hearthstones and winkles, of rotten rags and herrings.[46]

The valuation material gives no hint of these multifarious activities, and it may perhaps be assumed that such things were largely gone by 1910. The occasional reference to stables and workshops is all that may conceal something of the scene described by Dyos.

Despite this, the core area was still clearly a slum. The data are very detailed but need reinforcement from other material to maximise their usefulness. In particular, it has been emphasised that census data would be invaluable for assessing overcrowding, though local authority records may well survive and enable some assessment of this and other matters. The use of local authority housing and public health committee minutes would be particularly beneficial, as would the New Works and Plans subcommittee minutes available for Sheffield.[47] There are also schedules of property drawn up as Housing Confirmation Orders under the provisions of the Greenwood Act (1930) by London County Council relating to those properties which were to be compulsorily purchased by the LCC for demolition and rehousing of the working-class occupants. These contain descriptions of the property in terms of their use, the names of the owners, lessees and occupiers.[48]

Nevertheless, the data here certainly confirm Dyos' findings of complex

[45] PRO IR 58/78107. [46] Dyos, *Victorian Suburb*, 111.

[47] P. J. Aspinall, *Building Applications and the Building Industry in Nineteenth-Century Towns: The Scope for Statistical Analysis* (Research Memorandum 68, Centre for Urban and Regional Studies, University of Birmingham 1978).

[48] I am grateful to Mona Paton for information on Housing Confirmation Orders. Because of general retrenchment in the years immediately following the 1930 Act, slum clearance did not begin in earnest until 1933. All urban authorities with populations in excess of 20,000 were

systems of tenure, and that the overall inadequate living conditions described by him continued little changed into the twentieth century. Indeed, as late as 1932, a London School of Economics study described this area in some detail. It claimed that Wyndham and Avenue Roads, as well as the streets off Avenue Road, were inhabited by skilled workers. Toulon, Pitman and Bailey Streets were home to a mixture of skilled and unskilled workers. The inner core of streets was said to be home to a conglomeration of unskilled, very poor and criminal groups. Specifically, it stated that Sultan, Hollington, Bowyer and Crown Streets are 'poor and much overcrowded, and crime is not absent'.[49]

Conclusion

Whether one tackles urban social analysis through selecting streets on some pre-ordained basis, sampling streets, sampling hereditaments or taking a specific area for more detailed study, the Valuation material would seem to have a great deal to offer. For urban analysis the data on property ownership and tenure; residence of owners; the tenancy terms and lengths of rented property; levels of rent, rates and other outgoings; and the sizes, composition and state of repair of properties, are all generally available either from Valuation or Field Books. The four case studies thus share some of the same material, whilst other material is more localised in its interest, relevance or availability. Table 7.24 summarises this position for the four urban locations.

Because the case studies were designed to be illustrative, not all were focussed on as complete a coverage as might have been possible. In addition, some were based on complete spatial coverage of a limited area, and some on sampling frameworks, so the bases for the studies were not conceived in such a way as to present comparable material. Clearly topics such as house descriptions, ownership, rents, tenancy periods, and the size and condition of hereditaments were topics of a more general nature, and ones which can reasonably be expected to be covered in all the relevant 1910 material. But other material has a more localised interest, as with the involvement of the local authority in Camberwell's housing, or the study has specifically focussed on the issue, such as the availability of WCs in the Bradford area as an index of housing quality. In some cases the material was presented by the valuers as an extra item which was not strictly required, and this allowed the length of residence of occupiers in one Brighton street to be determined. Readers

required to submit proposals every five years for dealing with housing conditions in their areas. For an examination of the linkage of these records with the 1910 Valuation material for the Bethnal Green and Shoreditch areas of the East End of London see M. Paton, 'Urban property and the slum', University of London, unpubl. MA dissertation in London Studies 1989, esp. 28–42 and 53–60.

[49] London School of Economics, *The New Survey of London Life and Labour* vol. VI (Survey of Social Conditions) No. 2, The Western Area (London 1932).

Table 7.24 *Themes in the analysis of four Edwardian towns*

Theme	Brighton	Cromer	Bradford	Camberwell
House description	x	x	x	x
Tenure types	T	x	x	T
Number of owners	T	x	T	x
Owner/occupiers	T	x	x	–
Residence of owners	T	x	–	–
Tenancy periods	T	T	T	–
Rent levels	T	T	T	x
Responsibility for outgoings	T	–	–	–
Length of residence	T	–	–	–
Types of hereditament	T	–	–	–
Number of rooms	T	T	T	–
Condition of hereditament	T	–	x	x
Toilet provision	–	–	x	–
Local authority interest	–	–	–	x

Notes:
T Topic meriting a table in text
x Topic covered in text
– Topic not covered in text

should note that no comparative analysis has been attempted here, but only because this chapter is designed to illustrate what might be attempted, rather than to penetrate more deeply into the urban geography of Edwardian England *per se*.

Problems with the material have been noted, and may occur without warning for particular towns, but nevertheless the materials constitute a prime new source of urban data. And if we use them in conjunction with other materials such as Medical Officer of Health reports, local authority records, directories and other place-specific material, we should now be far better placed for the construction of an urban geography of Edwardian England and Wales.

8

Rural society and economy 1909–1914

Chapter 7 addressed the use of the 1910 documentation in the context of inter- and intra-urban differences in housing and the quality of life in the Edwardian town and city. This chapter now addresses issues which have a bearing on rural community life at the same period. To some extent there are points which need not be laboured, which are common to both town and country, and which can be expected from the data source irrespective of whether one is dealing with an inner city tenement or a large country estate. Ownership, estimates of rents or values, and comments, however skeletal in places, on housing conditions can all be expected. But only detailed local study will, of course, reveal the quality of the data. The aim here is to demonstrate the potential of the documents to reveal insights into the parish, into the estate, and into inter-parish contrasts.

The parish

By relating locally held and centrally held historical material together it is possible to build up a very good picture of life in the Edwardian countryside. An example will here be drawn from the Lincolnshire Wold parish of Binbrook, where documentation in fact appears to date from *c.* mid-1911 to late 1912, judging from dates given in the Field Books.[1] By relating the latter to the Valuation Books and Forms 37–Land held in the Lincolnshire Archives

[1] Much of the information upon which this section is based was originally published in Charles Rawding and Brian Short, 'Binbrook in 1910: the use of the Finance (1909–10) Act Records', *Lincolnshire History and Archaeology* 28 (1993), 58–65. For more comprehensive information on the parish, see Charles Rawding (ed.), *Binbrook in the Nineteenth Century* (Binbrook WEA 1989); Rawding (ed.), *Binbrook 1900–1939* (Binbrook WEA 1991); Rawding, 'To the Glory of God? The building of Binbrook St Mary and St Gabriel', *Lincolnshire History and Archaeology* 25 (1990), 41–6; R. J. Olney (ed.), *Labouring Life on the Lincolnshire Wolds: A Study of Binbrook in the Mid-Nineteenth Century* (Occasional Papers in Lincolnshire History and Archaeology 2, 1975).

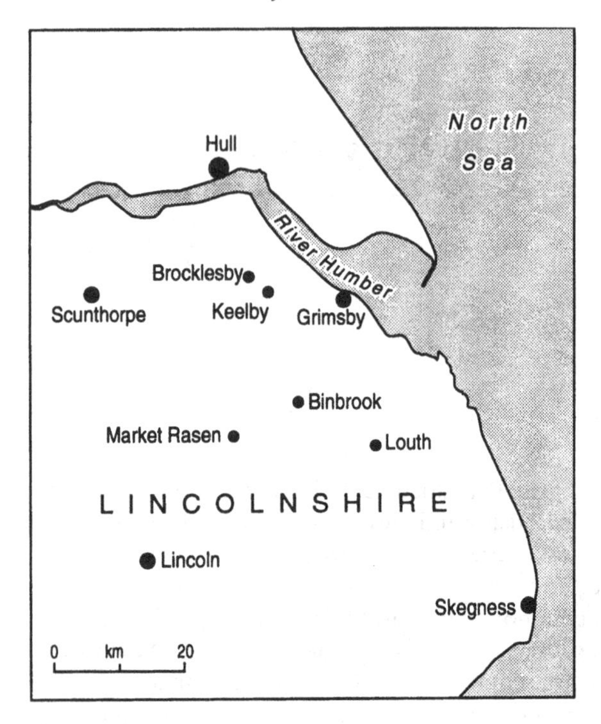

Fig. 8.1 Binbrook, Brocklesby and Keelby, north Lincolnshire.

Office, and the Record Sheet Plans, with the Field Books, at the PRO, an incisive amount of information can be gleaned.[2]

Lying between Grimsby and Market Rasen in the north Lincolnshire Wolds, Binbrook had a population of 874 in 1911 in 237 inhabited houses, a far cry from its 1861 peak of 1,344, and testimony to a steady rural depopulation compared with urban growth at nearby Grimsby and Scunthorpe. By the 1920s the number of local burials was exceeding the number of baptisms (Fig. 8.1).

At this time John Jennings was the largest landowner in the parish, with just over 1,155 acres at Binbrook Hill. The Turnors of Panton Hall, Wragby, owned just over 900 acres, whilst Woodthorpe Clarke, resident at The Manor, had 835 acres. The only other landholdings greater than 500 acres were those of William Dennis and Son of Boston, and W. C. Brocklehurst who owned Walk Farm. Table 8.1 and Fig. 8. 2. show the landowners with more than 125 acres in 1910. These ten owners held 93 per cent of the parish, whilst the five

[2] For Forms 37–Land covering Binbrook (though with gaps) see Lincolnshire Archives Office (LAO) 6 Tax/42/53; the 25 inch maps are at PRO IR 130/1/63. The Working Sheets for East Lindsey, covering Binbrook, somehow found their way into the District Planning Office.

Table 8.1 *Landownership in Binbrook, Lincolnshire c.1910*

Landowner	Acreage A	R	P	Annual value	Location
John Jennings	1155	3	21	£667 15s	Binbrook Hill/Top Farm
Christopher Turnor	916	3	25	£567 13s	Chestnuts/Spottle Hill/ Burkinshaws Top
Woodthorpe Clarke	835	3	0	£986 11s	The Manor/Binbrook Top/Sunnyside/ village properties
W.C. Brocklehurst	724	0	0	£572 6s	Walk Farm/Highfield Farm
William Dennis & Son	557	3	33	£436 0s	Binbrook Hall/Scallows
The Rector	351	0	27	£361 14s	The Poplars/allotments/ Tithe rent charge
Execs of J. Bingham	186	0	8	£163 17s	Bingham's Top
Thomas Sawyer	161	2	31	£141 0s	The Poplars
John Drakes	128	0	27	£108 8s	Low Farm
Ecclesiastical Commissioners	125	2	39	not given	Spottle Hill/Burkinshaws Top (pt sublet to Chris. Turnor)
Total (owners over 25 acres)	5143	1	11	£4004 4s	
Parish total	5518	3	17		

Notes:
A R P acres, roods, perches
Source: PRO IR 58/33575–8.

largest landowners alone held 76 per cent of the parish. Of these landowners, only John Jennings, Woodthorpe Clarke, the Rector and Thomas Sawyer were resident in the parish itself.

It is equally possible to determine the farmers within the parish (Table 8.2 and Fig. 8.3). The largest farm was Binbrook Hill at 945 acres, owned and farmed by the Jennings, which was in 'excellent repair'. In 1907 this large arable farm had stabling for twenty cart-horses,[3] the horse being a most important element in the farming regime on the Wolds, and a principal selling point was: 'a fine supply of water has been secured in a field some distance from the farm yard, and by means of a powerful mill, the water is forced into a New iron Tank (6,500 gallons) from which the yards and premises are all

[3] PRO IR 58/33576; LAO 2BD 7/136 Sale particulars of Binbrook Estate 12 July 1907.

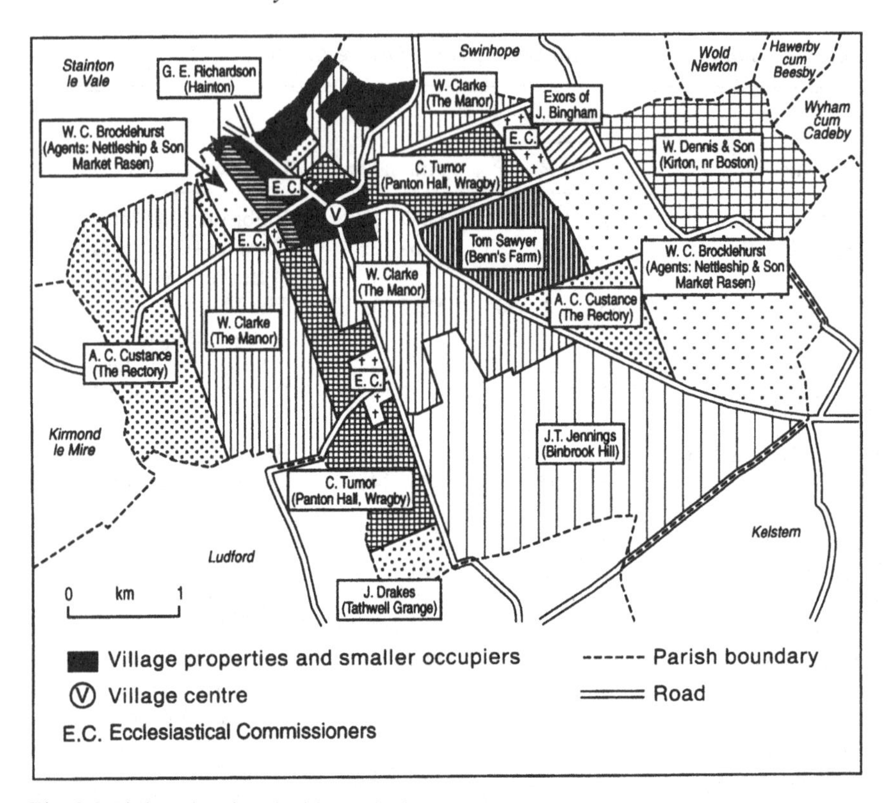

Fig. 8.2 Binbrook, Lincolnshire: principal landowners *c.*1910.
Source: PRO IR 58/33575–8.

supplied'.[4] Then came The Chestnuts at 852 acres, rented by Henry Burkinshaw from Christopher Turnor. Although he was resident at The Chestnuts, 'the bulk of the farm lies about a mile away'.[5]

Walk Farm of 576 acres was farmed by tenant William Drakes and was 'an extremely good arable farm, buildings occupied by Foreman. House and buildings in excellent repair, some wattle and daub outbuildings.' However, in common with several of the more distant farms such as Highfield Farm, it was poorly supplied with water. Much had to be brought from the pumps in the village, a considerable distance, although in later years an iron windpump was used to lift water from a well.[6] At Binbrook Hall, 'the house and all slated buildings [were] in good repair, all tiled buildings mostly require re-roofing'. This had been purchased in 1907 at the break-up of the Grimthorpe estate by the Dennis potato company from Kirton, near Boston, which held about 1,500 acres of potatoes across the county. The farm was then described as

[4] LAO 2BD 7/136. [5] PRO IR 58/33577. [6] PRO IR 58/33575.

Table 8.2 *Principal farms and farmers in Binbrook, Lincolnshire c.1910*

		Acreage		
Farmer	Farm	A	R	P
Henry Burkinshaw	Chestnuts	852	3	31
Woodthorpe Clarke	Manor House	214	0	17
William Dennis and Son	Binbrook Hall	557	3	33
John Drakes	Low Farm	128	0	27
William Drakes	Walk Farm	576	0	0
John Fieldsend	Binbrook Top/Rectory Farm	704	2	36
John Jennings	Binbrook Hill	945	0	0
Henry Odling	Lambcroft (Kelstern)	85	2	6
George Payne	Bingham's Top	100	2	2
Benjamin Penistone	Highfield Farm	148	0	0
Thomas Robinson	Top Farm (res. North House)	228	3	20
George Rushby	Jesmond Farm (Kirmond Road)	53	2	19
Thomas Sawyer	The Poplars	341	2	11
Tom Sawyer	Benn's Farm	212	1	27

Notes:
A R P acres, roods, perches
Source: PRO IR 58/33575–8.

'good wold land, well adapted for sheep farming and barley growing'. Dennis also purchased Scallows Hall at the same time, described in the sale catalogue as 'an ideal home for the family of a City or Town gentleman during the summer months'. But it had previously been empty for two years, and was described in the Field Book as requiring £200 to £300 to be spent on repairs to the roof and walls.[7]

One of the smaller farms, the 100 acre Bingham's Top which was run by George Payne who was also a carrier, was described as: 'A very compact little farm with a main road running on three sides of it. The house is very small and poor accommodation for a farm house. The farm is badly supplied with water and all drinking water is brought from Binbrook.'[8] John Fieldsend (1866–1964), a thrashing machine owner and contractor as well as a farmer, occupied Binbrook Top and Rectory Farm, covering much of the western end of the parish in 1910 (Fig. 8.3). This was rented from Christopher Turnor (150 acres), the Rector (156 acres) and Woodthorpe Clarke (398 acres).

The smaller landowners, those perhaps less likely to appear in directories or estate papers, are also represented fully in the 1910 material. Many are also tradespeople: carriers, for example, held land for horse paddocks, and

[7] PRO IR 58/33576; LAO 2BD 7/136. [8] PRO IR 58/33575.

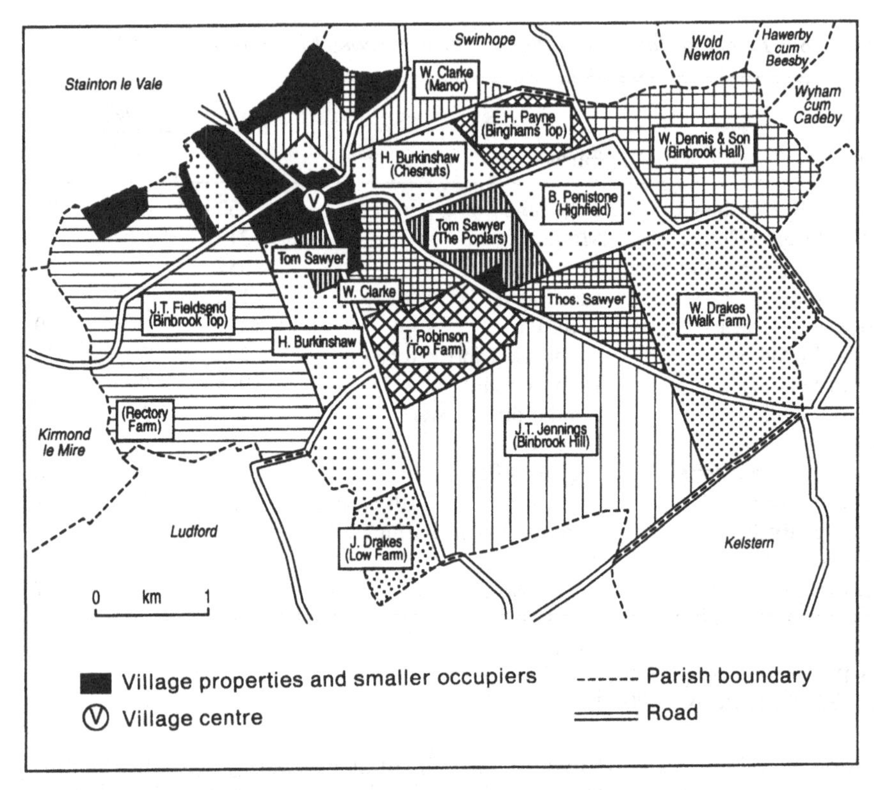

Fig. 8.3 Binbrook, Lincolnshire: principal farmers *c.* 1910.
Source: PRO IR 58/33575–8.

butchers for rearing and fattening animals. Charles Cook, a butcher, held 32 acres in total in nine separate hereditaments ranging in size from just over 1 acre to the largest field of 8 acres. His rented house and shop were in the Market Place.[9] John Maultby, a tailor, occupied 31 acres in three hereditaments, along with his tailor's shop on Ludford Road. A further eleven people occupied between 1 and 20 acres at this time, of whom seven were tradespeople in the village, whilst John Cook was listed as a cowkeeper. Mrs Rockliffe, listed as a smallholder in *Kelly's Directory* (1913), owned a half-acre plot in North Halls which included a house and two pigsties, and rented 3 acres of allotments from the rector along with just over 2 acres of land on Low Lane from Samuel Maughan, the chemist.[10]

In terms of the total acreage of the parish, these smaller units were perhaps insignificant, but there is no doubt that they provided a valuable supplement to the incomes of many of those involved in working this land. To reinforce

[9] PRO IR 58/33575–8. [10] *Kelly's Directory of Lincolnshire* (1913); PRO IR 58/33575–8.

Table 8.3 *Trade properties in Binbrook, Lincolnshire c.1910*

Description	Owner	Occupier	Occupation	Value
Marquis of Granby, Village	Market Rasen Brewery	Mrs Meannett	Publican	£34 17s
Plough Inn	W. Christian	Charles C. Reed	Publican	£32
Water Mill/House, Mill Lane	Marshall, Grimsby	Tom Topliss	Miller	£25 10s
House/shop, Market Place	Samuel Maughan (chemist)	J.W. Filsby	–	£24
House/shop, Market Place	T.B. Dawson	T.B.Dawson	Postmaster	£19
Shop/House, High Street	Tom Gelsthorpe	Tom Gelsthorpe	Blacksmith	£18 5s
Shop/House, High Street	Edwin Parr	Edwin Parr	Draper/ grocer	£18
Windmill, High Street	Mrs Short	Mrs Short	Miller	£18
Shop/House, High Street	William Surfleet	William Surfleet	Tailor	£17

Source: PRO IR 58/33575–8; *Kelly's Directory of Lincolnshire* (1905 and 1913).

further the notion of a community that attempted to be as self-sufficient as possible in food, forty-three properties (17 per cent of the total) in the village are listed as having pigsties.

Information was also supplied about the tradespeople of the parish and their premises (Table 8.3). The scale of their operations can be judged to some extent by the rental levels of their properties. Edwin Parr, for example, had a grocer and draper's shop in the High Street with a showroom behind, sitting room, pantry, backstairs, two rooms, a small lavatory place, four bedrooms, a brick and tile coal place, two store warehouses, stable and cart-shed with a drive in from the street.[11] On the other hand, Mr Barton's shop in the Market Place was obviously less prosperous, being described as 'a very old place. part barber's shop and part grocery shop: very old and dilapidated has been two old cottages'.[12] In common with most butchers at this time, Tom Sawyer's shop in the Market Place contained a slaughterhouse as well as a goose box, pigsty, cart-shed and stables.[13] Details of the now-demolished windmill are also provided: 'Windmill with sails complete and main shafting. 3 store rooms up above. 4 pairs of stoves, old engine house, brick and tile, pigsty. Wattle and thatch outhouses etc.'[14] The four-sail mill belonged to the Topliss family who also owned a bakehouse in the village which sold flour and yeast as well as

[11] PRO IR 58/33575. [12] PRO IR 58/33577. [13] PRO IR 58/33575. [14] PRO IR 58/33577.

Table 8.4 *Rental value of houses in Binbrook, Lincolnshire c.1910*

a. Properties above £15

Description	Owner	Occupier	Occupation	Value
The Manor	W. J. Clarke	W. J. Clarke	Farmer	£60
The Rectory	W. J. Clarke	A. C. Custance	Rector	£34
East House	W. J. Clarke	Dr B. Wilkinson	Doctor	£24
House, Grimsby Road	Thomas Robinson	Thomas Robinson	Farmer	£22
West House, Kirmond Road	J. T. Fieldsend	J. T. Fieldsend	Farmer	£18
House, Ludford Road	Tom Sawyer	Tom Sawyer	Farmer	£17.10s
House, Village	J. Tharratt	A. Keller	Teacher	£17
Hall Farm	William Dennis and Son	—	—	£16.10s
The Chestnuts	Christopher Turnor	H. Burkinshaw	Farmer	£15

b. Properties below £3

Description	Owner	Occupier	Occupation	Value
House, Ludford Road	Samuel Maughan	Jos Atkinson	not known	£2
House, High Street	J. Chapman	—	—	£2 7s
House, High Street	J. Chapman	—	—	£2 7s
House, Ludford Road	Samuel Maughan	J. T. Simpson	not known	£2 10s
House, Ludford Road	Samuel Maughan	—	—	£2 10s
House, Ludford Road	Samuel Maughan	—	—	£2 10s
House, Ludford Road	Samuel Maughan	H. Clayton	not known	£2 10s
House, Ludford Road	Samuel Maughan	J. Enderby	not known	£2 10s
House, Ludford Road	Samuel Maughan	J. Walsh	not known	£2 10s

Source: PRO IR 58/33575–8; *Kelly's Directory of Lincs* (1905 and 1913).

bread. The mill continued to grind corn beyond the 1930s although the amount handled decreased steadily.[15]

An analysis of rental values can also provide a good indication of housing quality (Tables 8.4a and 8.4b). A detailed description of the most expensive property in the village, The Manor, provides some insights into the domestic arrangements of the wealthy Clarke family: 'House very pleasantly situated in the centre of the village and well secluded by limes, beech and ornamental trees. The house and buildings are in excellent repair.' The house contained twenty-two rooms, listed as follows: 'Dining, drawing, breakfast, billiard, 2 kitchens, 2 dairys, pantry, wash house and cellar. 10 beds, boxroom, bath & W.C.'[16] In addition, all the uses of the numerous outbuildings are noted.

[15] Rawding, *Binbrook 1900–1939* (1991), 28. [16] PRO IR 58/33577.

The cheapest housing in the village provides a marked contrast within this parish community (Table 8.4b) Most are described as 'old cottages' on a 'very poor site'. Twenty-three cottages in Ranters Row were valued at £4. Nine were single cottages; five were 'two up, two down'; three more also had a privy, pigsty and shared a common garden; seven were listed as '4 up, 4 down' with 'out offices' and a common garden; and a further four were '5 up, 4 down'.[17] Out of 150 properties valued at under £5, forty-five were owned by non-resident small-scale owners of property, thirty-nine by farmers and/or landowners, and thirty-six by village tradespeople. However, if one analyses the village alone, leaving out the thirty farmworkers' cottages on the surrounding farms, then the importance of the tradespeople as landlords becomes clearer, with 30 per cent of all cheap rented accommodation falling within this category. The most significant of these landlords were Frank Hall with fourteen properties and Samuel Maughan with seven. Frank Hall (1844–1931) owned twelve cottages on Ranters Row and Mount Pleasant, and a further two in the centre of the village. He was listed as a blacksmith in 1910, although by 1922 (aged 78) he was recorded as the proprietor of the Temperance Hall, and by 1926 as a smallholder and proprietor as well. Living at Grant House (rental value £11), he also owned land on which he kept some cattle, and rented three 1-acre allotments on Kirmond Road from the rector and a further 3/4 acre from the Lord Chancellor. Samuel Maughan was a chemist in the village, living at West Villa, a 'good. well built house, built by the present occupier', valued at £10 10s. He also owned 15 acres of land in three lots as well as the cottage properties listed in Table 8.4b.[18]

One final insight into village life is provided by the 1910 material. Churches, chapels, schools, public houses and village halls are all scheduled. The Primitive Methodist chapel (a butcher's shop in the 1990s) is described as 'built in 1879 – pine seated with rostrum', and the Wesleyan chapel was 'exceedingly well built and well seated inside'.[19] The Free Methodist chapel is described as 'a chapel with gallery and school room', with reference also to changes made to the structure since the original inspection.[20] The Temperance Hall, one of the principal social venues in the village, is described as 'a lecture room with bottom portion made into two cottages'.[21] The village school, under the administration of the County Council since 1902, was not included in the valuation, although the playground, owned by the Ecclesiastical Commissioners, was included.[22]

In various ways, therefore, the 1910 material can shed light on the local economy and society of parish life. Much depends on the enthusiasm and dedication of the original surveyors, but in this Lincolnshire example of average quality the possibilities of the material, especially when used in conjunction with other contemporaneous material such as the directories as

[17] PRO IR 58/33575–7. [18] PRO IR 58/33577. [19] PRO IR 58/33578.
[20] PRO IR 58/33576. [21] PRO IR 58/33577. [22] PRO IR 58/33578.

used here, allow us to move towards a fuller representation of the village community than has hitherto been possible.

The landed estate

Well to the fore in the controversy surrounding the Liberal taxation procedures were the owners of large estates, and there is a correspondingly large amount of information available locally on the great and lesser estates at this time. One can be examined here, the Ashburnham estate in East Sussex, partly for its intrinsic interest and partly for methodological reasons of understanding the use of the documents in the analysis of the estates at a time of great uncertainty over their futures.[23]

Mention has been made of the Ashburnham estate in the Sussex Weald.[24] The estate can now be used to demonstrate the degree to which information can be obtained from the 1910 material on a large landed estate. It was centred on the parish of Ashburnham, where Bertram, 5th Earl of Ashburnham (1840–1913) owned almost all of the land. The year 1913 produced an 'occasion' for Increment Value Duty purposes, when Bertram died and was succeeded by his brother Thomas, the 6th (and last) Earl who died without issue in 1924.[25] However, the long-drawn-out nature of the assessment procedures are illustrated by the fact that Bertram's executors were only being informed of the position on Increment Value Duty on small plots in outlying parts of the estate in May 1917, four years after his death, by which time the 6th Earl had already been dealing in land transactions on the estate.[26]

The sources used included the Valuation Books in the East Sussex CRO (ESRO) and the Field Books in the PRO. Additional sources included the 1910 Record Sheet plans, and estate records which included rentals for this period and correspondence. There are also Rate Books for some relevant parishes in ESRO.[27]

[23] On the relationship at this time between government policy and the wellbeing of Dorset estates, see Janet Waymark, 'Landed estates in Dorset since 1870: their survival and influence', University of London, Ph.D. thesis, 1995, 53–77.

[24] See, for example, Chapter 6 where the problem of the 'hereditament parish' as opposed to the civil parish is discussed. The material on which this section is based was originally published as Brian Short, Mick Reed and William Caudwell, 'The county of Sussex in 1910: sources for a new analysis', *Sussex Archeological Collections*, 125 (1987), 199–224.

[25] Thomas was succeeded by Lady Catherine Ashburnham, daughter of the 5th Earl, and when she died in 1953 succession passed through the female line to the Rev. John Bickersteth.

[26] Form 181A–Land was used by the Land Values Branch of the Inland Revenue to inform Messrs Peake, Bird, Collins & Co. (the Ashburnham solicitors in Bedford Row, London) that no duty would be payable on the death of the 5th Earl for various cottages and small pieces of land in Battle, Pevensey, Herstmonceux and Warbleton (Chief Valuer's Library). Form 182–Land was used by the same office to advise that a part of High Holmstead Farm (Warbleton) which was sold in October 1914 was not liable to duty. (East Sussex Record Office uncatalogued Ashburnham Mss, dated 6 August 1918).

[27] PRO IR 58/29198–200 Field Books for Ashburnham parish; ESRO IRV 1/3 Ashburnham Valuation Book.

Table 8.5 *Number of hereditaments on the Ashburnham estate, East Sussex*

Parish	Total	'Missing'	%	Area of parish (acres)	Area owned by estate stated to be in parish*	% of whole
Ashburnham	206	15	7.3	4079	5330	131
Battle	54	3	5.6	8252	2476	30
Brightling	9			4901	599	12
Burwash	15			7452	1432	19
Catsfield	12			3018	623	21
Dallington	26	4	15.4	1941	1513	78
Hailsham	2			5330	46	1
Heathfield	3			8032	69	1
Herstmonceux	17			6507	513	8
Hooe	3			2473	570	23
Mountfield	8			3928	631	16
Ninfield	26	21	80.8	2619	528	20
Penhurst	27			1455	1049	72
Pevensey	20			4397	475	11
Warbleton	70	4	5.7	6226	948	15
Wartling	58	1	1.7	3287	2147	65
Willingdon	3			2572	80	3
Totals	559	48	8.6	76,469	19,029	25

Notes:
* As stated in the Field Books
Source: PRO IR 58/29198–769; East Sussex Record Office IRV 1/3.

The parishes in which the estate had land are indicated in Table 8.5 which also indicates the distribution of nearly 600 hereditaments covering the estate among these parishes.

The documents are of variable quality. The Valuation Books contain very little information other than that transcribed from the Rate Books. Few values are included and the figures for extent are almost all confined to the estimates from the Rate Books. The Field Books are certainly variable in quality. Those for Ninfield appear almost useless as far as estate-owned land is concerned. There are entries for only five of the twenty-six estate-owned hereditaments indicated in the Valuation Books. Entries are also missing for a few hereditaments in several parishes as indicated in Table 8.5. The reasons for this remain obscure. In some cases in Ashburnham parish, the 'missing' hereditaments were incorporated into others. This practice was common and was usually indicated in the Field Book, but in these cases it is necessary to examine the Valuation Books to establish this. For the remaining cases, it is possible that the hereditaments were simply not valued at all, perhaps because the war and/or the Scrutton judgement rendered the work impracticable.[28] Three areas in Ashburnham are indeed blank on the Record Sheet plans – Compass Wood, some small plots nearby, and an area around and including Northlands Wood – and this may account for three or more of the 'missing' hereditaments (Fig. 8.4).

Values are missing in almost all cases in the Valuation Books, and are only intermittently entered into the Field Books, where they are frequently limited to the Gross Value only. In most other respects though, the Field Books seem very useful. Property descriptions are usually very good, and detailed evaluation of property can be made.

Apart from owning most of Ashburnham parish itself, the estate had substantial interests in the neighbouring parishes. In all, some seventeen East Sussex parishes had an Ashburnham 'presence' at this time,[29] and Table 8.5 gives an estimate of the amount of land owned by the estate in each parish. Where possible, the extent determined by the valuer has been used, while in the absence of such a figure, the acreage in the Field Book that was ostensibly extracted from Form 4 has been used. Only in the absence of either has the estimate given in the Valuation Book been utilised. The figure for Ashburnham relates to the 'hereditament parish' because property primarily within the parish of Ashburnham, and returned as being under that address, overlapped into neighbouring parishes.[30]

The land owned by the estate was presumably mainly freehold. Unfortunately, only the Field Books for Ashburnham and Penhurst give this information for more than a tiny handful of hereditaments. In these two parishes almost all land was owned freehold, although the estate still had a few copyhold properties held of various manors. Copyhold tenure was abolished in 1928.

Land on the estate was also occupied in holdings of widely differing sizes as indicated in Table 8.6 and in Fig. 8.4. Overall, some 20 per cent of hereditaments were smaller than ¼ acre, being mainly cottages with gardens. The estate owned substantial numbers of these in Ashburnham and Wartling parishes, but few elsewhere. Almost as numerous were plots between ¼ acre and 1 acre. Again these were often cottages with larger gardens, or else were odd plots of land in awkward situations. These were also most prominent in Ashburnham and Wartling, but were a prominent feature of estate-owned land in Ninfield and Battle as well. Most noticeable overall is the finely graduated range of hereditament sizes, with substantial numbers between 5 and 50 acres. Of course, some of these hereditaments may have been held by the same occupier, and the data would enable assessment of this, but once again it

[28] See Chapters 2 and 4 for the implications of the Scrutton judgement.

[29] ESRO Ashburnham Mss, uncatalogued estate rental 1909.

[30] See Chapter 6, and Short *et al.*, 'The county of Sussex', 206, Fig. 2. The work on Ashley Walk (Hampshire) and Lockinge (Berkshire) revealed similar problems.

Table 8.6 *Hereditaments by unit size on the Ashburnham estate, East Sussex (acres)*

	A	B	C	D	E	F	G	H	J	K	T
Ashburnham	45	26	13	20	13	7	3	4	9	4	144
Battle	2	11	6	13	4	4	1	1	1	2	45
Brightling				3	2	1		2			8
Burwash					2	1		1	4	1	9
Catsfield	7				2	1	1			1	12
Dallington	6	2	1	4	4	2	1		3	1	24
Hailsham				1	1						2
Heathfield				1	2						3
Herstmonceux	7		1	1	2	3			1		15
Hooe				1		1				1	3
Mountfield				1			1		1	1	4
Ninfield		10	5	4	1	2	1		1		24
Penhurst		6		2	3	1	1	5	1		19
Pevensey		1	3	10	4	1	1				20
Warbleton		3	1	6	9	1		1	2		23
Wartling	24	16	3	5	3		1	1	2	3	58
Willingdon			1	1		1					3
Totals	91	75	34	77	50	25	10	15	24	17	418

A <0.25 acres; B 0.25 to 0.9 acres; C 1 to 4.9 acres; D 5 to 24.9 acres; E 25 to 49.9 acres; F 50 to 74.9 acres; G 75 to 99.9 acres; H 100 to 149.9 acres; J 150 to 299.9 acres; K >300 acres; T Total
Source: PRO IR 58/29198–769.

would be a fairly lengthy task. Almost certainly, multiple holding was not common, and the data suggest a strong existence of small- and medium-scale agriculture in this part of the Weald prior to the Great War. Overall, hereditaments in the parish of Ashburnham were usually compact, with all parts contiguous. Fragmented hereditaments were few, and mainly small.

Information on tenancy terms was only sporadically entered in the Field Books, either because it was not returned on Form 4, or because of selective transcription from the Form into the Field Book. In any event, information is available for less than half of the total number of hereditaments. These were let by the estate on either a weekly (in the case of cottages), or else on an annual basis. Six were let rent-free, and only four were let on any other basis. Table 8.7 shows this for each of the parishes on the estate.

Cottage rents seem to have been generally high, especially when compared to those on other English estates at this time. Information was again only sporadically entered into the Field Books, and it is possible that the rents featured in Table 8.8 may not be typical. Some 55 per cent of cottages were let at

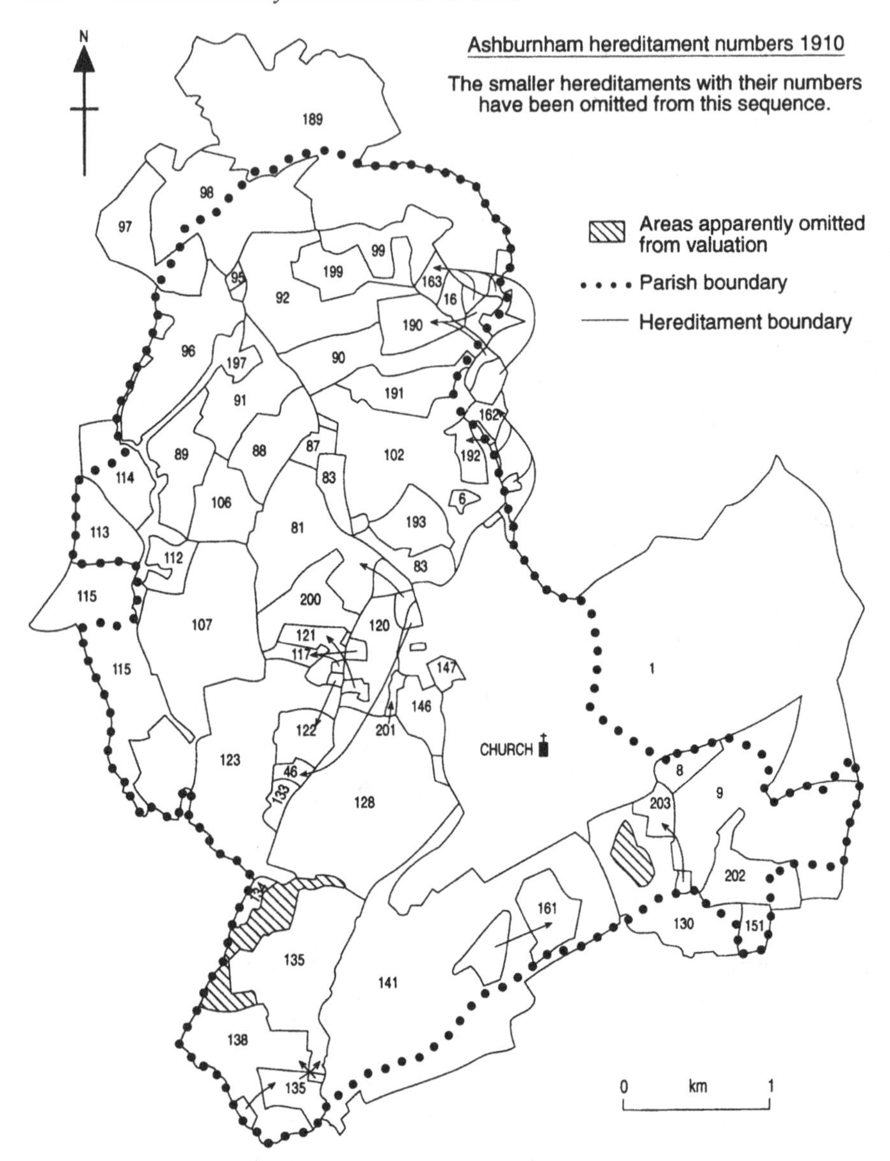

Fig. 8.4 Ashburnham parish, East Sussex: hereditaments 1910.
Source: PRO IR 58/29198–29200.

1. Ashburnham Place, Park, etc.
6. Brickyard (Court Lodge)
8. Agmerhurst Lodge
9. Agmerhurst farm
16. Ashburnham furnace (pt)

46. Bray's Hill, Beggar field and Barn field
81. Reedlands farm
83. Reedlands farm (pt), Court Lodge farm (pt) and Peltham farm (pt)
87. Thornden farm (pt)
88. Thornden farm (pt)
89. Lattenden farm (pt)
90. Buckwell farm (pt)
91. Thorndale farm
92. Buckwell farm
95. Olivers Hill
96. Sliverick's farm
97. Sliverick's farm (pt) and Olivers Hill
98. Herring's farm
99. Lakehurst farm
102. Court Lodge farm
106. Glyde's farm and Peltham farm (pt)
107. Brigden Hill, farm and Brownings (pt)
112. Farthings, Little Midge and Brigden Hill
113. Red Pale
114. Red Pale (pt)
115. Pear Tree farm (pt)
117. Brownbread Street
120. Street farm
121. Northlands farm
122. Bray's Hill, Brownbread Street (incl. Ash Tree Inn)
123. Frankwell farm
128. Lingham's farm
130. Corner House
133. Bray's Hill
134. Tillys farm (pt) and Henley Bridge
135. Wilson farm and Page farm
138. Gardners farm
141. Kitchenham farm
146. Pigknoll farm
147. Glebe
151. Marlpits
161. Luxford's and Wilding Woods
162. The Forge and Peens farm (pt)
163. Peens farm
189. Pannelridge Wood
190. Anderson's Wood
191. Furnace Wood
192. Malthouse Wood
193. Pontsgreen Wood
197. Wheeler's Wood
199. Buckwell Wood
200. Reed Wood
201. Pigknoll Shaw
202. Lower Freckly Wood
203. Lower Hurst Wood

Table 8.7 *Tenancy terms on the Ashburnham estate, East Sussex*

	Weekly	Monthly	Quarterly	6-monthly	Yearly
Ashburnham	53*		1		44
Battle	10		1~		19~
Brightling					2
Burwash					4
Catsfield	6				1
Dallington	7		1~		12
Hailsham					
Heathfield					
Herstmonceux					4
Hooe					3
Mountfield	2				2
Ninfield	1				2
Penhurst	8				10
Pevensey					8
Warbleton					13
Wartling	10			1	11
Willingdon					3
Total	97	–	3	1	138

Notes:
* Includes two hereditaments where house was let weekly and the land annually
~ Includes two hereditaments where house was let quarterly and the land yearly
Source: PRO IR 58/29198–769.

Table 8.8 *Cottage rents on the Ashburnham estate, East Sussex*

Rent(£)	No.
Rent-free	6
<4	11
4 to 5	11
5 to 6	54
6 to 7	13
>7	4
Total	99

Source: PRO IR 58/29198–769.

Table 8.9 *Living accommodation on the Ashburnham estate, East Sussex*

	Bedrooms						
	1	2	3	4	5	6+	Total
Living rooms (inc. kitchen)							
1	12	77	31	2	1		123
2		13	40	21	6		80
3			8	9	4	3	24
4					3	5	8
5					1	1	2
6+						1*	1
Total	12	90	79	32	15	10	238

Notes:
* Ashburnham Place
Source: PRO IR 58/29198–769.

between £5 and £6 per annum, compared to £4 which was the maximum on the Lockinge estate, Berkshire, which also had large numbers let at under £3 per year. On both estates, however, the owner almost invariably assumed responsibility for the costs of rates and repairs, as well as for tithes and land taxes where payable in connection with cottages.

The accommodation on the estate naturally varied. Once again, details are not always given, but information on the numbers and use of rooms is available for 238 houses. Most of these were small (Table 8.9), with 102 (43 per cent) having only one or two bedrooms. A number of the larger farmhouses, and of course Ashburnham Place itself, were at the opposite end of the spectrum, being very spacious. The mansion is very fully described in an old notebook, which was transcribed in August 1941 from the field notes taken by Mr H. S. Burt. The top floor, for example, was described as having two night nurseries and a day nursery; suites of blue, pink, green and orange rooms; a seamstresses' room, together with attics, WCs etc. The first floor contained bedrooms such as those 'occupied by the late Earl'; the state bedroom suite; a 'magnificent gallery' and a landing between Chinese rooms (white and green); blue and pink bedroom suites. On the ground floor were a large dining room; large and small dining rooms and libraries; 'the Earl's private entrance'; gun, billiard and smoke rooms and strong rooms. Outside were the dairy, brewery, coal house, a heated winter garden, with drinking water from the hill behind and washing water from the lake. The gardens were stated to cost £1,000 per annum. A large range of stables, eight lodges in the 140-acre grounds, a home farm and the walled-in garden described above completed the picture.

Fig. 8.5 Ashburnham Place, East Sussex *c.*1910.

The servants' quarters occupying three floors were said to be 'shut off' but a full description is also available.[31] Such an extraordinary level of detail allows the virtual reconstruction of the big house just prior to the First World War (Fig. 8.5).

Court Lodge farmhouse in Penhurst was described as follows:

Farmhouse A very old fashioned house part built of stone & tile & part of brick, weather-tile & tile. Well blt.
Accom:- porch, *Grd floor*. Kitchen, small Sitting Room, large front Room (very bad Repr!) Pantry & Scullery & Cellar underground.
1st floor. Landing, 1 front bed all oak pannelled, 4 bedrooms
Top. 2 attics (no use for bedrooms). Good Repair.
2 Cottages. A pair of semi-detached Cottages well built of Brick, Stucco & Tile.
Accom. of each Living Room & Washhouse & 3 bedrooms. Fair Repair

We are also informed that Court Lodge Farm comprised agricultural land, house and buildings and cottages; it was 109 acres 2 roods 30 perches in extent; the occupier was C. White sen.; the owner the Earl of Ashburnham; agent A.P. Ashburnham-Clement; the farm was freehold on a yearly tenancy at £90 rent; the tithe of £17 5s 4d was paid by the owner who also paid the insurance and was liable for repairs, the occupier paying other rates and taxes. An accompanying sketch map in the Field Book (Fig. 8.6) showed that the farm complex also included a brick, timber and tile cart lodge; a brick and tile three-stall stable; a brick and tile store room and granary over loose box and lean-to meal room and brick and tile oast kiln; a timber and corrugated iron lean-to cow

[31] PRO IR 58/29198 (Field Book description of Ashburnham Place).

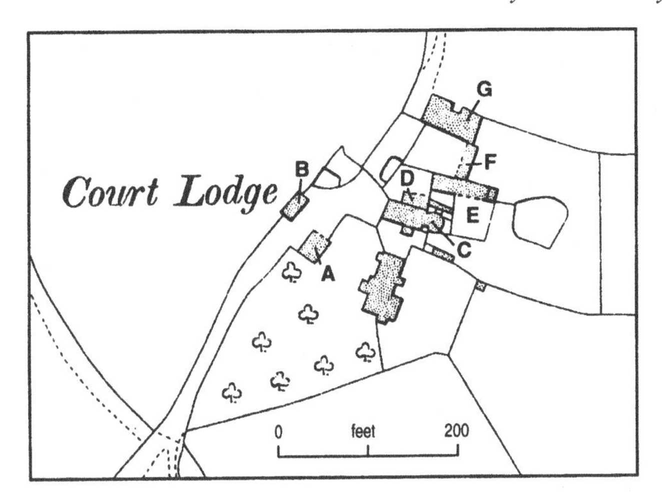

Buildings

A. Brick, timber and tile cart lodge
B. Brick and tile 3-stall stable
C. Brick and tile store room and granary over with loose box and lean-to meal rooms and brick and tile oast kiln
D. Timber and corrugated iron lean-to cow lodge and yard
E. Brick, timber and tile pig pound
F. Range of brick and tile open cow lodges and yards
G. Large brick, timber and tile barn and lean-to cow lodge
H. Brick and tile lean-to cart lode (location not marked on original sketch map)

Fig. 8.6 Valuer's sketch plan of Court Lodge farmstead, Penhurst, East Sussex *c.*1910.
Source: PRO IR 58/29199.

lodge and yard; a brick, timber and tile pig pound; a range of brick and tile open cow lodges and yards; a large brick, timber and tile barn and lean-to cow lodge; and a brick and tile lean-to cart lodge. All were referred to as in 'fair repair'.[32]

Students of garden history would have much to learn from looking at the Field Books for examples of garden design and content. The walled garden at Ashburnham Place, Sussex, for example, is described as having:

Range of brick and slate potting sheds – 2 ranges brick pits – range of brick pits – 2 ? sheds – Range of brick pits – Brick base 3/4/ tomato hse – Furnace Hse – h/t melon hse. T & T open potting shed – Range of lean-to brick-base grnhses & vinery – L/T

[32] PRO IR 58/29199.

Table 8.10 *Condition of housing on the Ashburnham estate, East Sussex*

'Shocking repair'	1
Bad/very poor	8
Poor	29
Fair	141
Good	30
Very good	2
Total	211

Source: PRO IR 58/29198–769.

vinery – L/T hot hse – Brick & ? hovel in ? – range of timber and tile bullock hovels in ditto. Small pig pound & timber.[33]

However, the condition of much of the other accommodation on the estate was not considered by the valuers to be particularly good. Table 8.10 shows the condition attributed to 211 houses.

The great majority of houses with details of condition were described as 'fair', but a considerable number were 'poor' or worse, and rather less were described as 'good' or 'very good'. Dampness was a fairly common feature, even of houses described as 'fair' by the valuer.

The descriptions of some 136 houses provide details of the water supply. Most (102), as would be expected, had access to a well, but nineteen (14 per cent) relied on springs for their water, and two had to obtain their supplies from a distance. Only thirteen had pumps. The bulk of houses, though, provide no details, and it cannot be assumed that these trends were repeated in the other cases. Toilet facilities largely consisted of outside earth closets. Water closets were fairly common in the larger homes, but were seldom found elsewhere.

There is still more information which could be exploited in this case study. Information on livestock accommodation, dairying and baking facilities, for example, is frequent in the property descriptions, but has not been analysed. The problems of allocating specific properties to the social situation of the occupier in the absence of access to the 1911 Census inhibits full exploitation of this material. For example, a number of cottages had dairies attached, yet there was apparently no land for keeping livestock. It may be that the dairy was disused; or that the occupant was solely engaged in dairying, buying all the milk required; but it could equally be that the cottage and its occupant were part of a larger unit, perhaps attached to a nearby farmstead, so that the dairy was part of the facilities attached to the latter. We certainly know that there were 231 dairy cows in the parish in 1909, a not inconsiderable number.[34]

[33] PRO IR 58/29198. [34] PRO MAF 68/ 2371 (June Returns for Ashburnham 1909).

This case study does serve to demonstrate some of the pitfalls in deriving conclusions about property size, landownership, and land occupation patterns in relation to the individual parish. Even after delineation of the ITP, and its constituent civil parish(es), the problems caused by what has been called the 'hereditament parish' are formidable, and although these can mostly be resolved by reference to the Record Sheet Plans, one must exercise caution in assuming that the documents 'innocently' reveal all the information about the parish, or indeed that they do not contain information relating to contiguous parishes.

There will be many other estates throughout the country which demonstrate similar arrays of documentation, and for which a thoroughgoing reconstruction at this time of landed crisis can be undertaken. The possibility arises, for example, that land sales will have generated sale catalogues or other material related to the sale of parts or whole of estates in response to some degree to governmental pressure. In Dorset, for example, the period 1910–14 saw the appearance of just under 5 per cent of the county's total acreage of large (500 acres and over) estates on the market, eighteen in all, totalling nearly 20,000 acres. A busy Dorset land market in the years leading up to 1910 is caught by the Field Books which register the new owners. Thus the 1906 sale of the parish of Poyntington in north Dorset by Sherborne Castle Estates resulted in its ownership in 1910 by two absentee owners from the rapidly expanding Bournemouth. The Netherbury Field Book shows the similarly rapid acquisition of farms and small holdings by Thomas Colfox.[35]

To catch an estate at the moment of sale would be very revealing. The Newport estate in Almeley, a parish in the Wye valley, Herefordshire, for example, was sold on 28 September 1909 with parts being acquired by sitting tenants.[36] This raised the possibility of comparing ownership and other data in the Valuation Books, which as transcripts of the 1909–10 Rate Book should give the pre-sale position, with the information contained in the Field Books which were mostly compiled between 1911 and 1914. An examination of changes that were involved in the process of dissolution of this landed estate might then be possible. Form 4, which was sent out to owners in August 1910, would have been sent to the new owners so the Field Books would have unambiguously provided data of change of ownership when compared to the Valuation Books.

The Valuation Books were basically well compiled with substantial cross-referencing, but with considerable alteration, due no doubt to the 1909 sale. This made them difficult to read and to analyse. This 'messiness' is also a feature of the Field Books. As well as the Valuation Book, a sale catalogue

[35] Waymark, 'Landed estates in Dorset', 70.

[36] A. Whitehead, 'Social fields and social networks in an English rural area, with special reference to stratification', University of Wales, unpubl. Ph.D. thesis, 1971, 76.

exists in the Hereford and Worcester Record Office giving some assistance in unravelling the much-altered detail in the Valuation Book. The estate, while centred on Almeley, also extended over neighbouring Eardisley, Kinnersley, Lyonshall and Kington.[37]

In 1909 there were 175 hereditaments in the Almeley Valuation Book. There were many, mainly small, owners. The Newport estate, in the person of Mrs Gurney Pease, owned thirty-eight hereditaments. As a result of the sale and other reasons, several hereditaments were sub-divided and additional numbers allocated bringing the total to 193. Of the extra eighteen hereditaments, twelve had been part of the Newport estate, bringing the total number relevant to this case study to fifty.

The greatest problem for this study was that the Field Books did not indicate a full change of ownership from that shown in the Valuation Book. In the latter the owner is shown as Mrs Gurney Pease, and this relates of course to April 1909. In the former, the owners of the relevant hereditaments are shown as the Trustees of Mrs Gurney Pease. This information was extracted from Form 4, completed by the owners usually between late August 1910 and the end of that year. This presumably indicates that even by late 1910, the sale had not actually been finalised. Since the Field Book entries were not altered, we must assume that the position was still not finalised by the time the provisional valuation was issued, or that the valuers were not concerned about altering the Field Book entry. Access to Form 37 would help resolve this, but although they have been retained by Hereford CRO, they are unlisted. In any event, the problem does reinforce the potential significance of Form 37 for certain purposes, and once more highlights the possible damage done by their widespread destruction.

The difficulties of the data may best be illustrated by listing the relevant entries in Table 8.11. The occupier in the Valuation Book is the person in occupation at 1 April 1909, while the person indicated in the Field Book is taken from Form 4 and presumably relates to the situation after August 1910. From the point of view of the valuation procedure, it is the later occupier who is relevant. The Act (sec. 26.1) required every piece of land under separate occupation to be separately valued. But this permitted pieces of land in the same ownership and occupation, but not contiguous, to be valued together, and to be considered as one hereditament. *Instructions to Valuers* Pt1 (para. 142) indicates this when it states that detached portions of hereditaments should be 'braced together' on the Record Sheet Plans. Under this provision,

[37] Herefordshire Record Office (Hereford and Worcs.) AG9/91 Valuation Book for Almeley in Weobley Division. Eardisley is AG9/96, Kinnersley AG9/97, Lyonshall AG9/43 and Kington (Urban and Rural) AG9/37–8. The latter, together with Lyonshall, are in Kington Division. The Field Books for Almeley parish are PRO IR 58/38519–20. The Newport estate sale catalogue is Hereford CRO K10/10

Table 8.11 *Newport estate hereditaments in Almeley, Herefordshire*

Hereditament no.		Occupier		Area		
Main	Components	In VB	In FB	(A	R	P)
24	24	J. Mainwaring	Same	338	1	4
	5	T. Vaughan	Same	0	0	36
	89	W. Pinnock	Same	n.s.		
	144	F. D. Price	Same	31	2	0
30	30	E. Williams	Same	0	1	27
	81	E. Williams	Same	n.s.		
64	64	E. J. Davis	F. E. Davis	300	3	27
	18	S. Barnett	F .E. Davis	0	1	39
	65	E. J. Davis	F. E. Davis	n.s.		
	75	Wm. Michael	F. E. Davis	0	3	31
	124	Albert Williams	F. E. Davis	0	3	6
73	73	Geo. Lewis	Same	314	0	3
	28	J. C. Jones	Same	45	3	21
	122	C. Griffiths	Geo. Lewis	1	2	31
	154	H. W. Smith	Same	6	1	19
	Kinnersley 25	No other information				
80	80	S. Price	T. Price	81	1	10
	Lyonshall 86					
97	97	S. Powell	S. Powell	273	3	38
	40	M. Isaac	S. Powell	n.s.		
108	108	n.s.	T. Hughes	n.s.		
	35	T. Hughes	Same	236	1	3
123	123	A. L. Williams	Same	2	3	27
	153	J. Mainwaring	Same	16	1	5
	161	A. L. Williams	Same	10	2	19
151	151	A. Williams	Same	152	2	29
	6	W. Joseph	A. Williams	0	2	7
	41	A. Hards	A. Williams	n.s.		
	53	M. Webb	A. Williams	0	1	0
152	152	R. Vigors *et al.*	Same	33	0	6
	32	Mrs G. Pease	Same			
	136	Mrs G. Pease	Same	186	2	1
	15	R. Vigors	Same	22	2	31

Notes:
NB The 'component' hereditaments in the table are those stated in the documents to be included in the main hereditament for valuation purposes.
A R P Acres, roods, perches
n.s. not stated
Source: Herefordshire RO AG9/91; PRO IR 58/38519–20.

therefore, it would appear that only some of the hereditaments (nos. 30, 64, 97, 108 and 151) should have been combined with their component hereditaments. However, by the Revenue Act 1911 sec. 5, this procedure was amended, so that the owner could request the valuing together of contiguous pieces of land, not exceeding 100 acres in aggregate extent, and in separate occupation, although the valuers were not obliged to comply with any such request.

The table does suggest that if any such request were received, the valuers did not necessarily comply with the requirements of sec. 5 of the Revenue Act 1911. The hereditaments in separate occupation that were merged with nos.24, 73 and 152 totalled well over 100 acres in each case.

The fate of the various hereditaments after the sale is even less clear. The estate was auctioned in twenty-nine lots of varying sizes. The owners of each hereditament following the sale are not indicated in the Field Books, although they may be included on Form 37. The data examined only indicate that one J. C. Mason bought Lot 1, by far the largest lot in the sale, and occupied by the Rev. Vigors and others. The material indicates that Lot 1 included, presumably *inter alia*, part of a hereditament no.24 (excluding those merged with it) totalling just over 57 acres in extent. This portion was allocated a new number 177 and was said still to be in the occupation of James Mainwaring. This procedure seems to have been at variance with the instructions issued to Valuers, which states:

> In the event of any original unit of valuation being subsequently sub-divided, or the original valuation apportioned, every part of the original unit should be given the same number with the addition of a subordinate number. Thus, assume a property valued in the parish of Leeds and originally numbered 1485. – This property would throughout be identified as 1485 Leeds, and would be so recorded and referred to in all books, forms, plans, correspondence, &c. If that property were subsequently sub-divided, the parts lopped off might be referred to as 1485/1 Leeds, 1485/2 Leeds and so on. Any further sub-division should then be referred to as 1485/1/1 Leeds and so on.[38]

Had the valuers consistently adhered to this practice, then the task in hand would have been both much easier and more rewarding.

Similarly part of no.73 was sold off to J.C. Mason in Lot 1. This amounted to 20 acres out of the total 314 acres of no.73. The part sold was allocated the number 178 and it apparently remained in the occupation of George Lewis. We have seen that part of no.24 was sold in Lot 1. Part was also sold in Lot 6 and allocated the new number 183. The new owner is not known, but it appears to have still been occupied by James Mainwaring.

The data are still more confusing in the case of Hereditament no.89 occupied by W. Pinnock according to both the Valuation and Field Books, and incorporated with no.24 for valuation purposes. In fact this is said to be the

[38] Inland Revenue Valuation Department, *Instructions to Valuers* Part I (HMSO, London 1910), para. 123 'Subsequent subdivisions of the Unit of Valuation' (p. 25).

case in the Valuation Book, whereas the Field Book included it in no.5 which was in its turn included in no.24. The Field Book states that no.89 formed part of Lot 3, occupied by J.Mainwaring according to the sale particulars. There is no indication against these entries of any new number being allocated to the part sold. However when the new numbers are themselves examined, no.184 (23 acres), occupied by T. E. Davis, is said in the Valuation Book to be included in no.24 occupied by J. Mainwaring. Unfortunately we are also directed to see 64 (inc. 65) both of which were occupied by Davis. The Field Book only adds to the confusion by stating that no. 184 was included in no. 144 occcupied by F. D. Price, and itself included in no. 24. In summary, it seems as if no. 24 was merged for valuation purposes with three other hereditaments, part of which together formed parts of at least three lots. Hereditament boundaries were not adhered to when drawing up lots for sale purposes. Similar problems occur with other hereditaments but it would not serve our purposes to list them all.

The detail given above may seem disproportionate in the light of the limited findings which are yielded. The important conclusion, however, is that the readily available data are of little use on their own for examining the breakup of this landed estate in the Wye valley. The complexities are enormous and the ambiguities often equally so. Once again we are forced to recognise the fundamental unity of the various categories of material and to regret their dispersal and frequent destruction. It does seem very clear that for detailed local study of ownership and holding structures, which are the two least complicated groups of data, Form 37 is the only document that can be linked fairly straightforwardly with the Record Sheet Plans. The two together would provide a highly desirable framework from which to begin. Given a relatively unambiguous set of data and by working backwards, it seems probable that the ambiguities of the Valuation and Field Books would be less formidable.

Nevertheless, these two estates – one in process of dissolution and the other subject to an 'occasion' on the death of its owner – do demonstrate the validity of the documents in helping to reach an understanding of the stresses facing the landowning elite on the eve of the Great War.

Inter-parish contrasts

Two examples, taken respectively from the north Lincolnshire Wolds and from the Fens, will demonstrate how the 1910 documents can be used to throw light on parish contrasts. They will also show the varying degrees of use to which the records can be put, given the constraints of data entries in some cases.

The north Lincolnshire study concentrated on two adjacent parishes in the north Lincolnshire Wolds, and sought to examine contrasts between an 'open' and a 'close' parish. Keelby was selected as an example of the former, while Brocklesby – the seat of the Earl of Yarborough – was chosen as a 'close' example. The Earl was by far the largest landowner on the Lincolnshire Wolds at this time, owning perhaps one-third of the Wolds by value in 1831, and over 55,000 acres in the 1873 Return. He was the patron of both parish churches (Fig.8.1).[39]

Brocklesby was the only civil parish within the ITP of that name, and contained 128 hereditaments. The ITP of Keelby included Keelby civil parish as well as Riby. Keelby hereditaments were numbered 1 to 293 inclusive within the ITP. The Valuation Books in the Lincolnshire CRO contained no information at all, beyond that transcribed from the Rate Books, except for the hereditament numbers which were added in the customary way at the beginning of the valuation process.

The Field Books in the PRO were far from satisfactory in the case of Brocklesby. Only forty-nine of the 128 hereditaments (38 per cent) had any descriptive material at all. Most of the descriptions were limited to the odd word such as 'wood', 'useful grass field', 'rough grass', or slightly more helpfully 'all grass coarse and wet' (of a 7 acre smallholding) and so on. The 190 acres of woodland had a fuller description of small oaks, good elms and some clear ash, with mixed coppice and both young and full-grown larch.[40]

Only nine houses or cottages had any descriptive material. Hereditament no.112 was a brick and thatch cottage (two up, two down and a pantry) and a smallholding of just under 11 acres with two old sheds, all in very poor condition, and since it was in between two lines running into Ulceby station, it was stated to be in a 'bad situation'.[41] The reasons for the otherwise poor amount of detail are clear. The great house – seat of Lord Yarborough – was merged for valuation purposes with fifty-one other Brocklesby hereditaments as well as three from adjoining parishes. The details of all these were said to be in the 'schedule in file'. The amount of detail necessary for such a vast collection of hereditaments could simply not be entered in the limited space available in the Field Books. A number of other hereditaments were also merged with others in adjacent parishes. The Field Books for Keelby were more complete, and adequate for most purposes.

The Valuation Books make it clear that, with the exception of the railway line, owned by the Great Central Railway Company, the whole of Brocklesby was owned by Lord Yarborough. Ownership in Keelby was more diffused. The seventy-four different owners included Yarborough of course, and also Ernest George Pretyman of Ipswich, the Unionist MP, ardent opponent of the valuation and the driving force behind that most successful oppositional organisation, the Land Union.[42] He and Yarborough owned together 113 of the 293

[39] Detailed local knowledge was most helpfully supplied by Dr Charles Rawding of Market Rasen, Lincolnshire. See also Charles Rawding, 'The iconography of churches: a case study of landownership and power in nineteenth-century Lincolnshire', *Journal of Historical Geography* 16(2) (1990), Table 1, 160. For a more detailed discussion of the 'open' nature of Keelby, see C. K. Rawding (ed.), *Keelby Parish and People, 1831–1881* (Keelby 1987), esp. Chapter 3. [40] PRO IR 58/33584 Hereditament no. 54. [41] PRO IR 58/33584.

[42] See Chapter 1.

Table 8.12 *Number of hereditaments in single ownership in Keelby, Lincolnshire*

No. of hereditaments	No. of owners
1	33
2	17
3	8
4	3
5	5
6	1
7	1
8	2
9	1
10 and over	3*
Total	74

Notes:
* Ellen Warmesley owned 13 hereditaments; E. G. Pretyman owned 44 (inc. 14 allotment plots); and Lord Yarborough owned 69 (inc. 45 allotment plots).
Source: Lincs. Archives Office (uncatalogued Valuation Book) and PRO IR 58/33925–7.

hereditaments in the parish (38.6 per cent), although many of these were allotment plots. Table 8.12 shows the numbers of hereditaments owned by different people in Keelby.

Two-thirds of property owners in Keelby owned only one or two hereditaments. Even excluding the allotment plots (mainly 0.25 to 0.5 acres each), Pretyman and Yarborough owned many more hereditaments than anyone else. Accurate acreage figures are not given consistently, so that detailed assessment of the proportions of the parish acreage owned by the different owners is not easily made. It is clear though that Pretyman and Yarborough owned all the larger hereditaments. Of the eight known to exceed 50 acres, these two men owned seven. The Rev. Neville Bourne owned a hereditament of about 53 acres; Pretyman owned five hereditaments totalling some 925 acres, including the 499 acre Manor Farm ('lightish arable land . . . no milk trade. Timber rough'), while Yarborough owned two large farms, including Keelby Grange with its house suiting a 'good gentleman farmer' of about 598 acres.[43]

Apart from one hereditament in Keelby, freehold was universal in the parish. The exception was a small plot with sheds thereon, held copyright from

[43] PRO IR 58/33925. Keelby Grange was returned as consisting of 357 acres in the Valuation Book (taken from the Rate Book?) but the Field Book shows it to have been over 598 acres according to the valuer's survey. For Manor Farm see PRO IR 58/33926.

Table 8.13 *Tenancy periods in Brocklesby and Keelby, Lincolnshire*

	No. of hereditaments	
Period	Brocklesby	Keelby
Weekly		20
Monthly	14	6
Quarterly	1	3
Six-monthly		6
Yearly	21	129
Over one year		1*
Not stated	82	128
Total	118	293

Notes:
* Leased for 10 years
Source: PRO IR 58/33925–7 (Keelby), IR58/33584–5 (Brocklesby).

the manor of the Earl of Yarborough. To judge from those hereditaments with appropriate information, freehold was universal in Brocklesby.

Tenancies were predominantly yearly in Keelby, even for cottagers. Information is sparse for Brocklesby, but annual tenancies were also quite common there. In the latter parish, it is likely that many cottages were tied. Table 8.13 gives details of tenancy periods. The difficulties of the data may best be illustrated by listing the relevant entries in Table 8.13. Many of those hereditaments without information in Keelby were in fact allotments, which would presumably have been let on an annual basis.

By correlating the information in the Valuation Books with that from the Field Books, it is possible to establish rents for some cottages and houses in Brocklesby but by no means all. These are set out in Table 8.14, but the problems arising from the separation of the Valuation and Field Books are once again demonstrated, since it was only by reference to both that the table could be constructed at all. The figures are confined to those homes with less than an acre of ground.

The substantial numbers of 'rent-free' houses in Brocklesby were presumably tied cottages let to estate workers. Apart from this, Brocklesby houses were let at between £4 and £6 a year. A majority of Keelby houses were let at between £4 and £7 a year. In all, seventy-nine of those with information (58 per cent) were let in this range. In addition, Keelby had a smattering of much cheaper cottages as well as a fairly substantial number of properties with higher rents. The village would seem to have been socially heterogeneous *c.* 1910 compared to Brocklesby.

Table 8.15 gives details of living accommodation in each parish.

Table 8.14 *House rents in Brocklesby and Keelby, Lincolnshire*

	No. of homes	
Rent (£)	Brocklesby	Keelby
Free	14	
1 to 1 19s 11d		1
2 to 2 19s 11d		2
3 to 3 19s 11d		5
4 to 4 19s 11d	11	24
5 to 5 19s 11d	11	27
6 to 6 19s 11d		28
7 to 7 19s 11d		17
8 to 9 19s 11d		14
10 to 12 9s 11d		4
12 10s 0d to 14 19 11d		4
15 and over		10
Total	36	136

Source: PRO IR 58/33925–7 (Keelby), IR58/33584–5 (Brocklesby).

Table 8.15 *Number of rooms: Brocklesby and Keelby houses, Lincolnshire*

	No. of houses	
No. of rooms	Brocklesby	Keelby
2		6
3	1	25
4	4	72
5	1	22
6	3	15
7		8
8		8
9		4
10 and over		5
Total	9	165

Source: PRO IR 58/33925–7 (Keelby), IR58/33584–5 (Brocklesby).

Unfortunately, information for Brocklesby is so sparse – with details for only nine houses – that no valid conclusions can be made. Almost two-thirds of Keelby homes had fewer than five rooms. Two-up and two-down was the most frequent type of house in the parish, although there were a substantial number of three-room cottages (known locally as 'parlour cottages'). Nearly all of these smaller houses had piggeries attached and several had dairies. The cottager's pig was clearly still a very real feature of village life in this part of Lincolnshire.

Comments on the condition of properties were infrequent and permit no realistic statements to be made. Only the one property in Brocklesby was singled out. Hereditament no.112 – already noted above as being in 'very poor' condition, and badly situated between the railway lines. In Keelby information was almost as sparse in this respect. Keelby Grange, again as noted above, which included land in adjoining parishes and owned by Yarborough, had a house and buildings in 'excellent condition'. Only seven other properties had comments about condition. Five of these were owned by Pretyman, and ranged in condition from 'very old and poor', to 'very nice and good'. No. 163, a 118 acre farm owned by Pretyman, was 'very badly done, full of rubbish'. Only two other properties, both owned by small proprietors, have comments, and both were described as 'poor'.

Keelby was well endowed with allotments, though whether they were solely for parishioners is not known. Pretyman and Yarborough let out fifty-nine allotment plots. All those owned by Yarborough were 0.25 acres in extent, and were let for 7s 6d a year. Pretyman's ranged upwards from 0.25 acres, with rents of 7s 6d a year to 27s 6d.

The inadequacies in the completion of the Field Books precluded any real comparison between the two communities in terms of their 'open' and 'close' characteristics. However, the fuller information for Keelby does enable a number of useful statements to be made about the parish. The paucity of information for Brocklesby parallels other estate villages such as Lockinge (Berkshire). Where hereditaments were valued together, presumably at the owners' request, the resultant units are likely to be too large for the relevant details to have been entered into the Field Books in their entirety, so that the 'hereditament files' – known to have existed for each hereditament – were the most likely repository for the information. Once again the decisions and practices of individual DVOs and valuers have rendered the documents less valuable than one might reasonably have expected.

Turning from the Lincolnshire Wolds, a second study of parish contrasts is taken from the Fenland of Cambridgeshire. Both parishes selected for this case study had been the subjects of previous work: Isleham was chosen, to examine some of the comments made by Mary Chamberlain in *Fenwomen: A Portrait of Women in an English Village* where the parish was disguised as 'Gislea', and to assess whether those comments could be substantiated or modified by the valuation documents. The second parish, Willingham, had featured in Margaret Spufford's *Contrasting Communities: English Villagers in the Sixteenth and Seventeenth Centuries.* In particular it was hoped that the documentation would enable an investigation to see whether there remained any vestiges of the landholding structures that had featured during the seventeenth century.[44]

The documents were well compiled in both cases. The Field Books for Isleham were especially fully completed, and even had field-by-field land use schedules for most hereditaments. Both parishes contained large numbers of hereditaments. Isleham, which was an ITP in its own right, had 788 hereditaments. Willingham was even larger, with almost 850 hereditaments. The latter parish was in the ITP of Over, along with the civil parish of that name and Rampton. Willingham comprised hereditament nos.634 to 1478 within Over ITP. In addition, there were 140 unrated hereditaments for which the precise parish in which they were situated is not clear. They have therefore been excluded from the following analysis. For the purposes of this analysis, a sample of every seventh hereditament was examined, giving a sample size of almost 15 per cent. In all, 112 hereditaments were studied for Isleham, and 120 for Willingham (Fig. 8.7)

There were significant differences in landownership structures between the two parishes. Chamberlain indicates that before the 1930s, the village of Isleham was largely controlled by the Coatesworth family who owned virtually all the land.[45] Certainly the Coatesworths are mentioned by most of her older informants as employers and large farmers. They resided at the Hall.[46] Presumably Chamberlain disguised the people in the same way that she disguised the village itself – by changing names. If the Coatesworths really lived at the Hall, then their correct identity was the Robins family. R.T. Robins was the occupier of Isleham Hall in 1910.

The Robins family was certainly the most prominent landowner in the parish. Three members of the family owned a total of twenty-three hereditaments in the sample (20.5 per cent). The sampling method is not really suitable for gauging the proportion of land owned by the family within the parish, and in any event, areas of many hereditaments are not given. Neither could any social or political dominance be discerned from the documents. Overall though, the sample does not support Chamberlain's assertion of landowning dominance by a single family. The 112 hereditaments in the survey were in fact owned by eighty-two individuals, including the three members of the Robins family. Of these, twenty-five were owner-occupiers.

[44] Mary Chamberlain, *Fenwomen: A Portrait of Women in an English Village* (London 1975); Margaret Spufford, *Contrasting Communities: English Villagers in the Sixteenth and Seventeenth Centuries* (Cambridge 1974). Information was also obtained from Cambridgeshire County Record Office 470/0105 (Valuation Book for Over, including Rampton and Willingham) and 470/0106 (Valuation Book for Isleham); and PRO IR 58/16043–50 (Isleham); 16648–56 (Willingham Field Books). [45] Chamberlain, *Fenwomen*, 20. [46] *Ibid.*, 91.

(a)

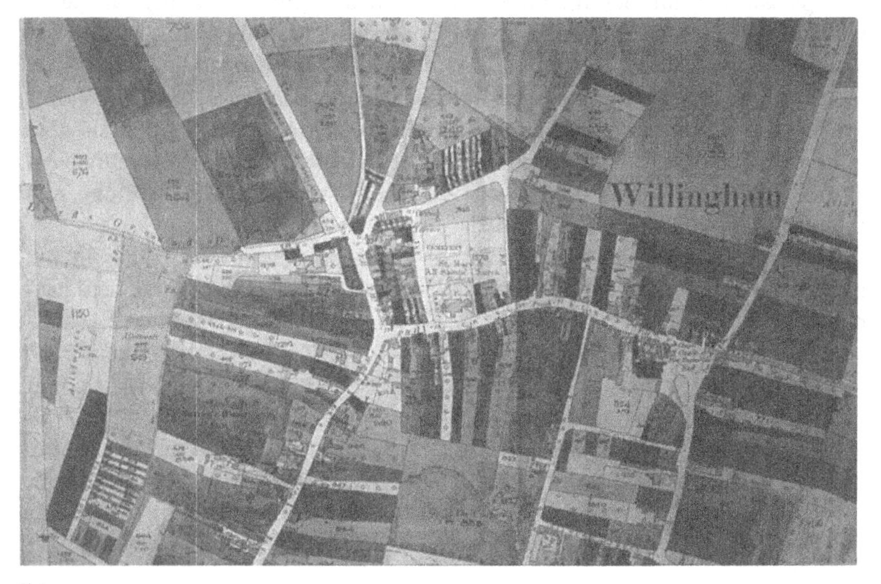

(b)

Fig. 8.7 Fenland villages *c.* 1910: (a) Isleham, (b) Willingham, Cambridgeshire, as illustrated on the Record Sheet Plans.
Source: PRO IR 127/2/120 and IR 127/2/134.

Table 8.16 *Hereditament size in Isleham and Willingham, Cambridgeshire*

	Number	
Size (acres)	Isleham	Willingham
<1	40	55
1–4.9	17	31
5–24.9	19	22
25–49.9	1	2
50–74.9	2	
75–99.9	2	
100–149.9		3
150–299.9		1
>300		
Not stated	29	8
Total	110	122

Source: PRO IR58/16043–50 (Isleham); 16648–56 (Willingham).

In Willingham, eighty-three different people owned the 114 hereditaments for which ownership is given. This was almost the same proportion as in Isleham, but Willingham had no owners as prominent as the Robins. As in Isleham, freehold was almost universal, and only nine hereditaments were wholly or partly owned by copyhold, and none were leasehold. In Isleham, leasehold was also absent, while ten hereditaments were held by copyhold, at least in part.

Smallholdings were a significant feature in both parishes, as Table 8.16 shows. In Isleham, over 80 per cent of hereditaments larger than an acre were in the 1–24.9 acre size-groups. To what extent this can be extrapolated to assess the numbers of occupiers is not clear. Sampling reduces the likelihood of detecting multiple holding by individuals. In the sample, the thirty-six hereditaments in this size-group were owned by thirty-one individuals. Three people occupied more than one hereditament in the sample.

Although the same problem occurs with the Willingham data, there are significant differences. Willingham is notable for a far greater preponderance of smallholdings. Spufford makes the point that between the sixteenth and early eighteenth centuries, 'the large farm never emerged as the dominant unit . . . whilst holdings in general got smaller and smaller'.[47] From comparison with her data it is clear that some two centuries after the close of Spufford's study, landholding in Willingham was still unusually fragmented, with no heredita-

[47] Spufford, *Contrasting Communities*, 158.

Table 8.17 *Land use on Willingham hereditaments, Cambridgeshire*

Land use	No.
Arable	16
Pasture	11
Arable/pasture	3
'Garden ground'	9
Fruit	6
Total	45

Source: PRO IR58/16648–56.

Table 8.18 *Number of rooms in houses: Isleham and Willingham, Cambridgeshire*

	No. of houses	
Rooms	Isleham	Willingham
2	5	3
3	8	3
4	30	22
5	3	7
6	5	13
7	4	9
8	1	3
9	6	4
10+	5	3
Not stated	5	2
Total	72	69

Source: PRO IR58/16043–50 (Isleham); 16648–56 (Willingham).

ments in the sample being as large as 75 acres. This was virtually identical with the position in 1720. The large Fenland-edge farm had no place here, in either the seventeenth or early twentieth centuries.

Somewhat unusually, field-by-field land use is available for many Isleham hereditaments, and this would certainly merit detailed study. And in Willingham land use was indicated for the holding as a whole in forty-five cases, again an unusual but extremely interesting addition (Table 8.17). Asparagus was said to be grown on three plots, while several others were devoted to soft fruit. Unfortunately, though, the crops are seldom mentioned.

Housing is not always fully described. Numbers of rooms are usually given, but the use of rooms is much less frequently available (Table 8.18). In Isleham, the great majority of the houses sampled had fewer than five rooms, with four-room cottages (usually two-up and two-down) being by far the largest single group, a point made also by Chamberlain. In the early years of the century, there was a tremendous social divide between the 'Uptown' parts of the village and the 'East End' together with 'the Pits', where 'the unemployed and poachers' lived.[48] Several of her informants made this point, and while the documents can give us little information on such social issues and perceptions, they can shed light on the polarised differences in housing quality within Isleham.

Seventeen of the sampled hereditaments were definitely situated in the 'less respectable' parts of the village. Fourteen of these were very small cottages with fewer than five rooms. This comprised almost half of the total in this size-group in the sample, although the number of houses in the district was less than one-quarter of the total sample. Housing was clearly smaller in general than elsewhere in the village. Few other details of these houses are available. Two were reported to be built of clunch, and several had earth closets (often shared). Ambrose Cole from Ely owned an old cottage at the Pits, occupied by A. Ball, constructed from clunch and with a tile roof over a three-up, one-down structure with an earth closet. Such statements confirm Chamberlain's comments in reporting the use of clunch as a building material, as well as 'Mary Coe's' exclamation relating to the Pits during her childhood, 'And the earth toilets!'[49]

Housing tended to be a little larger in Willingham. In Isleham, some 60 per cent of houses had fewer than five rooms, but in Willingham the figure was only 40 per cent. Although, as in Isleham, homes with four living rooms formed the largest single group, they comprised only a third of the houses with relevant information, compared to 45 per cent in Isleham. Earth closets were almost ubiquitous in Willingham, being specifically referred to in fifty-two houses.

Houses and most hereditaments were let on an annual basis in both parishes. Only a handful were let for shorter periods, though in Willingham leases for up to fifteen years were more frequent than in Isleham. As always, house rents were variable. Table 8.19 gives rents for houses with less than an acre of ground. Over 60 per cent of Isleham house occupiers paid less than £5 a year in rent. In Willingham, rents were somewhat higher, commensurate with the fact that houses tended to be larger. Here only 43 per cent of sampled houses were let at less than £5 per annum. In both parishes rates and other outgoings were generally paid by the owners.

In conclusion, there were marked distinctions between and within these two Fenland parishes. Smallholdings were a feature of both, as was a fragmented

[48] Chamberlain, *Fenwomen*, 18–20.

[49] PRO IR 58/16046, Hereditament no. 399; Chamberlain, *Fenwomen*, 18–27. Clunch is a poor building material deriving from chalk.

Table 8.19 *Annual house rents: Isleham and Willingham, Cambridgeshire*

	No. of houses	
Rent (£)	Isleham	Willingham
Free	1	
<3	6	3
3 to 3 19s 11d	21	7
4 to 4 19s 11d	7	9
5 to 5 19s 11d	2	7
6 to 6 19s 11d	4	6
7 to 7 19s 11d	6	3
8 to 9 19s 11d	1	4
10 to 12 9s 11d		1
>15	6	4
Total	54	44

Source: PRO IR58/16043–50 (Isleham); 16648–56 (Willingham).

landownership structure. In Willingham, though, these features were far more marked. To judge from the sample, there were no large landowners and no large occupiers in this parish. Housing seems to have been less cramped in Willingham, and rents were higher. Superficially, the village had fewer marked divisions between the different inhabitants, with no observable concentrated areas of deprivation. Isleham seems to have been a village of contrasts, as suggested by Chamberlain. Although poor housing was to be found all over the village, it was obviously found in greater concentration in the Pits and the East End.

Such studies of contrasting parishes clearly depend upon the initial conceptualisation of the issues to be tackled as well as upon the quality of the available documentation. And such studies do not have only to take a cross-sectional view. If the data are available, for example, they would be of great interest to trace and compare subsequent or antecedent change between the two (or more) communities. Certainly the availability of good cross-sectional materials *c.* 1840, *c.* 1910 and 1941–3 opens up a wide, 100-year field of study in this respect. The Lincolnshire Wolds and Fenland studies here are rural in every respect, but it would also be of great interest to tackle parishes which were becoming suburbanised, or in other ways changing quite fundamentally. The next chapter also demonstrates that such studies need not be those based purely around agricultural communities, since the countryside *c.* 1910 was also a place with significant amounts of non-agricultural employment.

9

Rural industrial communities on the eve of the Great War

Although the transference of much of Britain's industrial production from countryside to town had long since taken place, rural communities were by no means devoid of non-agricultural employment. Service industries had grown alongside the more traditional craft and workshop employments or by-employments as the gentrification of villages began towards the end of Victoria's reign in lowland England. But in upland Britain and the western counties, a great variety of primary and secondary activity still prevailed. In this chapter the 1910 documentation will be examined in order to analyse the information available about such activities. Those examined will include the slate quarrying industry of North Wales; coal mining in the Rhondda valley; peat digging in the Somerset Levels; and the industrial village of Cromford, Derbyshire.

Penrhyn and Bethesda: a quarrying community

It is significant to take the case of slate quarrying at Bethesda on the Penrhyn estate first, since the quarries and the attitudes of the local landowner, the Englishman George Douglas-Pennant, 2nd Baron Penrhyn, had been important influences in forging Lloyd George's attitudes to landownership that culminated in the land clauses of the Finance Act. In November 1910 Lloyd George published a collection of his speeches as *The People's Will*, with Hodder and Stoughton also publishing a *Caernarfon Herald* edition. In the chapter headed 'Rural Intimidation' Lloyd George wrote of his memory of living 'in the blackest Tory parish in the land' (presumably Llanystumdwy, some 30 miles away from Bethesda) and of the time in 1868 when the Liberals dared to field a candidate to run for Caernarfon against the 'country squires'' choice of Douglas-Pennant – apparently a quite unheard-of audacity. The eviction of tenants who dared to vote Liberal from Lord Willoughby de Eresby's estate was also engraved on his mind. A loathing for 'landlordism' stayed with him throughout his political career, and indeed ensured that his concerns were 'more for the class above him, "the dukes", rather than those below'.[1]

The Penrhyn estate was very large – 26,278 acres at this time – and included the slate quarries which had yielded £150,000 per annum in good years. The hugely impressive terraced Penrhyn quarry in the Cambrian slates was reputedly the largest in the world, with massive injections of capital into its infrastructure having been undertaken over the previous century, and with narrow-gauge railway links to Port Penrhyn, Bangor for the export of the slates. But in 1896–8 and 1900–3 bitter strikes involving lockouts and the employment of non-union men at the Bethesda quarries alienated Lord Penrhyn still further from many sectors of the local community. However, he was 'a tory of the old school' who 'managed his affairs in the feudal spirit', and who 'scorned popularity and played a detached part in public affairs'. Perhaps not surprisingly he was founder and chairman of the North Wales Property Defence Association![2] Lloyd George was the local MP and acted as the lawyer for the quarrymen, using the case to bring overtones of nationalist and radical fervour to his politics. Thus in 1897, following the refusal of Lord Penrhyn to let the dispute go to arbitration at the Board of Trade, he asserted:

> Lord Penrhyn . . . says, 'This is my private affair, it is my business'. The Board of Trade says that it is not his business alone, where the rights of three thousand men are concerned. It is not his business when there is nothing between ten thousand people and famine, but the charity of their sympathetic countrymen. [3]

At the May Day carnival at Caernarfon in 1897 he emphasised 'the right of the people to the land, to the mountains, and to the resources of the earth', and in another speech he spoke of the land before the Penrhyns stole it: 'The free mountain has become Lord Penrhyn's quarry, and three thousand Welshmen, the descendants, many of them, of the simple folk who knew not how to rob in legal fashion, have become his hewers of wood and drawers of water . . .' And when the 1899–1903 strike became an important issue Lloyd George appealed at the 1902 TUC meeting for union support for the quarrymen against Penrhyn, so that 'while they were fighting their children should be placed beyond the reach of starvation'.[4]

[1] D. Lloyd George, *The People's Will* (London 1910), 89–94; B. B. Gilbert, 'David Lloyd George: land, the budget, and social reform', *American Historical Review* 81(5) (1976), 1059.

[2] *Dictionary of National Biography* (Twentieth-century supplement 1901–1911, ed. Sir Sidney Lee, Oxford 1912), Vol. I, 517–18. The *DNB* noted cautiously that 'he acted throughout in accordance with what he believed to be stern equity and from a wish to obtain justice for non-union men'. In 1907, just before his death, he 'generously accorded' his workmen a 10 per cent bonus on their wages because of bad weather. See also his *Penrhyn Quarry* (privately printed April 1903, cited in K. O. Morgan, *Wales in British Politics 1868–1922* (Cardiff 1970), 212 fn. 1).

[3] C. J. Wrigley, *David Lloyd George and the British Labour Movement: Peace and War* (Hassocks 1976), 13.

[4] Wrigley, *Lloyd George*, 15. Both the Land Nationalisation Society and the English Land Restoration League were actively campaigning on behalf of the unemployed and their families (S.Ward, 'Land reform in England 1880–1914', University of Reading, unpubl. Ph.D. thesis, 1976, 309, 356).

This case study was formulated to examine the linkages between the three quarries of Penrhyn, Pantdraenog and Moel Faban on the one hand, and the village of Bethesda, as revealed by the valuation documents. Bethesda, emerging as a settlement from the grey cottages strung along the Holyhead Road near the Penrhyn quarry, and named after the chapel founded in 1820, was the largest civil parish within the ITP of Bethesda. The others were Llanllechid and Llandegai – the latter being Lord Penrhyn's attempt at a model village – and together they contained over 3,000 hereditaments. Bethesda itself comprised over 1,700 of these, together with an unknown proportion of the 160 unrated hereditaments appended at the end of the Valuation Book (Fig. 9.1).

Analysis was confined to the rated hereditaments and a sample of these was taken by examining every tenth hereditament in Bethesda in the Field Books. Additional details were extracted on landownership by the largest owners, and details of the slate quarries themselves. The 10 per cent sample resulted in a data base of 170 hereditaments plus two entries for the slate quarries – 172 in all. Of these, six had no information of any kind, and one was merged with another hereditament. Thus the following refers to 165 hereditaments. However, since two of these are slate quarries that were not in the original sample, percentages are based on the 163 hereditaments that did form part of the sample. The Valuation Books held by Archifau ac Amgueddfeydd at Caernarfon were well compiled, and this was also the case with the Field Books in the PRO.[5]

Landownership was fairly complex. As expected, Lord Penrhyn was the largest owner. He held the freehold of twenty-four hereditaments in the sample (14.7 per cent), and was the superior interest in a further thirty-two (19.6 per cent). In all then, Lord Penrhyn had an interest in about one-third of all hereditaments in the Bethesda sample. Details of the numbers of all freehold hereditaments owned by Lord Penrhyn had also been taken from the Valuation Books, and it was found that he owned 243 (14.1 per cent) in all, thus supporting the impression gained from the Field Book sample. Hence we have the 'small community of 10,000 which was, and remains today, intensely religious and closely knit, ranged against a great autocratic landowner, who refused equally to grant the workmen a minimum wage or to grant them chapel leases'.[6]

Another significant interest in the parish was the Cefnfaes estate

[5] Archifau ac Amgueddfeydd (Gwynedd Archives and Museums Service), Valuation Books: XLTD/11 (2 vols.); PRO IR 58/2524–7, 2529–41 (Field Books). The then Gwynedd Archives Service accepted but destroyed all Forms 37–Land in accordance with PRO guidance.

[6] Edward Sholto Douglas-Pennant, Lord Penrhyn, owner of the neo-Norman fantasy of Penrhyn Castle in Llandegain parish (Hereditament no. 2364), became the Third Baron Penrhyn in 1907 on the death of his controversial father (Morgan, *Wales*, 212). Penrhyn was acquired by the National Trust in 1951.

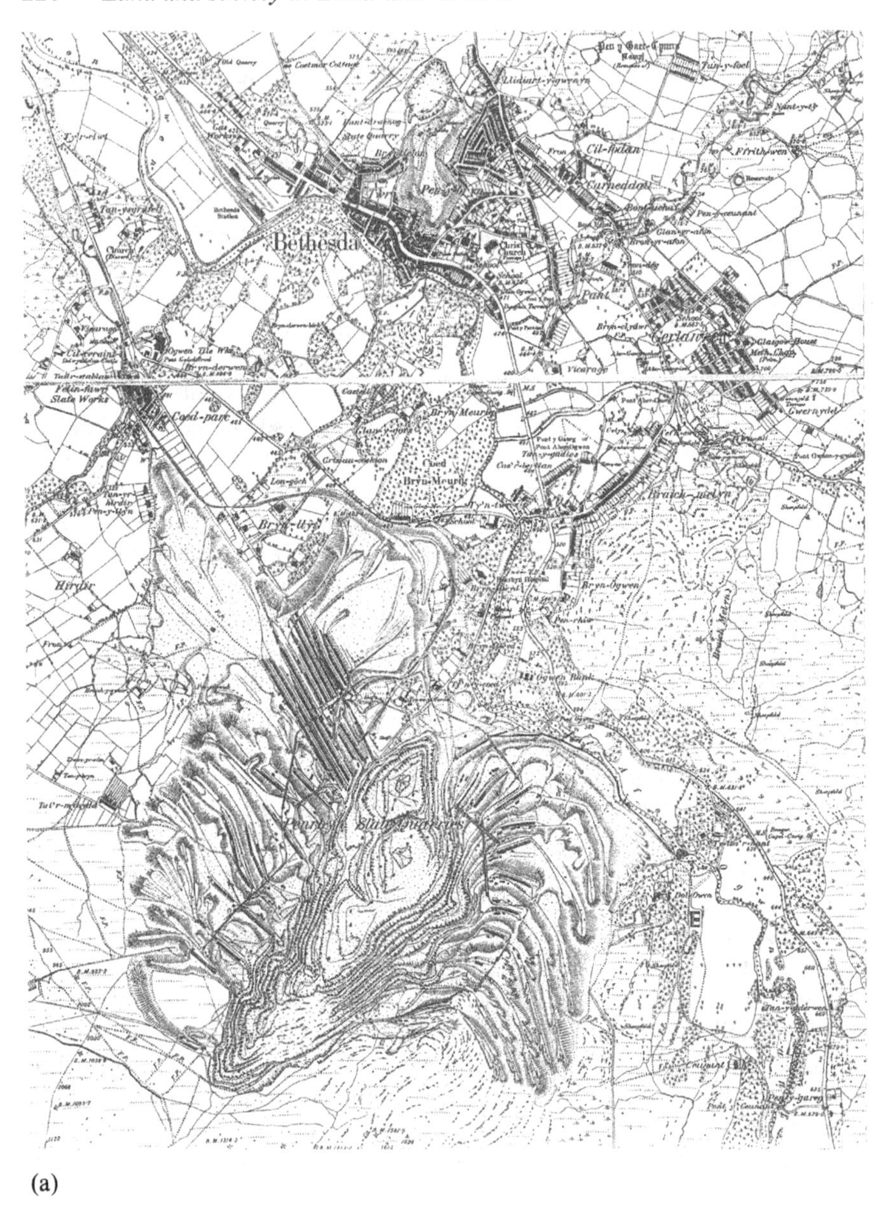

(a)

Fig. 9.1 Bethesda, Caernarfonshire: (a) Bethesda and the Penrhyn quarry 1889: an extract from the Ordnance Survey 1:10560 map; (b) Bethesda on the Record Sheet Plan *c.* 1910.
Source: PRO IR 131/2/44.

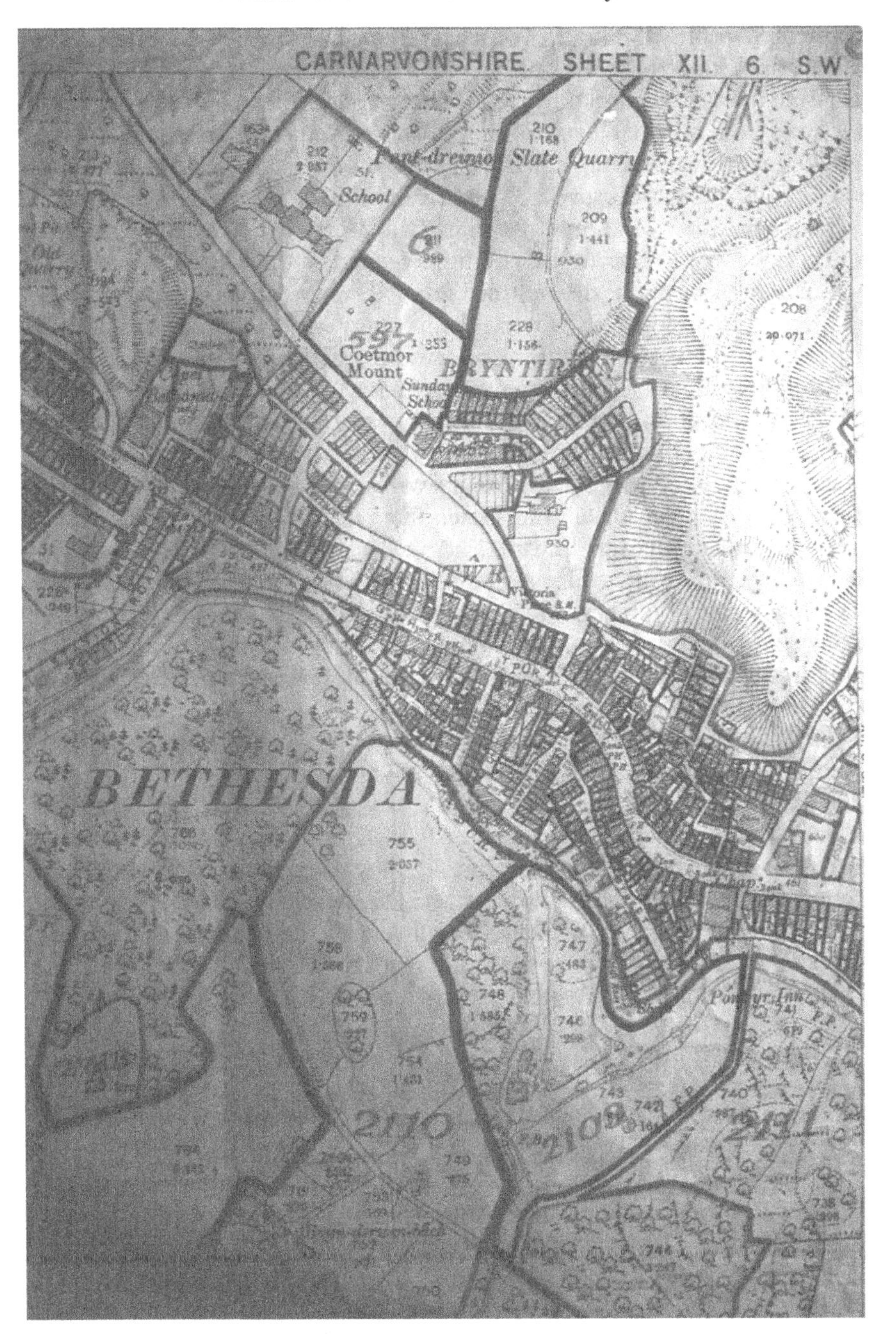
CARNARVONSHIRE SHEET XII. 6 S.W.
Slate Quarry
School
Coetmor
Mount
Sunday
School
BRYNTIRION
BETHESDA
2110
2109

(b)

represented locally by its agent W. J. Parry, 'the Quarryman's Champion'.[7] The estate owned the freehold of only five hereditaments in the sample (3.1 per cent), but it was the lessor of no fewer than forty hereditaments (24.5 per cent). It follows from this that tenure in the parish was frequently leasehold. Sixteen hereditaments in the sample had no information on the type of ownership, and of the remainder, seventy were freehold and seventy-nine leasehold. Owner-occupation was not uncommon. In all, forty-one hereditaments (25.2 per cent) were owner-occupied with twenty-seven of these being leasehold.

A national list of quarries returned in a 1907 report showed that quarrying was present to some extent in every Welsh county, but that in Caernarfonshire there were 120 quarries employing no fewer than 10,412 people, about 60 per cent of the total Welsh employment in quarrying.[8] The industry was thus extremely important economically and socially within the county. But in examining the 1910 documents one finds that, apart from the two quarries of Pantdraenog and Moel Faban, which were both owned by North Wales Quarries of 40 Deansgate, Manchester, there is little evidence of any direct link between the quarry industry and the village generally. The company owned only three other properties in the sample (1.8 per cent). Of these two were unoccupied cottages, and the third was the small farm of Ty Mawr. Building societies were not uncommonly owners of housing in industrial villages at this time, but in Bethesda there were only five sampled hereditaments which were owned by two different societies.

The actual ownership of the two slate quarries is in fact unclear from the documents. Moel Faban (Hereditament no.47) was owned by North Wales Quarries Ltd, but all other details as to the type of ownership etc. were said to be entered against Hereditament no.92, which was unfortunately not the case. Pantdraenog (Hereditament no.44) was 'owned' on a fifty-year lease from May 1904 from the Cefnfaes estate. The ownership of the estate is not revealed in the documents. There is no link with the Penrhyn estate, which owned the Penrhyn quarry to the south, but the difficulty of establishing the true ownership of a property from these records is certainly well illustrated.[9]

[7] On 13 March 1903 a libel action had been taken out by Lord Penrhyn against the Cefnfaes agent, W. J. Parry, because of an article in *The North Wales Clarion* accusing him of cruelty to his workmen. He was awarded £500 damages with costs (*DNB*, 518). The W. J. Parry Papers are in the National Library of Wales (see, for example, NLW MSS 8733–67).

[8] D. Morgan Rees, *The Industrial Archaeology of Wales* (Newton Abbot 1975), 160–1.

[9] The Cefnfaes estate ownership was not in the hands of Lord Penrhyn. In the two volumes of Penrhyn quarry records in the Gwynedd Archives Service there is much material on the industrial disputes 1885–1905, correspondence on the unions and a benefit club, and many other additional quarry papers. In a letter of 16 November 1967 from the Penrhyn Quarries to the County Archivist there is a request for a 30-year rule to apply to all records except those relating to the strikes. There are some Pantdraenog records among the Coetmor papers at University College, Bangor, where all other Penrhyn estate papers other than those relating to the quarry are held. I am most grateful to Mr Gareth H.Williams, Principal Archivist and Museums Officer at Gwynedd County Council, for information on these matters.

Table 9.1 *Number of rooms in Bethesda houses, Caernarfonshire*

No. of rooms	No. of houses
1	
2	20
3	21
4	37
5	26
6	15
7	8
8	1
9	
10+	2
Total	130

Source: PRO IR 58/2524–41.

Information on Pantdraenog is sparse, with most details said to be in the 'file'. The quarry had been worked partly as a source of employment for those quarrymen who set their face against Lord Penrhyn, but in 1910 the only comment was: 'This quarry is at present idle. The formation is the Cambrian bed of slate, and is of good quality. It is worked on the open pit system and entails hauling & pumping, with a revival in the Bldg trade this quarry possibly will start again.'[10] This suggests high unemployment in the parish at that time, but no other information on this is available, although another remark suggests the general air of depression that must have prevailed. The 'Douglas Arms' (Hereditament no.80) was a large hotel in the centre of the village, run by James Mitchell, but: 'the ppty has long since lost its trade & connection as a "Hotel" & is now practically nothing more than a "drinking" place'.[11]

Information on housing is inconsistent and indeed is frequently absent. Stone and slate unsurprisingly make up the great majority of building materials noted. Details of rooms are inconsistently given, and the use of each room is not always stated. Thus there is likely to be a degree of inaccuracy in Table 9.1 concerning the number of rooms in the 130 houses for which any information is available, although any discrepency is unlikely to be great.

The figure for those houses with only two rooms may be especially overstated. Several of these are said to have a 'taflod' or loft. This compares with the use of the English word 'attic' used in connection with some larger houses. It may be that 'taflod' has a rather specific meaning and that sleeping accommodation may have been situated there. In those houses with a 'taflod', there is often no mention of another room being used for sleeping, or else it is

[10] PRO IR 58/2524, Hereditament no. 44. [11] PRO IR 58/2524.

Table 9.2 *Condition of housing in the Bethesda sample, Caernarfonshire*

Good	26
Fair	52
Poor	27
Very poor	2
Derelict/insanitary	2
Total	109

Source: PRO IR 58/2524–41.

Table 9.3 *Period of tenancies in the Bethesda sample, Caernarfonshire*

Weekly	2
Monthly	46
Quarterly	2
Half-yearly	2
Yearly	23
Total	75

Source: PRO IR 58/2524–41.

clear that all living space is on one level. Thus the resident (we are given the name of W. H. Parry as the occupier, which is unlikely as he was the agent for the Penrhyn estate) of No. 2 Brynhedydd Street (Hereditament no.260) was living in two rooms on the ground floor with a taflod over. A coal shed and WC were outside, together with a large garden.[12]

Even given this problem, it is clear that houses were frequently small in Bethesda. Sixty per cent of the houses in the sample for which information is available had four or fewer rooms. The condition ascribed to this housing by the valuers was variable, as Table 9.2 indicates. Overall, almost three-quarters of houses in the sample were in 'fair' or better condition. However, one of these, occupied by William Griffiths at No. 5 Brynhedydd Street (Hereditament no. 1140), had a living room on the ground floor, two rooms above, and an outside privy, and was said to be in 'fair repair but of poor "type" – no thro' ventilation'.[13]

Tenanted property was let for various periods. Information on tenancy terms is available for seventy-five hereditaments of which most were let monthly as Table 9.3 shows, with a substantial number being let by the year. House rents (including estimates for owner-occupied properties) are available for 111 houses in the sample, and are tabulated in Table 9.4.

[12] PRO IR 58/2526. [13] PRO IR 58/2535.

Table 9.4 *Annual house rents in the Bethesda sample, Caernarfonshire*

Rent (£)	No. of houses
<3	2
3 to 3 19s 11d	18
4 to 4 19s 11d	14
5 to 5 19s 11d	23
6 to 6 19s 11d	21
7 to 7 19s 11d	13
8 to 9 19s 11d	13
10 to 12 9s 11d	3
12 10s to 14 19s 11d	2
>15	2
Total	111

Source: PRO IR 58/2524–41.

The 1910 documentation, particularly that part relating to the ownership of land, is concerned with legal issues only, and this case study demonstrates the fact that it is necessary to delve more deeply to find out who owned the corporate bodies like North Wales Quarries, or the Cefnfaes estate. Accordingly, any examination of the relationship between the slate industry and the village in general is not greatly assisted by the 1910 material by itself. The sources herein examined point to a substantial Penrhyn interest in the parish, but, superficially at least, do not support a Penrhyn interest in the quarries other than the vast Penrhyn quarry itself. However, this certainly highlights the need to look below the surface of landownership patterns as revealed merely by the owners' names. This is, of course, particularly significant with the increased numbers of corporate bodies as landowners.

Beyond this, the documents reveal a good standard level of information on rents, housing and so on, though the entries are inconsistently completed. It suggests that the material can be used relatively unproblematically for 'straightforward' information such as the extent of leasehold, freehold and other forms of tenure; to derive rent levels and tenancy periods; and to illuminate housing conditions and accommodation. But the precise details of landownership remain tantalisingly obscure.

Maerdy: a pit village in the South Wales Coalfield

Moving southwards into Glamorgan, Maerdy, known as 'Little Moscow' for its strong socialist views and postwar industrial confrontation, and with the last working pit in the Rhonddas from 1969 to its closure in December 1990, was selected for study.[14] In the valley of the Rhondda Fach, Maerdy was producing steam coal commercially by 1878, and by 1884 output had exceeded 160,000 tons. A major new colliery had been recently opened by 1910, and the village was chosen in order to examine relationships between the colliery company and the village as a whole. Maerdy is one of a number of mining villages in Rhondda Fach and Rhondda Fawr that were within the Sanitary District (until 1894), Urban District (until 1897), and civil parish of Ystradwfodwg. Development had been generally rapid after 1860 as coal mining boomed, and the large number of villages were grouped together in the ITP of Rhondda by 1910.[15]

The ITP comprised only a single civil parish, but it contained over 29,000 hereditaments. It was possible to isolate Maerdy within the Valuation Books mainly by reference to street names. The village contained over 900 hereditaments (nos. 24161 to 25070 inclusive, as well as a handful of later numbers). The finding aid in the PRO, somewhat unusually, gives an indication of the numbers relating to Maerdy within the ITP, but is not entirely accurate.[16]

Limited information on owner-occupation and company house-ownership was extracted from the Valuation Books, but most work was done on the Field Books. The large number of hereditaments involved using a sampling procedure, and accordingly every tenth hereditament was sampled commencing at no.24170. The Field Books were for the most part very fully compiled, though they contained no information on the colliery itself (no.26331), presumably owing either to its ownership by a statutory company, or to a lack of space in the Field Book.[17]

Property ownership in Maerdy was fairly widespread, although over one-third of all the houses in the village were owned by four companies. Information on company ownership was taken from the Valuation Books and for this part of the study, the figures are based on the entire population of books rather than just the sample. The Maerdy Cottage Company was the largest single owner with 228 houses, with large concentrations in the south of

[14] I am grateful for the assistance of Mrs Annette M. Burton, Glamorgan Archivist (Glamorgan Record Office), with my queries on Maerdy, and also for the local knowledge of Mr Andrew Bartholemew. See also K. O. Morgan, *Rebirth of a Nation: Wales 1880–1980* (Oxford 1981), 60–1, 175, 318.

[15] The population of the Rhondda Valleys rose from 3,025 in 1861 to 113,735 in 1901. See W. K. D. Davies, 'Toward an integrated study of central places', in H. Carter and W. K. D. Davies (eds.), *Urban Essays: Studies in the Geography of Wales* (London 1970), 193–227; D.Morgan Rees, *The Industrial Archaeology of Wales*, 97–8.

[16] Glamorgan Archives Service, Valuation Books D/D PRO/VAL/1/134; PRO IR 58/67600–8; 67622 (Field Books). Forms 37 also exist for all parts of the county except the Port Talbot area, where they were destroyed. The forms for the Rhonddas comprise about thirty archive boxes and have been neither sorted nor listed, making the retrieval of the forms for Maerdy an extremely time-consuming operation.

[17] See Chapter 6 for more detail on the Field Book contents for the interiors of properties in Maerdy.

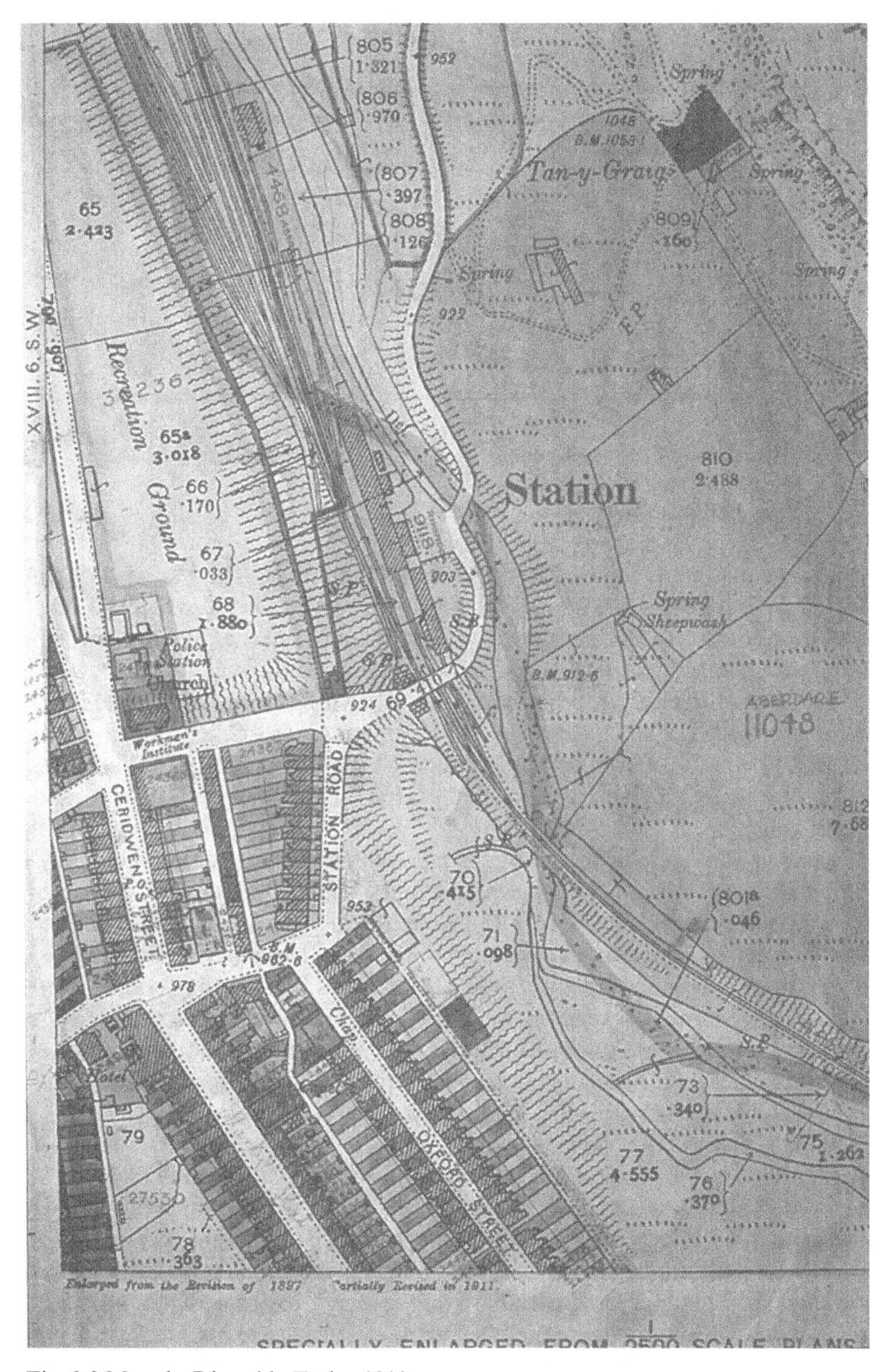

Fig. 9.2 Maerdy, Rhondda Fach *c.*1910.
Source: PRO IR 131/8/26.

Table 9.5 *Hereditament ownership in Maerdy, Glamorgan (sample)*

No. of hereditaments	No. of owners
1	29
2	2
3	
4	1
5	1
>6	2*
Total	35

Notes:
* The Maerdy Cottage Company owned nine hereditaments, and E. Davis owned thirteen.
Source: PRO IR 58/67600–8, 67622.

the village in Oxford Street, Pentre Road and Hill Street. The Maerdy (1904) Building Club owned 103 houses, all in Miles Street and Edward Street, while the Rhondda Cottage Company, based in Station Terrace, Maerdy, owned thirty-eight houses scattered across several streets, although half of them were in Ceridwen Street. The other company to own housing was the colliery company itself: Locketts Merthyr Coal Company (alias Locketts Steam Coal Co. Ltd) owned just two houses in Wood Street, although it also had subordinate interests in properties. In all, these four companies owned 371 houses in the village.

The companies though were under-represented in the sample, where only the Maerdy and Rhondda Cottage Companies feature at all. According to the sample, companies owned under 15 per cent of houses, less than half of the known actual proportion, as calculated in full from the Valuation Books. This discrepancy is because most of the houses in Miles Street and Edward Street – where the companies were especially prominent as owners – have no ownership details entered into the Field Books. In short, we have another case here where the Valuation Books are a superior source of information on basic landownership detail than are the Field Books.

Table 9.5 shows that most owners of property in the village (just under 83 per cent) were operating on a very small scale. There is a possibility that multiple ownership in the sample is also overstated, since it is assumed that where the same initials and surnames are given, they refer to the same person. This is always a problem of course, but it is particularly so in Maerdy where Welsh surnames such as Jones and Evans are very common. On the other hand, sampling can tend to understate the phenomenon of multiple holdings, since an owner may have other properties that do not feature in the sample. The understatement of company ownership has already been referred to. Apart from the companies, E. Davis (assuming this is one person) appears as a substantial private property owner in the village.

Owner-occupiers were fairly numerous. Once again, details for the whole village were extracted from the Valuation Books, and 124 owner-occupiers were noted. Of the 909 houses in the village, 13.6 per cent were owner-occupied, which seems a relatively high proportion for the Edwardian period, and compares with around 5 per cent in the Bradford area and under 12 per cent in the Brighton study reported above.[18] Once again the sample seems to understate this. Of the sixty-two houses in the sample for which suitable information is available, only 9.7 per cent were owner-occupied.

The representativeness of sampling procedures could be improved, of course, for a more detailed study. But questions of spatial bias in sampling from the Valuation or Field Books is a further methodological question to be raised, although one which could in practical terms be resolved by the coordinated use of either the Record Sheet Plans or the Working Plans (if complete). This would circumvent the problem of any spatial bias in the valuation of Field Books being transferred into the sample results by the imposition of an independent spatially valid framework onto the maps using a spatial grid of some kind. Without this use of the maps, the aspatial sampling frameworks, as employed primarily within this text, cannot be expected to yield results which can be subjected to more intense spatial scrutiny. The Maerdy study, then, highlights another potential difficulty for users of the 1910 material.

Information on types of tenure in Maerdy is only sporadically given. For the thirty-six hereditaments with information on this point, all were leasehold (usually 99 years). The lessor in twenty-nine of these cases was the Maerdy estate, while the executors of Messrs. Jones and Cobby were lessors of a further six. No.74 Maerdy Road (Hereditament no.24260) was owned by the Maerdy Cottage Company (whether freehold or leasehold is unknown) and leased to the colliery company who, in turn, let it to D. A. Thomas, the occupier. Whether this single example is the tip of an iceberg of colliery involvement as direct landlords in the village cannot be assessed from the data, which apart from this instance only show the colliery company as owning two houses.[19]

Rents for these houses seem high when compared to other case studies, and are detailed in Table 9.6, for those houses for which the relevant information is available. In addition, rents for three shops are given which ranged from £30 to £48 for the shop and living accommodation. Almost all homes for which information is available were let at between £15 and £17 a year, or from 5s 9d

[18] See Chapter 7. Only Cromer appears, from the sample studies undertaken, to have had a higher proportion of owner-occupied hereditaments.

[19] For the dominance of leasehold tenure in the towns of South Wales, and their less congested 'through' type construction with yards at the rear, see H. Carter and C. Roy Lewis, *An Urban Geography of England and Wales in the Nineteenth Century* (London 1990), 160.

Table 9.6 *Annual house rents in Maerdy, Glamorgan (sample)*

Rent (£)	No. of houses
10 to 10 19s 11d	3
14 to 14 19s 11d	1
15 to 15 19s 11d	26
17 to 17 19s 11d	1
Total	31

Source: PRO IR 58/67600–8, 67622.

Table 9.7 *Rooms in Maerdy houses, Glamorgan (sample)*

No. of rooms	No. of houses
4	5
5	44
6	11
7	7
8	2
9 and over	1
Total	70

Source: PRO IR 58/67600–8, 67622.

to 6s 6d a week. This is far higher than working-class house rents in other case study areas, though without knowledge of real regionally adjusted wages, meaningful comparisons are difficult. Living accommodation (including kitchens) is detailed in Table 9.7.

Five-roomed houses were the most typical in Maerdy, usually containing a kitchen, a living room, and three bedrooms. There were apparently no really small houses like those commonly found in rural and deprived urban areas.

Houses were virtually all built of stone, referred to as 'rockfaced masonry', with slate roofs. Some of the better ones also had brick dressings. The state of repair is only sporadically given. Only thirty-one houses in the sample had comments on condition and of these, twenty-five were described as 'fair'. Only two were considered 'poor' or 'indiff[erent]'. Dampness was mentioned in connection with eight properties, almost 10 per cent of the sample. The valuer was particularly impressed with No.1 Brook Street, a six-roomed stone-built house which he considered to be a 'good wkg class ppty'. Six-roomed houses though were not common in Maerdy.

It is, of course, a severe limitation that the colliery at Maerdy has no information entered in the Field Book. This is so for all collieries in the Rhondda valleys, though in the entry for Wattstown colliery (Hereditament no.26331), it was noted: 'No valuation made. For schedules of buildings and Machinery see Form IV cover.' This probably also applied to Maerdy, where the amount of information necessary to describe such a large industrial undertaking was almost certainly more than the Field Book had space for.

The documents are clearly once again very useful for assessing aspects of landownership and housing conditions in the pit village of Maerdy. However, any examination of the influence of the colliery company in the village generally could not be pursued through the 1910 material. Superficially it would seem that the company had little role as landlord in the village. However, the extent to which there was any involvement by the company in the cottage and building companies that were major house-owners in Maerdy could not be ascertained. The role of the Maerdy estate as the superior interest for many houses is clear, but its involvement in the mining industry, if any, cannot be determined from these data. However, the agent for the estate in Maerdy, Hugh Caldwell of Blackwood, Monmouthshire, was by profession a mining engineer, which does perhaps raise the possibility of some estate involvement in the local mining industry.[20]

Peat diggings in the Somerset Levels

The Somerset peat diggings have been concentrated to the north of the Polden Hills, between Bridgwater and Glastonbury. The land is generally too wet in winter for it to be useful for arable farming, or indeed for peat cutting, but excellent summer grazings are available in this highly distinctive landscape of peat strips separated by drainage channels (rhynes), fringed by pollarded willows, and droves.[21] Small proprietors of land and cottages fringed the peat beds, and by 1921 it was estimated that the industry employed about 1,400 men in Somerset. Fitzrandolph and Hay describe how:

> The beds here are largely worked by independent men, who rent a piece of land for a few years and cut the turf, and there were about one hundred men working beds in this way in the neighbourhood of Meare, Ashcott, Shapwick, Burtle, and Edington. The land from which the turf has been cut is also of a certain amount of use. One firm allows each man to have a piece of it rent free for three years to clean and crop it with potatoes. The rent for the peat beds before the war was £5 per acre, and in 1921 it had risen to £10 an acre.[22]

20 PRO IR 58/67603, Hereditament no. 24510.

21 For more detail on the landscape of the Somerset Levels see M. Williams, *The Draining of the Somerset Levels* (Cambridge 1970).

22 Helen E. Fitzrandolph and M. Doriel Hay, *The Rural Industries of England and Wales* vol. II (1926, 1977 reprint, Wakefield), 114.

What light can be shed by the Valuation Records upon the Somerset peat digging industry in the period before the Great War? The parishes chosen for study were those indicated by Fitzrandolph and Hay as being the centre of the peat industry, namely Ashcott, Catcott, Edington, Godney, Meare and Shapwick.[23]

For this analysis, the primary interest was in holding sizes, rents and length of tenure, so only this information was extracted for the parishes as a whole. In addition, any reference to the peat industry was also extracted. The relevant hereditament numbers were obtained from the Valuation Books, together with figures for acreage. The Field Books in the PRO were examined for details of tenure and rent, and to fill in any gaps in the acreage figures. The Field Books were also the main source for information on peat cutting. The 1910 Record Sheet Plans were also consulted.[24]

The Valuation Books were fairly well compiled except for the problem of entries other than those made by the Land Valuation Officer from the Rate Book. The status and date of these are often unclear and are particularly so in these Somerset examples. The book for Ashcott and Shapwick parishes was especially poor, and the last fifty or so hereditaments were entered in pencil with extremely sparse information in both Valuation and Field Books.[25] This was unfortunate since a number of earlier hereditaments were said to have been 'split up' and the information on them included in the later hereditament numbers. Thus there is very little information on some hereditaments in these parishes, including several with peat connections. Equally problematic was the book for the Income Tax Parish of Catcott. This included Catcott itself, together with Chilton Polden and Edington, though the first of these was not included in this study. Some forty hereditaments (all unrated property) were added at the end of the book, but to locate them in a specific parish is impossible from the documents consulted. Similarly, seventy-two hereditaments in Godney and Meare could not be assigned to their specific parish.

Apart from the above problems, the best and most informative documents were the Field Books for Godney and Meare. These were in the Weston-super-Mare District, unlike the other parishes, and were especially informative in that field-by-field land use was indicated for the majority of holdings, though this material was not analysed in this particular instance.

The material was informative on holding size, though not every hereditament had the valuers' acreage entered in either the Valuation or the Field

[23] *Ibid.*, 112–14.

[24] Somerset Record Office DD/IR W 46/3; T 2/2; 8/4 (Valuation Books); PRO IR 58/82077, 82227, 92005 (Field Books); PRO IR (Record Sheet Plans). There appear to be no Forms 37–Land for Somerset, and they were said by the Somerset County Archivist to have been 'destroyed locally', although it is not known whether they were destroyed by the Record Office on advice from the PRO, or by the Valuation Office, or through some other factor.

[25] Somerset Record Office DD/IR T 2/2.

Table 9.8 *Hereditament sizes in the Somerset Levels*

	Ashcott		Shapwick		Godney		Meare		Catcott		Edington	
Size (in acres)	No.	%	No.	%	No.	%	No.	%	No.	%	No.	%
<0.25	61	25.4	48	22.5	22	7.3	99	20.0	47	19.9	45	20.1
0.25–0.9	55	22.9	36	16.9	44	14.7	74	14.9	32	13.6	28	12.5
1.0–4.9	61	25.4	48	22.5	73	24.3	114	23.0	75	31.8	52	23.2
5.0–24.9	34	14.2	42	19.7	135	45.0	127	25.7	57	24.2	73	32.6
25–49.9	7	2.9	11	5.2	13	4.3	26	5.3	5	2.1	9	4.0
50–99.0	10	4.2	9	4.2	6	2.0	18	3.6	10	4.2	6	2.7
100–149.9	4	1.7	4	1.9	2	0.7	8	1.6	1	0.4	1	0.4
150–299.9	2	0.8	7	3.3	2	0.7	2	0.4	1	0.4	2	0.9
>300			1	0.5								
ns	6	2.5	7	3.3	3	1.0	27	5.5	8	3.4	8	3.6
Totals	240		213		300		495		236		224	

Notes:
ns not stated
Source: PRO IR 58/82077, 82227, 92005; Somerset RO DD/IR W46/3; T2/2; 8/4.

Books. Moreover, the civil parish boundary is not coterminous with the hereditament boundaries in many cases, so that it is impossible without extremely detailed work on the Record Sheet Plans to link hereditaments precisely and exclusively to a single civil parish.[26]

Nevertheless, the main trends are very clear. Large holdings were infrequent in this area, as indicated in Table 9.8. Only Shapwick had a farm larger than 300 acres, and the same parish also had seven holdings between 150 and 300 acres. The remaining five parishes had only nine holdings in the latter size group between them. Notable were the large numbers of holdings between 1 and 25 acres. In Godney these comprised almost 70 per cent of the total number of holdings. Holdings of this size were least common in Ashcott and Shapwick, where they comprised just over 40 per cent of the total in each parish.

Many of these smaller holdings were let to, and worked by, peat cutters, and were frequently no more than an acre or so in extent. Most of the holdings over 5 acres seem to have been small farms rather than peat holdings, though the latter might form part of a small farm holding. Those holdings smaller than a quarter of an acre were cottages and gardens and were present in substantial numbers in every parish except Godney. The reasons for this are not

[26] For an analogous problem, see Chapter 10 for difficulties in dealing with material from the South Hams, Devon.

Table 9.9 *Length of tenure in the Somerset Levels*

	Ashcott		Shapwick		Godney*		Meare*		Catcott+		Edington+	
Period	No.	%	No.	%	No.	%	No.	%	No.	%	No.	%
Weekly	13	5.4	2	0.9	3	1.0	10	2.1	2	0.8	6	2.9
Monthly			2	0.9								
Quarterly	15	6.3	6	2.7			12	2.6	7	2.9	4	1.9
Six-monthly	12	5.0			2	0.7	9	1.9	2	0.8	10	4.8
Yearly	87	36.6	80	36.5	186	62.2	202	43.3	162	68.1	102	49.3
Two-yearly												
>Three Years	7	2.9	42	19.2			27	5.8	11	4.6	11	5.3
Void/tied			11	5.0			1	0.2	1	0.4	6	2.9
ns/ambig.	103	43.2	66	30.1	108	36.1	204	43.8	53	22.3	68	32.9
Owner/occ.	1	0.4					1	0.2				
Totals	238		219		299		466		238		207	

Notes:
*Meare is in Godney ITP. Hereditaments 747 to 819 have been excluded from these figures, as the civil parish in which they are situated is not clear.
+Edington is in Catcott ITP, as is Chilton Polden (not analysed). Hereditaments 612 to 643 are excluded from these figures, since their civil parish is not clearly designated in the sources. The detail is in any case sparse.
Source: PRO IR 58/82077, 82227, 92005; Somerset RO DD/IR W46/3; T2/2; 8/4.

yet apparent. It is not possible to comment here on the existence or non-existence of multiple holding by occupiers, since information about occupiers was not extracted in this case. A more detailed locally based study would, of course, be able to extract such information.

Holdings were let for various terms. A large proportion (36 per cent) unfortunately have no information in the records, or are ambiguous, on this point. This problem was greatest in Ashcott and Meare, where some 43 per cent of holdings are without this information. Nevertheless, almost two-thirds of holdings overall do have information on tenancy periods, and it is clear that most were let on a yearly basis, as Table 9.9 shows. Only a few in each parish were let for a shorter period. A number of holdings were let on leases of three years or more, though there were none in Godney, where there was also no mention of peat being worked. Many peat diggings were in this category. Fitzrandolph and Hay claimed that it was customary in Somerset to let peat beds for a ten-year period.[27] In fact no turf land was let for this time according to the documentation. Leases varied from four to twenty years, with substantial numbers being let for six, fifteen and twenty years. Only one lease

[27] Fitzrandolph and Hay, *Rural Industries*, 115.

approximated to Fitzrandolph and Hay's figure. Meare hereditaments numbered 630 to 636 (totalling just under 6 acres in area) were leased to Edward Rogers for eleven years, though the lease expired in 1909.[28]

Data on rent are available for most hereditaments, but are largely meaningless unless the type of holding or its land use is known. For example, the rent for a holding of about an acre varied enormously. In Godney, Hereditament no.23 (area just over 1 acre) was let for £2 a year, while Hereditament no.5, almost exactly the same size, was let at £8 10s. Accordingly, only holdings said to be peat workings, referred to in the Valuation Books as 'Turf in cutting x years', have been analysed in respect of rents.

Rent on peat diggings was frequently stated, but not always unambiguously so. It seems to have been normal for a high rent to be charged for about 80 per cent of the period of the lease, followed by a few years on a low rent when the holding was exhausted. Edward Rogers, on the hereditaments already referred to at Meare, paid £7 for the first nine years of his eleven-year lease, and a 'peppercorn' rent thereafter. He actually leased seven hereditaments with a total acreage of just under 6 acres. He also leased several peat plots in Meare parish, each of just over 0.25 acre at £2 10s per year (i.e. £6 12s per acre). Other hereditaments were let at around £7 per acre, e.g. Meare nos. 734 and 735, so it seems probable that the rent for turf land was between £6 and £8 per acre during its productive life. Hereditaments nos.507 to 521 in Meare were let for seven years at rents of £8 per acre for five years and rent-free thereafter. Hereditaments nos. 592–3 in the same parish were let for fifteen years at a rent of £70 for the first twelve years and 5s per annum for the remaining three. These hereditaments totalled an acre in extent, so that this figure was presumably for the entire twelve-year period. In Shapwick, a number of hereditaments were let at £7 per acre on leases of twenty years. Once again the data conflict with Fitzrandolph and Hay who claimed that rent before the [Great] War was £5 an acre, which would now appear to have been an underestimate.[29]

Once the peat beds were exhausted, rents fell dramatically. The low rents payable for the last years of a lease indicate this. Presumably only the most inferior pasture could be had from an exhausted peat digging, and the documents frequently refer to holdings as being 'exhausted turbary'. These were let

[28] Several generations of the Rogers family have been involved with peat cutting, with traditional skills being passed down between the generations. Thus we find Mrs Harriet Rogers at Catcott occupying a 4 acre holding of pasture land on Sedgemoor (PRO IR 58/82227), Sidney Rogers on 11 acres at West Heath, Shapwick (PRO IR 58/82077), Thomas Rogers on 5 acres at Godney (including exhausted turbary), and Charles Rogers on 4.25 acres of land on Sedgemoor in Catcott parish (Somerset RO DD/IR W 46/3, Hereditament nos. 195–6; T 8/4, Hereditament no. 196). In 1976 the family opened a small garden shop to supply those increasing numbers of gardeners coming in search of peat. Within ten years it had grown into 'The Willows', a large garden centre, with Somerset Levels museum, information centre, demonstration hall and tea rooms in the former cottage of Edmund George Rogers.

[29] Fitzrandolph and Hay, *Rural Industries*, 114.

at only a few shillings per acre. Meare Hereditament no.329 was typical. For a rent of £3 a year, the tenant acquired a holding of 5.25 acres. In the same parish, Hereditament no.425 was nearly 17 acres in extent, yet was rented out for only £4. At Meare the valuer recorded: 'Turbary land at Westhay Moor let to various occupiers 4 ½ acres let from 1908 at £8 per acre for 5 years last year rent free. Remainder let from 1908 on same terms.'[30] At Edington we also find a typical entry for an exhausted turbary as 'No rent paid'. This was Hereditament no.501, and no.500 (a 20 acre holding) actually had the rare entry for such a large holding of 'I cannot say' in the Valuation Book against its rent. There were several smaller slips of land for which the same statement was entered.[31]

Other miscellaneous material on the peat workings can be gleaned. At Godney, Hereditament no.310 was a peat works, and 57 acres of land, nearly all turbary, were connected to the works by a tram line. But 'the land lies low and is under water during wet seasons'. Timber had been cut off the land during the three years prior to the valuation.[32] It would be possible in the parishes of Godney and Meare, where the Field Books give field-by-field information as to land use (including exhausted turf land), to examine how the diggings shifted over time, and perhaps, in conjunction with other sources, to investigate more thoroughly other aspects of the management of this finite resource.

Other information, much as available in the other case studies, is also available in the documents. It could be interesting, for example, to undertake work on the brick and cob cottages on the levels, and perhaps to follow up Fitzrandolph and Hay's statement to the effect that: 'Peat-cutting is not considered unhealthy work, the open air life in some cases even counteracting the bad effects of poor houses, a number of which are on the "droves" across the marsh, on very spongy ground.'[33] It would also be possible with the aid of the Record Sheet Plans to plot the precise location of diggings in the pre-Great War years (Fig. 9.3).

A declining industrial village: Cromford, Derbyshire

Cromford had been the home of Sir Richard Arkwright (1732–92), who built his cotton-spinning mills there in the 1770s and 1780s. His first mill was erected there in 1771, a second in 1777 (destroyed by fire in the 1890s) and a third in 1780, so that the Hon. John Byng could complain of the Derwent

[30] Somerset Record Office, Valuation Book for Meare, Hereditament no. 513.

[31] Somerset Record Office DD/IR T 8/4 (Hereditament nos. 500 and 501) in Edington parish within Catcott ITP. [32] PRO IR 58/92005.

[33] Fitzrandolph and Hay, *Rural Industries*, 116. The professionalism of the valuers in dealing with peat cuttings in Somerset was in marked contrast to the situation in Ireland where the valuers were urban specialists who found great difficulty in assessing peat cuttings (see Chapter 11).

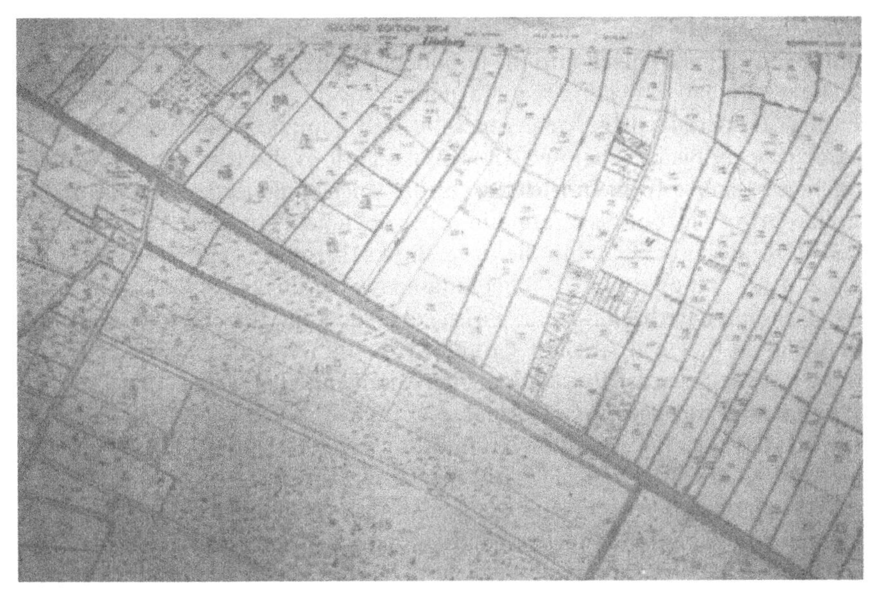

(a)

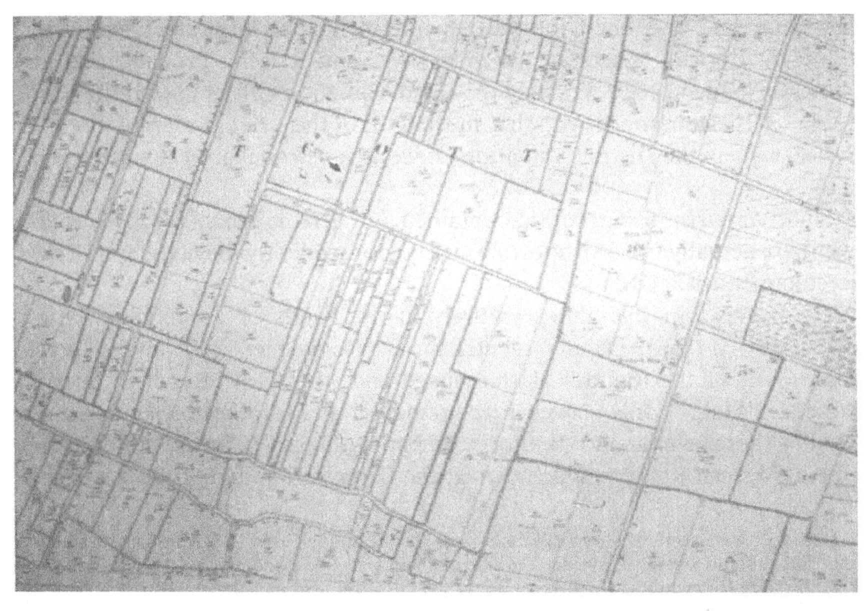

(b)

Fig. 9.3 Peat diggings in the Somerset Levels: an extract from the Record Sheet Plans: (a) Godney, and (b) Catcott, Somerset.
Source: PRO IR 128/9/564.

valley that 'Every rural sound is sunk in the clamours of cotton works, the simple peasant is changed into the impudent mechanic.'[34]

The documents providing the information for this study of an eighteenth-century industrial village were generally well compiled, and presented no major new problems. There was however a curious departure from normal practice, presumably due to a clerical error, which could lead to confusion. A similar 'error' has not been found in other documents to date. According to the PRO Finding Aid, and to the Field Books themselves, Cromford was the name of an ITP of some 3,500 hereditaments. On this basis, Cromford civil parish should have been the first parish covered within the ITP, and the hereditaments in the civil parish of Cromford should have been numbered from no.1 onwards. In order to isolate the hereditaments that were actually in the civil parish, it was necessary as usual to examine the Valuation Books. These showed that the civil parish of Cromford actually comprised Hereditament nos. 2957 to 3283 within the ITP. But they also gave the ITP name as Matlock, and yet Matlock hereditaments comprised nos.1126 to 2956. In fact the first parish in the Valuation Book was Matlock Bath, which would suggest that the ITP in which Cromford was situated should have been termed Matlock Bath, rather than Matlock.[35]

The departure from normal procedure presented no real problems in using the documents, once the Cromford hereditaments had been identified in the Valuation Books – a necessary step in any event. It seems probable that an error was made in 1910. Possibly the Land Valuation Officer copied the Rate Books of the component parishes in the ITP in the wrong order, or else the clerical staff in the District Valuation Office assigned an incorrect name to the ITP.

The civil parish of Cromford contained 326 hereditaments of which fourteen were actually tithes or sporting rights. Consequently the number of 'real' hereditaments was 312.

Landownership was almost entirely in the hands of Frederick Charles Arkwright (b.1853), Deputy Lieutenant of Derbyshire, a JP and a direct descendant of the founder of the village and who lived himself at nearby Willersley Hall.[36] Only sixteen hereditaments were in other ownership, and most of these belonged to statutory companies, in particular the London and North-Western Railway which owned the Cromford and High Peak railway

[34] Cited in J. V. Beckett, *The East Midlands from 1000* (London 1988), 288, with the second half of the quotation from the abridged edition of the diaries by F. Andrews in 1954.

[35] Derbyshire Record Office, Valuation Books D595R/4/1/73 (from Matlock District Valuation Office); PRO IR 58/55858–61, Field Books. The Valuation Book has a street index. There are also Working Sheet maps, including some 1912 Special Editions, marked up for the Matlock District (Derbyshire RO D595R 4/2/1–359). The map numbers retain the original Valuation Office numbering.

[36] W. T. Pike, *Nottinghamshire and Derbyshire at the Opening of the 20th Century: Contemporary Biographies* (London 1901), 283, 235.

Table 9.10 *Hereditament size in Cromford, Derbyshire (in acres)*

Area (in acres)	No. of hereditaments
<1	38
1 to 4.9	17
5 to 24.9	22
25 to 49.9	9
50 to 74.9	4
75 to 149.9	9
150 to 299.9	1
Over 300	
Total	100

Source: PRO IR 58/55858–61.

line and associated works, as well as five cottages. The LNWR also leased several cottages from the Arkwright estate, which were in turn sub-let to their own tenants at 1s 8d per week. Only three hereditaments – two houses and a small factory – were owned by private individuals and firms, who in each case were also the occupiers. Of these, Henry Cooper owned his house from the Arkwright estate on a 999-year lease.[37]

It follows that almost all land was tenanted, but unfortunately virtually no information on tenancy periods is available. Information on holding size is given sporadically, though the majority of those hereditaments without information on this point are houses with gardens, workshops and so on, and were presumably small. Table 9.10 gives a breakdown by unit size of those hereditaments with the necessary information.

Despite the inadequacies of the data, it would seem that holdings were small, all except one being less than 75 acres. What little information is given suggests that agricultural holdings were under grass for the most part.

Cotton manufacture had finally stopped in 1891, by which time part of the mill had been leased to a brewery, while part had become a laundry. These buildings were described in the Field Books. Hereditament no. 3241 was part of the old mill site, and was rented at £15 per year.[38] The ground floor was used as a steam laundry, but the remainder was 'unusable' for its declared purpose as a mess room. The adjoining mill building was in ruins, without a roof and with trees growing in it. Ancillary buildings were in disuse, or were used for storage. That part of the mill which had become a brewery was still being actively used, with various offices and storerooms, but the brewery plant was considered by the valuer to be very old and out of date. The buildings were

[37] Cooper's house was a three-bedroomed, bay-windowed detached property (PRO IR 58/55861, Hereditament no. 3261). [38] PRO IR 58/55861, Hereditament no. 3241.

very substantial but in much need of repair.[39] In all, an impression of near-dereliction is engendered by the descriptions.

There are other references to industrial activity in the documents. A number of workshops are mentioned, including one in the Market Place, the central core of Arkwright's planned village dated to 1790, but with no indication of usage; and there was an office at the coal wharf. Several quarries, some derelict, are noted. Hereditament nos. 3287–90 were said to be quarries let by the estate to the township of Cromford and to Mrs Walker respectively, but interestingly, a note is appended to the entry in the Field Book denying that any quarries were let to these people.[40] Only the old brickyard (no. 3193) was specifically identified, but even this had been disused for many years. The kiln was almost in ruins and there was 'no machinery as bricks were made by hand'.[41] At the large stone and slate estate workshops there was a sawmill on the ground floor worked by an undershot wheel, with a large joiner's shop overhead.[42]

Service activities are referred to in the shape of a large double-fronted refreshment room, a tearoom and shop, a blacksmith's shop with forge, a joiner's shop, slaughterhouse with two hereditaments designated as shambles, a Post Office, a barber's and several other shops of unspecified usage. A corn mill (Hereditament no.3208) was in good repair with the miller's house adjoining. The mill was described in some detail: the waterwheel and gearing outside; the first-floor grinding room with four pairs of stones, a storeroom and loading room; a second-floor malt house now used as a store, with drying kiln and more storerooms; and a third floor with another large storeroom. Outside was a six-stall stable with loft over, a cart-shed and gardens.[43]

Housing was much more fully described, although, as already indicated, letting periods are unknown. Rents are usually given however, and are shown in Table 9.11 for those houses without land attached.

Over half the houses in Cromford were let at less than £5 a year, though the occupier was usually responsible for payment of rates, but not repairs. Apart from these cheaper types of property, there were several more substantial houses in the village let at over £15 per annum, many of which were accommodation plus shops on the ground floor.[44]

The condition and repair of houses was almost always described as 'good', an unusually consistent position compared to other areas examined. A typical

[39] *Ibid.*, Hereditament no. 3242. The old mill buildings still exist but on three rather than five storeys as when it was painted by Joseph Wright *c.* 1783, the top two being removed after a fire in 1930 (Beckett, *The East Midlands*, 287, Plate 9.3). The mills were later purchased by the Arkwright Society for conservation.

[40] PRO IR 58/55861, Hereditament nos. 3279–80.

[41] PRO IR 58/55860, Hereditament no. 3193.

[42] PRO IR 58/55858, Hereditament no. 2973.

[43] PRO IR 58/55861.

[44] See below for a description of e.g. Refreshment rooms, Post office etc. Not all shops were identified in the Field Books, but cross-checking with an appropriate *Kelly's Directory* would certainly elucidate the situation.

Table 9.11 *Annual house rents in Cromford, Derbyshire*

Rent (£)	No. of houses
< 3	8
3 to 3 19s 11d	27
4 to 4 19s 11d	73
5 to 5 19s 11d	19
6 to 6 19s 11d	12
7 to 7 19s 11d	7
8 to 9 19s 11d	8
10 to 12 9s 11d	7
12 10s 0d to 14 19s 11d	7
15 and over	12
Total	180

Source: PRO IR 58/55858–61.

house rented at just over £4 a year had two rooms upstairs and two down, an earth closet, gas, but no mains water. Sometimes the earth closets were shared with neighbours in a common yard. Pigsties were a common feature. Three-storey cottages had been a feature of Arkwright's original plan for the village, with cellar, living room, bedroom and an attic weaving room for the frame-work knitters.[45] These can be seen in the North Street houses, built 1771–6, with their elongated windows under the eaves. Thus, attics are frequently mentioned in the Field Books as a result. A typical entry, Hereditament no.3130, therefore reads: 'Hse/gdn, 2 up 2 down & attic. Earth closet, gdn.'[46]

At the other extreme was 'Rock House', belonging to Sir Richard Arkwright in the late eighteenth century. In 1910 it was let furnished at £250 with 5 acres of ground, although it is unclear whether this sum was the annual rent, or whether it referred to the five-year period of the lease. The ground floor had an entrance hall, a smoke room, dining room, drawing room, study, and a large library (27′×21′). There was also a kitchen, servants' hall, house-keeper's room, and other utility rooms. On the first floor was a very large (36′×21′) drawing room, four bedrooms, four maids' rooms, two bathrooms, and various closets. On the top floor was a bathroom, drawing room, seven bedrooms, and a WC. The 5 acres of grounds included a kitchen garden with two greenhouses and a potting shed, plus lawns, tennis lawns, stables, a motor house, a carriage house, saddle rooms, and so on.[47]

Mrs J. C. Arkwright's house was also fully described. It was said to be about 130 years old – which would have put it back to the very beginnings of

[45] Beckett, *The East Midlands*, 287–8. [46] PRO IR 58/55860.
[47] PRO IR 58/55858, Hereditament no. 2964.

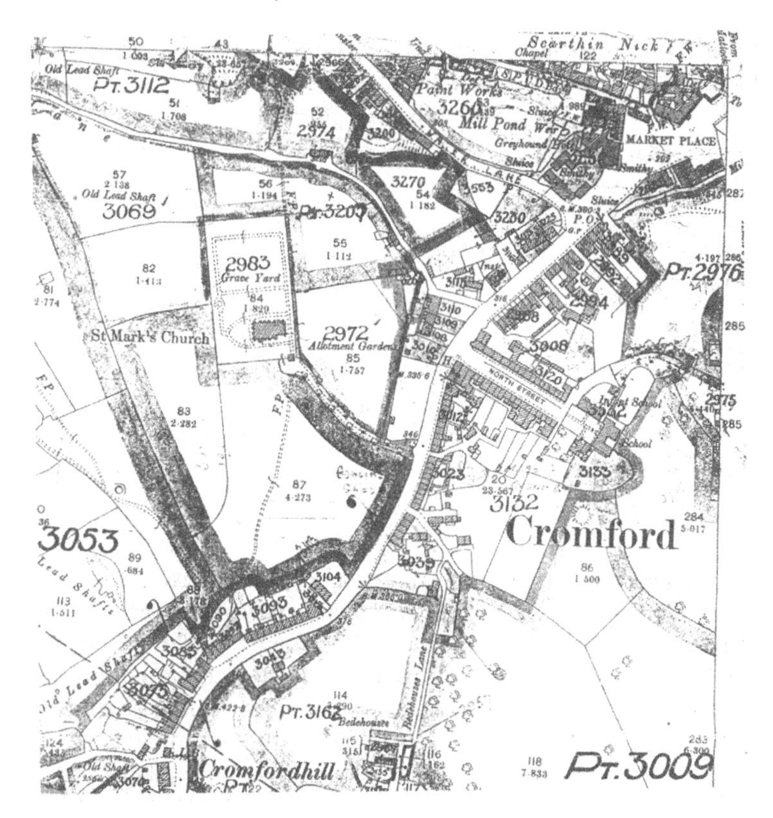

Fig. 9.4 Cromford, Derbyshire: extracts from the Record Sheet Plans.
Source: PRO IR 130/8/378–9.

Arkwright's original village. Built of stone and slate on two floors, it was a very comfortable residence. There were five bedrooms, a dressing room, drawing room, WC, servants' hall, maid's room, butler's dressing room, another room for three maids, a footman's room and other small rooms on the first floor. On the ground floor (with all measurements given in the Field Book) were the drawing room, dining room, scullery, pantry, larder and servants' hall, coal room and WC. Outside were a stabling, harness and carriage room; a knifehouse; two looseboxes; a potting shed and cokehouse. Elm, sycamore and larch mixed with greenhouse and sheds in the small area of garden.[48]

The village of Cromford was provided with football and cricket grounds on a 59 acre hereditament belonging to Arkwright which was otherwise meadow and pasture with elm, oak and fir trees in fair quantity. The tenants were to

[48] PRO IR 58/55858, Hereditament no. 2975.

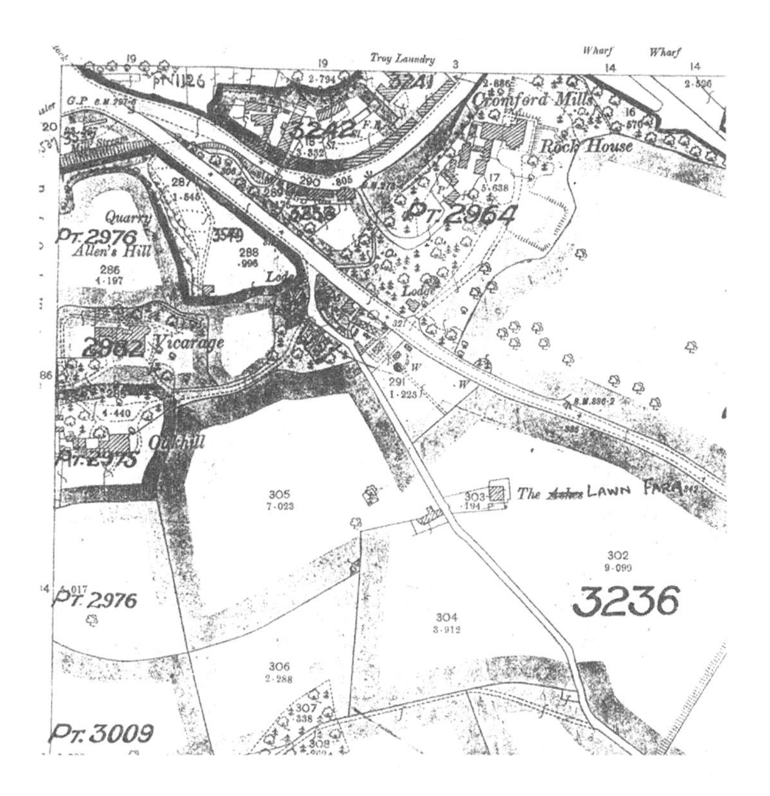

allow Mrs Hoasley the pasturage.[49] Fishing rights on the river were leased by Arkwright to the Matlock Angling Association, and there was a bowling green (taken out of a large meadow after 1909) and allotments. There were also three inns at this time: the Railway Inn; the Greyhound Hotel (built by Arkwright in 1788) with its coach house, club room, smoke room with three-quarter-size billiard table, wine and beer cellars etc. and a seven-day full licence; and the '6 days only freehouse' Bell Inn which purveyed one barrel per week together with 1½ gallons of spirits.[50] There was also a reading and billiard room, and a room for 'boys games' sited over the 'Pay station'.[51] St Mary's Church, refashioned in stone and slate in the Victorian period, accommodated between 180 and 200 people in its congregation.

The documents suggest a village of good-quality housing, albeit frequently a little cramped, let at reasonable rents, and in good repair. The condition of

[49] PRO IR 58/55860, Hereditament no. 3199.

[50] PRO IR 58/55859, Hereditament no. 3010 (Bell Inn); 55860, Hereditament no. 3194 (Railway Inn); 55861, Hereditament no. 3230 (Greyhound Hotel).

[51] PRO IR 58/55861, Hereditament no. 3265, but included with 3283.

several industrial premises was less good and a sense of decay is manifest. The population of Cromford was therefore smaller in 1901 than it had been in 1801.

Generally the quality of descriptive material is good, and in conjunction with the Record Sheet Plans, it would be possible positively to identify each building. Given the historical importance attached to Arkwright's industrial village and the educational significance of the museum centred there, it would seem that the material offers considerable scope for industrial archaeologists and others interested in industrial history. Such a study would also be readily applicable elsewhere, and work could be undertaken on the fortunes of similar industrial 'new' villages of the nineteenth century, such as Port Sunlight, Bourneville and Saltaire.

Conclusion

It is obviously possible to use the 1910 material to study countrysides other than those of the agricultural English lowlands. Here four different communities have been taken to demonstrate how the documents can aid in an investigation of industrial rural communities. But hints of difficulties have been given throughout the chapter, and it will therefore be important to end by drawing some of these difficulties together, in order to elucidate some further important methodological concerns about the material.

Four issues can be addressed briefly. The first relates to the fact that we are here dealing with more complex undertakings than agricultural concerns, and that such complexities, involving capital and management decisions, frequently resulted in equally complex ownership and tenurial relationships between people and property. Teasing out the ownership of large industrial undertakings such as collieries or quarries will often require more information than would normally be present in the 1910 material. Corporate ownership has already been mentioned earlier, but the translation of land ownership into political and social power – if that is the aim of a particular analysis – will be complicated by company ownership. The Bethesda example illustrates this very well, with the documents unable by themselves to throw much light upon the ultimate sources of control over the quarries, and thereby to trace more particularly the relationships of power between estate and village.

Secondly, the documents frequently fall just short of providing the necessary amount of information on the industrial concerns and their encompassing communities because they are ultimately concerned to establish ownership and thereby liability for taxation, rather than to investigate any structures above ground which do not contribute directly to the value of the hereditament. And certainly they are unconcerned with quantitative or qualitative measures such as employment figures, the various factors of production, or the social and cultural role of the industries. The attempt to trace the relationship between the Penrhyn ownership of the slate quarries at Bethesda and the local community could therefore only be deemed a partial success for this reason.

Lack of information can be compounded by a third issue: that of the physical size of the Field Books when valuers came up against large and complicated hereditaments. Too often we find a note to the effect that we should see the file or are otherwise directed some other long-lost document. The situation is similar to that faced when dealing with large agricultural estates or even large country houses, but it is no less frustrating. There is however the added complication that we may be dealing with a public utility company or other large public undertaking that was actually excluded from taxation in the immediate future. Although such undertakings were to be valued eventually, the rush to finish the valuation across the country inevitably meant that such undertakings were left until the end – and that could mean, because of the history of the valuation, that they were not valued at all.

Finally, the sampling problem has been touched upon in this chapter. Again, sampling will be an obvious alternative in any study of a large town or large tract of countryside, especially if a thematic approach to a study dictates that no individual settlement requires in-depth treatment. But in attempting to assess patterns of ownership, sampling is fraught with difficulties. The Valuation Books or Field Books represent an obvious sampling framework for the researcher, who might take every tenth or fifth etc. hereditament to make up the sample. But the relationship of the order of hereditaments as set out in the books to any actual spatial pattern on the ground requires prior consideration, and will vary from one locality to another. It is somewhat analogous to the relationship between entries in the Census Enumerators' Books and the actual spatial contiguity of the properties visited by the enumerator. This difficulty does not, of course, preclude satisfactory reconstructions of the community, but it certainly makes them very much more difficult. If the extent of spatial contiguity of hereditament ownership *c.*1910 is unknown, then to generalise from a 10 per cent sample to statements about land and property ownership could be a dubious exercise. One really does need to examine all hereditaments before making such a statement. In the case of a factory or other industrial structure set amidst countryside or village, the tracing of its 'tenurial tentacles' should be approached therefore with great care.

10

Contrasts and comparisons

It is possible to effect comparisons for particular locations between the 1910 Valuation material and contemporary, or near-contemporary, sources, whether they be documents or other types of qualitative description or analysis. It is also possible to hold the location constant but to investigate comparisons between cross-sections drawn through time by correlating information for *c.* 1910 with that of either a later or earlier period.

The following chapter attempts to outline some potential projects of this kind, and in so doing also sets out some of the pitfalls which lie waiting for the unwary or over-enthusiastic researcher! Three main case studies demonstrate the possibilities in different ways. First the linkages which can be made through time by comparing a group of parishes in the South Hams of Devon in 1840 and 1910, with a short diversion on Cumbria; and secondly the linkages between documents of the same date by a comparison between the 1910 material and the 4 June 1910 Returns for parishes in Wharfedale, West Yorkshire. Thirdly a comparison is undertaken with a previously published book on Corsley, Wiltshire, written almost contemporaneously with the valuation survey.

The South Hams of Devon: a comparison of the tithe manuscripts and the 1910 material, with a note on Cumbria

The possibility of using the Valuation data within a sequence of source materials for the longitudinal study of a region or locality is an exciting one for many historical geographers. Its insertion into a run of land use data which might begin with the tithe material and continue through the 1870s 25 inch Ordnance Survey field books and progress onto the first and second edition Land Use surveys, and currently end with the 1941–3 National Farm Survey, is a tantalising prospect.[1]

[1] See B. Short and C. Watkins, 'The National Farm Survey of England and Wales 1941–43', *Area* 26(3) (1994), 288–93; and P. Barnwell, 'The National Farm Survey, 1941–43', *Journal of the Historic Farm Buildings Group* 7 (1993), 12–19.

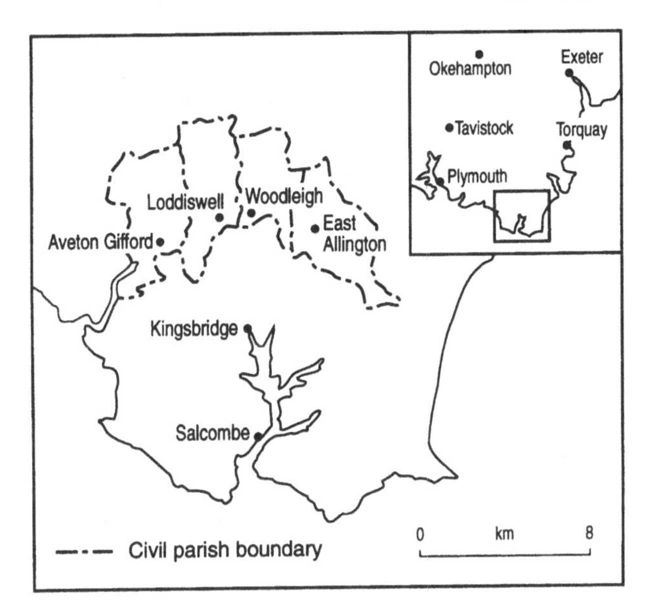

Fig. 10.1. South Hams, Devon: the case study parishes in 1910.

In this section attention will be confined to an obvious comparison: with the tithe surveys of about seventy years earlier. A case study drawn from the South Hams of Devon will be used to show both the possibilities and difficulties of such comparative work. The study area comprised the four parishes of Aveton Gifford, Loddiswell, Woodleigh and East Allington – all just to the north of Kingsbridge (Fig. 10.1).

The tithe awards presented no problems other than those generally recognised for this type of document, and although this material has its own intrinsic interest, it will not be specifically evaluated further in this section, other than to comment on the findings derived from it in comparison with the Valuation material.[2] The latter, in the shape of the Valuation Books, was as well compiled as any others seen. Most hereditaments had their extent entered in the Valuation Books, though this was not always the valuers' determination of the area. Most of the figures omitted from the Valuation Books, at Devon County Record Office, Exeter, were later obtained from the Field Books in the PRO.[3] Thus only a small proportion of hereditaments have no figure at all for

[2] For detailed analyses of the tithe material see R. Kain and H. Prince, *The Tithe Surveys of England and Wales* (Cambridge 1985); R. Kain, R. Fry and H. Holt, *An Atlas and Index to the Tithe Files of Mid Nineteenth-Century England and Wales* (Cambridge 1986); R. Kain and R. Oliver, *The Tithe Maps of England and Wales* (Cambridge 1995).

[3] Devon Record Office (Exeter) Tithe apportionments for each parish. Valuation Books for each parish (uncat.); PRO IR 58/83383–5 (Field Books).

extent. The problem is the familiar one of having to use a mixture of 'actual' areas determined by the valuers from the Record Plans, estimated areas taken from the Rate Books, and owners' statements returned on Form 4 and available in the Field Books. Precision is often impossible therefore at the local Record Office, without access to the Record Plans from which areas could be calculated for each hereditament. This alone makes comparison with the tithe material risky if not impossible, but there are other more profound difficulties inherent in the material, that can only be resolved by the most intensive and detailed work.

These problems originate in the ways that the data were compiled, and are present in most work involving the Valuation material to some degree. Broadly speaking, the impact of this on the South Hams study was that the unit of analysis – the 'parish' – is not the same, either theoretically or practically, in the tithe survey as in the 1910 valuation. Thus comparison is being made between two unlike units.

The tithe surveys give rise to several well-known problems, of which two are of particular relevance here. The first is that land not subject to tithe was not always surveyed and therefore no details relating to that land are available. This amounted to some 87 acres in Loddiswell, and 121 acres in East Allington, although there is no mention of exempt land in the other two parishes. The second problem is that the tithe survey was usually confined to a single ecclesiastical parish, or other unit. Holdings that crossed the parish boundary are therefore not included in full in that parish. Thus the tithe survey is likely to understate the size of holdings that cross the parish boundary, and moreover, they do not indicate that this is the case. The numbers of smaller holdings in a parish are therefore likely to be overstated.

The reverse is the case with the 1910 data, and the problem is rather more complex. The unit of valuation in 1910 was the Income Tax Parish (ITP). Where the ITP was synonymous with a civil parish, the problems may be reduced, though they are still formidable. In the parishes of this study, only East Allington did not form an ITP on its own, since it included the civil parish of Slapton. Where an ITP was identical with a civil parish, it is quite likely that the boundaries of the civil parish are different from that of the ecclesiastical parish of the same name that was the basis for the tithe survey. Accurate comparison requires adjustment to take account of this. Moreover the instructions to the valuers required that where a hereditament crossed the boundary of an ITP, it was to be included in the ITP in which the largest part was situated. The problem is obvious. The invisible 'hereditament parish' noted for the Ashburnham estate and parish example features again, and is identical neither to a civil parish nor to an ITP.[4]

To reiterate the situation, hereditaments in any given parish that cross

[4] See Chapter 6.

parish boundaries may (a) have a larger area than that actually within the parish boundaries because of the inclusion of land from an adjoining parish or parishes, or (b) be excluded from the parish entirely because the portion in an adjoining parish is of greater extent so that the whole is included in the adjoining parish. Thus land may be 'imported' into or 'exported' from a parish for the purposes of valuation, without this being indicated in a precise way in the documents.

In fact in a properly compiled Valuation or Field Book, some indication should be found, though accurate areas are unlikely to be present. It must be remembered that the Valuation Book was a transcript of the Rate Book so that every piece of rateable property ought to be indicated in the Valuation Book, and it ought also to have been given a hereditament number. Thus if a hereditament was included in a larger unit that crosses the ITP boundary, then ideally the entry in the Valuation Book should indicate that it has been included in an identified hereditament in a neighbouring parish, or else that it contains land from a neighbouring parish. The accurate area of the combined hereditaments should be found against the entry for the largest component hereditament but the area of each component will, at best, be that estimated by the rating officers and entered in the Rate Book. In practice of course, even this procedurally correct situation is frequently not found.

The same problem occurs when an ITP comprises more than one civil parish, but there is an additional difficulty in this case. Information for a particular civil parish can only be extracted – in the first instance – from the Valuation Books. This is because hereditaments in an ITP were assigned the name of the ITP and not the name of their civil parish. Thus a hereditament in Slapton civil parish might be identified as East Allington 150, since Slapton was in East Allington ITP. The Field Books and Form 37–Land only identified hereditaments in this way.

To isolate hereditaments in a civil parish, it is necessary to go first to the Valuation Books. These, it will be recalled, were a transcription of the Rate Books for the component civil parishes in the ITP, and once transcribed the hereditaments were numbered consecutively throughout the book (including unrated hereditaments which were entered later). It is therefore possible to identify the numbers that relate to a component civil parish and these numbers can then be examined in the Field Books or on Form 37. However, if a hereditament crosses the boundary of the component civil parishes, but remains wholly within the ITP, there may be no obvious indication of this in the Field Books. Only in the Valuation Book is a clue likely to be found. In cases of this kind, part of a hereditament should be found in the Rate Book for each component civil parish, and if it is included with others, or includes others, then this should be stated. If the hereditament numbers it is included with are those relating to a different civil parish, then it is possible that the hereditament crossed the civil parish boundary, but once again it will be difficult to ascertain what the areas of the component parts were without recourse to the Record Plans. Moreover in this case, there was no need for the valuers to have reference to the parish in which the larger part of the hereditament was situated. In fact, since Valuation Books were often compiled inadequately in this respect, these indications are often absent. Information is usually fuller in the Field Books, but even here it is often incomplete. The Record Plans are indispensable for the fully accurate demarcation of hereditaments that cross parochial boundaries whether within an ITP or between ITPs.

Other problems are also likely to be encountered. Under the 1910 Act, section 26(1), each piece of land under separate occupation was to be separately valued, and the owner could require each piece of land to be subdivided and valued separately. This could create an impression of the existence of smaller holdings than was in fact the case. And more significantly from the present point of view, it should be remembered that section 5 of the Revenue Act 1911 amended Section 26(1) of the original Act so that the owner could request pieces of land under separate occupation that were contiguous, and which did not exceed 100 acres in aggregate extent, to be combined for valuation purposes. This of course could create, and has created, an impression of fewer and larger holdings than actually existed. The original hereditament should be indicated in the Valuation Book, but all references to the area – and indeed to other information relating to it – would be combined, and, needless to say, this is not always indicated in the documents.

Another relevant problem is that most Valuation Books, including all those in the present study, have a number of additions that were derived from sources other than the Rate Books. Hereditaments in this category include unrated property and many entries of uncertain status. Sometimes these relate to the kinds of instance just referred to, where an owner has requested a hereditament to be subdivided for valuation purposes, or where a hereditament has been subdivided as a result of a sale, or has been let in separate parcels and so on. Unfortunately, it is often the case that there is no indication of the reasons for the additional numbers, and it is almost invariably the case that information relating to these hereditaments is extremely sparse. Indeed it is not always obvious when these additions were made, and therefore whether they relate to the period around 1910 at all.

It will be clear that the problems are of considerable complexity, and although most can be unravelled and the gains from so doing are considerable, this requires a great deal of work, and in particular it does need access to the Record Plans. Any comparison between the tithe surveys and the 1910 data that relies on Valuation and Field Books alone is only possible within probably substantial margins of error.

These general theoretical and practical problems make precise comparison between the tithe and valuation data difficult. A study of the parochial data gives some idea of the problems involved.

Table 10.1 *South Hams, Devon: acreage by tithe and valuation documents compared with the 1911 Census*

	Tithe*			Valuation			
Parish	A	R	P	A	R	P	1911 Census
East Allington	3646	2	38	3658	0	22	3702
Loddiswell	3658	1	13	3687	1	22	3597
Woodleigh	2319	0	21	2836	3	33	2655
Aveton Gifford	3952	2	39	3955	1	00	3946

Notes:
A R P acres, roods, perches.
*Includes non-titheable areas.
Source: Devon CRO, Tithe Apportionments; PRO, Field Books; Census Abstracts 1911.

In no case does the documentation provide figures that, when added together, equal the area of the parish. Table 10.1 gives the relevant figures for size of parishes as revealed by three different data sources. Discrepancies between the area according to the tithe survey and that of the 1911 Census are probably due to the fact that the tithe is based on the ecclesiastical parish, while the 1911 figure is for the civil parish. In three cases the 1910 valuation documents give figures that exceed the actual area of the parish, and in one case they are less than the total area. In every case, though, there is only a small discrepancy between the figures, and it may be thought that this enables a fairly reliable analysis of landholding in each parish. This may indeed be the case, but there are several reasons for doubting it.

First the figures for area in 1910 are an aggregation of 'accurate' figures taken by the valuers from the Ordnance Survey sheets; of estimates in the Rate Books; and of owners' statements returned on Form 4, and transcribed into the Field Books. In addition some hereditaments have no figures for area at all. It will be clear then that the figures for total area are extremely unlikely to be accurate.

However, the greater problem is likely to be the inability to determine the boundaries of the 'hereditament parish' without recourse to the Record Plans, necessitating the combination of locally based tithe data with PRO-based maps. Some insight into this problem can be gleaned from Table 10.2

Table 10.2 suggests a great increase in the numbers of hereditaments in each parish between 1840 and 1910, a point which will be taken up later. For the moment, however, it is only necessary to refer to those hereditaments 'imported' and 'exported'. Each parish features this. In East Allington, a single hereditament (no.135, area about 2.5 acres) was 'exported' to Slapton,

Table 10.2 *South Hams, Devon: numbers of hereditaments (excluding tithes, sporting rights etc.)*

			1910 hereditaments	
Parish	*c.* 1840	1910	'Imported'	'Exported'
East Allington	89	139	1	1
Loddiswell	135*	285	4	7
Woodleigh	28+	75	5	2
Aveton Gifford	146	263	1	6

Notes:
*Loddiswell included about 87 acres of non-titheable land for which no details are available.
+Woodleigh included about 121 acres of non-titheable land for which no details are available.
Source: Devon CRO, Tithe Apportionments; PRO, Field Books.

while no.42 included several other hereditaments from within the parish, as well as the 'imported' hereditament Stokenham 315. The total area of the combined hereditaments was some 223.5 acres, but from the material looked at, it is not possible to ascertain how much actually lay in Stokenham ITP.

Loddiswell had a 'net loss' of three hereditaments, and Aveton Gifford of five, while Woodleigh had a 'net gain' of three. The areas of these hereditaments cannot be determined from the collected data, and the overall effect upon the boundaries of the 'hereditament parish' and the deviation from those of the civil parish remains unclear. What is obvious is that comparison between the tithe parish and the 1910 Valuation parish is between two different units, and that the boundaries of one can only be ascertained precisely with much detailed investigation. So only generalised statements can be made, though these may well stimulate questions that can be pursued in more detail with additional research.

Table 10.3 compares the holding size in each parish *c.* 1840 and 1910. All of the foregoing cautions should be borne in mind when studying the figures.

These figures are compiled by a straightforward extraction of the information entered against each hereditament on the tithe schedule, and in the Valuation and/or Field Books. Using this method, only the tithe schedule gives figures for area entirely derived from a survey. The 1910 figures are an amalgam of surveyed figures, Rate Book estimates and owners' statements. So even at this level there are numerous uncertainties as to the accuracy of the figures.

Moreover the practice of 'merging' hereditaments in the Valuation survey affects the figures seriously – indeed since the necessary cross-referencing by

Table 10.3 *South Hams, Devon: holding sizes c.1840 and c.1910*

Parish	Unit size (acres)	No. of holdings c.1840	No. of holdings c.1910
Woodleigh	<1	1	20
	1 to 4.9	3	6
	5 to 24.9	2	9
	25 to 49.9	5	2
	50 to 74.9	5	6
	75 to 99.9	4	2
	100 to 149.9	5	5
	150 to 299.9	3	3
	>300		2
	Not stated		18
	Total	28	73

NB The Woodleigh tithe survey excluded 121 acres of non-titheable land, which probably included many cottages and doubtless other small parcels.

Parish	Unit size (acres)	No. of holdings c.1840	No. of holdings c.1910
Loddiswell	<1	59	86
	1 to 4.9	11	33
	5 to 24.9	14	6
	25 to 49.9	14	6
	50 to 74.9	8	8
	75 to 99.9	6	3
	100 to 149.9	8	8
	150 to 299.9	1	5
	>300	1	1
	Not stated		102
	Total	122	258

NB The Loddiswell tithe survey excluded nearly 88 acres of non-titheable land. The effect of this on holding size is unknown.

Parish	Unit size (acres)	No. of holdings c.1840	No. of holdings c.1910
East Allington	<1	29	47
	1 to 4.9	10	30
	5 to 24.9	13	9
	25 to 49.9	8	6
	50 to 74.9	11	3
	75 to 99.9	6	3
	100 to 149.9	8	7
	150 to 299.9	3	8
	>300	1	1
	Not stated		25
	Total	89	139

Table 10.3 (*cont.*)

Parish	Unit size (acres)	No. of holdings	
		c. 1840	*c.* 1910
Aveton Gifford	<1	53	90
	1 to 4.9	15	16
	5 to 24.9	20	12
	25 to 49.9	20	12
	50 to 74.9	8	7
	75 to 99.9	4	6
	100 to 149.9	5	6
	150 to 299.9	5	6
	>300		
	Not stated		90
	Total	130	245

Source: Devon CRO, Tithe Apportionments and Valuation Books; PRO, Field Books.

which merging can be identified is not always entered, it may do so even more seriously than is discernible without recourse to the Record Plans. Many of the hereditaments for which there is no statement as to area have in fact been merged with others. Thus the area of many hereditaments actually includes the area of some without a stated acreage. In Table 10.3, then, these have been double-entered. Only when they are merged with hereditaments within the parish, can any assessment be made of the effects. Even here the practice is seldom clear-cut. An analysis of the East Allington data shows that the areas of fifteen hereditaments actually included other hereditaments, but that this was either completely or partially omitted from the entries of the fifteen involved. It has been possible to deduce this since the entries for the component hereditaments did indicate that they had been included in one of the original fifteen. This information had to be gleaned from both the Valuation and Field Books as it was not fully available in either. Thus we can see that the information needed for this exercise is only intermittently entered into any specific category of document, and this raises the unfortunate possibility that there are instances when the information is not entered at all in any of the available documents.

Some examples from East Allington follow. Hereditament no. 9 was said to have an area of 252 acres and 3 roods, and the entry gave no indication that other hereditaments were included. However, the entries for other hereditaments did indicate that they were included in no. 9. These were nos. 92 and 93 (no area stated), no. 122 (area 5 acres 1 rood), and no. 126 (12 acres 2 roods).

Even worse was no. 25 with an area of 342 acres 1 rood that apparently related to it alone. In fact it included no. 26 (area 159 acres 1 rood), nos. 75 and 91 (no area stated), no. 124 (10 acres 1 rood), and no. 129 (3 acres 3 roods). Moreover it is not clear whether the area of no. 25 (342 acres 1 rood) referred to that holding alone, or whether it included the areas of the component hereditaments. Thus it is not known whether double-counting exists or not.

A case where an assessment can be made is that of no. 42, where the area, as determined by the valuer, is given as 223 acres 2 roods and 2 perches. This entry states that the following hereditaments were included in this: no.2 (area 39 acres 3 roods), no.46 (area not stated), no.121 (1 acre 2 roods), no.134 (3 acres 1 rood), and Stokenham 315 (area not available). Also included but not indicated in entry 42, were nos.115 and 116 (areas not stated). It is clear that in this case no. 42 had an area before merging of at least 45.5 acres less than that stated (assuming the areas of the other numbers are correct), and that the discrepancy is undoubtedly greater than this since the area of four of the component hereditaments are not known. In this instance, the problem may not affect Table 10.3 since if we reduce the area of no. 42 by the known areas of the component hereditaments, we arrive at a figure of some 178 acres. This would not alter the placing of the hereditament within the size-group it is already assigned to, but it is easy to see that this may not always be the case.

The problems are such as possibly to cancel each other out to some extent, but we are working with any number of imponderables, that can only be resolved by detailed and intensive work on all the appropriate documents. It is therefore most unfortunate that it is impossible to work on the Field Books and Valuation Books together, since it is now clear that they actually complement each other in many instances.

The exclusion of some 120 acres from the Woodleigh tithe survey, and 88 from that of Loddiswell, adds to the difficulties of comparison. It seems unlikely that Woodleigh would have had only a single cottage/garden with less than an acre of land *c.*1840, so that the discrepancy between that and the twenty shown in 1910 is probably due wholly or partly to the unsurveyed area. It may also contribute to the apparently substantial increase in the number of smallholdings (5 to 24.9 acres) between the two dates. The most striking feature of the Woodleigh data that cannot be attributed to this is the apparent emergence of two large holdings over 300 acres in extent, while the numbers of those between 100 and 299.9 acres remained constant, implying that a great many smaller holdings had disappeared. But even this is not all it seems. Hereditament no. 33 (area 370 acres 3 roods 18 perches) actually included nos. 37 (65 acres 1 rood), 38 (10 acres), 39 (not stated) and 41 (10 acres), so that it should probably be included in the size group below. Similarly no. 55 (321 acres 3 roods 22 perches) included hereditament no.64 – in different occupation – though no area is given for this.

The most obvious difference between the two sets of data for Loddiswell

Table 10.4 *South Hams, Devon: hereditaments by size-group c.1840 and c.1910*

	No. of hereditaments	
Area (acres)	*c.* 1840	*c.* 1910
1 to 24.9	88	121
25 to 99.9	99	64
100 to 149.9	26	26
150 and over	14	26

Source: Devon CRO, Tithe Apportionments and Valuation Books; PRO, Field Books.

Table 10.5 *South Hams, Devon: owner-occupied hereditaments c.1840 and c.1910*

	No. of hereditaments	
Parish	*c.* 1840	*c.* 1910
Woodleigh	5	16
East Allington	6	27
Loddiswell	53	60
Aveton Gifford	51	52

Source: Devon CRO, Tithe Apportionments and Valuation Books; PRO, Field Books.

that are clearly not due to the 88 acres unsurveyed in 1840 is the sharp fall in the numbers of holdings in the 25 to 49.9 acre size-group, and a similarly sharp rise in the numbers in the 150 to 299.9 acre group. A sharp rise in the numbers in the latter size-group is also found in East Allington. To what extent this is the result of merging of hereditaments is seldom clear. Valuers' figures for area are infrequent for East Allington, and whether double counting has occurred in Table 10.3 with any effect is not apparent. In Loddiswell, cross-referencing is infrequent, but whether this indicates that merging was also uncommon, or whether the books are simply poorly compiled in this respect, is unknown.

Overall, and with considerable reservation, the data support the view that between *c.* 1840 and 1910, there was a 'squeeze' upon occupiers of medium holdings (25 to 99.9 acres), with increases in the numbers of those smaller (1 to 24.9 acres), and larger (especially those over 150 acres). Table 10.4 makes this point for the four parishes together.

Owner-occupation was not uncommon in two of the parishes, as Table 10.5

Table 10.6 *South Hams, Devon: landowners c.1840 and c.1910*

	No. of landowners	
Parish	*c.*1840	*c.*1910
Woodleigh	10	17
East Allington	34	15
Loddiswell	86	90
Aveton Gifford	79	81

Source: Devon CRO, Tithe Apportionments and Valuation Books; PRO, Field Books.

indicates. The table suggests that owner-occupation was proportionately far more significant *c.*1840 in Loddiswell and Aveton Gifford than in the other two parishes. In the two former parishes, they constituted over one-third of all hereditaments at this time, whereas they comprised only a handful of the total in Woodleigh and East Allington. In absolute terms, the numbers varied little by 1910 in Loddiswell and Aveton Gifford, but they apparently comprised a smaller proportion of the total. An absolute and relative increase is more noticeable in Woodleigh and East Allington by 1910, though in the case of Woodleigh the unsurveyed portion in 1840 may contribute to this.

Landownership was dispersed in each parish, as Table 10.6 indicates. From this it can be seen that the numbers of owners remained fairly constant in Loddiswell and Aveton Gifford, and increased somewhat in Woodleigh, though once again this may be due to the unsurveyed area *c.*1840. Only in East Allington is there a sharp drop in the number of owners during the period.

The large numbers of landowners suggest at first sight that none of the parishes could be considered 'close' at either date. In the cases of Loddiswell and Aveton Gifford, this is clearly the case. In neither parish, at either date, were there owners who owned substantial proportions of each parish. But in Woodleigh and East Allington the situation was rather different. In the former parish *c.*1840, the three largest owners owned together 69.5 per cent of the area surveyed for tithe purposes. In 1910 – and these figures may be less reliable – the three largest owners held together 71.4 per cent (Table 10.7).

In Woodleigh then, there was almost no change in the proportion of the acreage owned by the three largest owners. Yet the numbers of owners had increased from ten to seventeen. The increase – if it is not an illusion – must have been as a result of the fragmentation of some smaller estates. In East Allington a very different picture emerges and the scale of change is so great that the problems of the data can only qualify it in some degree, but there can surely be no doubt as to the trend. Table 10.6 has shown that the numbers of

Table 10.7 *South Hams, Devon: land owned in Woodleigh by the three largest owners*

Name	Area			
c. 1840	A	R	P	% of total
J. B. Sweet	810	0	26	34.9
John Luscombe	495	3	21	21.4
John Netherton	305	1	37	13.2
c. 1910				
C. W. Tayler	1108	2	31	39.1
Lord Ashcombe	471	1	15	16.6
Rev. Olliphant	446	0	18	15.7

Notes:
ARP acres, roods, perches
Source: Devon CRO, Tithe Apportionments and Valuation Books.

owners fell by more than half – from thirty-four to fifteen – between c. 1840 and 1910. In this parish at each date, there was only one substantial landowner. In c. 1840 it was W. Blundell Fortescue, who owned 1,461.5 acres – 40.1 per cent of the total area of the parish. Seventy years later, Lord Ashcombe owned just over 2,352 acres (these figures are not likely to be precise), or 64 per cent of the parish. The estate undoubtedly dramatically expanded its acreage during the period at the expense of lesser owners.

The South Hams study highlights many substantial problems of the 1910 data. It makes clear that analyses based on that favourite unit of geographers and historians – the ecclesiastical and/or civil parish – is not usually possible in any straightforward way. To continue to use the parish as the unit of study requires some effort and knowledge on the part of the researcher, including an examination of material relating to surrounding parishes. The problems are theoretically fairly straightforward, but in practice they may be less so given the dispersal of the sources and the inconsistent ways in which the data were compiled. If the Record Sheets, the Valuation and Field Books are to hand, then the task of identifying the boundaries of the civil and ecclesiastical parish and its relation to the ITP will be easier, as will the identification of the practice of merging, and the assessment of acreages. Without this, comparison with other sources such as the tithe can only be carried out with substantial margins of error.

These findings seem extremely positive since it is now clear what the researcher needs to do in order to utilise the documents fully for the study of changes in landholding and ownership structures. Another positive finding is that it is possible, even given the problems, to identify trends in an area, providing they are sufficiently pronounced, and providing claims to full precision are avoided.[5]

In the light of these findings, it is necessary to warn future scholars attempting to use the material for such comparisons by means of one further example. In a stimulating study of Cumbria from the eighteenth to the twentieth centuries, Searle was one of the first scholars to make use of the Valuation material after its entry into the public domain. His use of the documents was confined to a study of changes in landownership structures in a number of Cumbrian parishes chosen to be representative of several major regions.[6] He compared landownership structures at the time of the 1910 valuation with those at the time of the tithe surveys some seventy years earlier, and concluded that small owners declined in numbers and in the proportion of the land owned. In an appendix, he provides the raw figures upon which his conclusions are based, together with a statement regarding the importance of the valuation material. Although he refers in this appendix to the 'Valuation Books', he is in fact referring to the Field Books deposited in the PRO.[7] Here the re-examination will concentrate on the 1910 data, rather than on the changes found by Searle, since it is vital to correct his findings and to demonstrate the methodology for future uses of the data for time comparisons.

Searle selected six regions in Cumberland and one in Westmorland for his analysis, using data – ostensibly – from thirty parishes. Here an examination is made only of those regions and parishes that had been in Cumberland.[8]

Given the fact that the Field Books were based on Income Tax Parishes rather than upon civil parishes, it is unlikely that Searle could have isolated the hereditaments situated in a particular civil parish from the Field Books alone. An initial examination of the Valuation Books in the Cumbria CRO at Carlisle – a source not used by Searle – was made in order to determine the constituent civil parishes of each ITP.[9] The Valuation Books were generally well compiled and it was relatively easy to identify the hereditaments in each

[5] For an analogous study of the integration of Tithe and Valuation material (together with 1941–3 National Farm Survey data) see J. D. Godfrey, 'A century of agriculture on the South Downs *c.*1840–1940', University of Sussex, unpubl. D.Phil. thesis, forthcoming; for an evaluation of the three sources as they relate to one parish, that of Hamsey, East Sussex, see Hazel Lintott, 'Mapping rural landownership. An evaluation of three national surveys as cartographic databases', University of Sussex, unpubl. MA term-paper, April 1995.

[6] C. Searle, 'The odd corner of England: a case study of a rural social formation in transition, Cumbria, *c.*1700–*c.*1914', University of Essex, unpubl. Ph.D. thesis, 1983. I am extremely grateful to Dr Searle for his ready cooperation in this re-evaluation of this part of his wider work.

[7] *Ibid.*, Appendix C, 396–8.

[8] B. Short and M. Reed, *Landownership and Society in Edwardian England: The Finance (1909–10) Act 1910 Records* (University of Sussex 1987), 40–3.

[9] Cumbria CRO, Carlisle: TIR 4/4, 20, 22, 26, 28, 33, 44, 50, 53, 63, 78, 79, 81, 86, 88, 93, 105 (Valuation Books); PRO IR 58/18671–2, 18802–4, 18821–4, 19015, 19039–44, 19047–50, 19082–3, 31624–30, 31641, 31698–700 (Field Books). Unit size categories are those used by Searle. For further information on the availability of 1910 material in Searle's study area see Chapter 5, Figs 5.2 and 5.3.

civil parish, and also their extent. Figures for extent were usually the Rate Book estimate, as the valuers' figures were normally absent.

In his study, Searle appears to have assumed that the parishes named in the Field Books were civil parishes when in reality the names are actually those of ITPs. These could comprise several civil parishes. So, to what extent did Searle's assumptions lead him to include other unidentified parishes in his study? Table 10.8 shows the relevant ITPs and their constituent civil parishes.

This analysis demonstrated that in the six Cumberland regions studied by Searle, his analysis of twenty-one parishes in the Field Books in fact included at least thirty-three civil parishes. Closer precision is impossible since the Valuation Books for Ennerdale and Kinniside were missing, and as a result of this, his group of parishes on the Lake Dome could not be included in this re-evaluation. The analysis here is therefore confined to a study of the remaining five regions, comprising according to Searle seventeen parishes, but in reality totalling twenty-nine. It appears therefore that Searle was dealing with a substantially larger area than he had believed. His failure to isolate the civil parishes within each ITP is further illustrated by his use of Field Books that do not in fact relate to the civil parish of his choice. For example, in his study of the Foothill Belt, Searle used fourteen volumes of the Field Books to examine the four parishes, but in fact only nine of these actually related to the civil parishes, the remaining five referring to other civil parishes within the ITPs. These five parishes appear to have been inadvertently included in Searle's analysis.

Searle's comparison of the valuation data with those of the tithe is fraught with difficulties, some of which are outlined above for the South Hams. It is clear that Searle had no knowledge of the problems caused by the 'hereditament parish'. To illustrate the problem further, a comparison may be made between the figures derived from this re-examination of the data and Searle's figures of unit size. The assumption – almost certainly incorrect – had to be made that all the hereditaments lay entirely within the parishes under study. Thus one cannot claim any precision, and the exercise should only be seen as a preliminary one to highlight the problems of using the material without full knowledge of the way in which it was compiled. The results are shown in Table 10.9. The unit size categories are those used by Searle.

There is a clear distinction between the two sets of figures. Those derived from the re-examination suggest that Searle overstated the numbers of larger properties in his chosen area, while seriously understating the numbers of smaller properties. Even given the lack of precision in the re-examination, the discrepancies are too stark to be completely erroneous. The reasons are clear enough. By assuming the Field Books dealt with civil parishes rather than ITPs, Searle included a number of other parishes in his analysis and was therefore dealing with a larger geographical area. This explains the fact that there now appear to be fewer large properties than Searle found. However by the

Table 10.8 *Cumbria: Income Tax and civil parishes*

Income Tax Parish		Civil parish
	Coalfield Area	
Lowside Quarter		Lowside Quarter
Gilcrux		Gilcrux
Dearham		Dearham
Harrington		Harrington
Moresby		Moresby Parton
	Foothill Belt	
Ireby High		Ireby High
Ireby Low		Ireby Low
Lamplugh		Lamplugh, Salter and Eskett
Bridekirk		Bridekirk, Dovenby, Tallentire
Caldbeck		Caldbeck
	Eden Valley	
Crosby-on-Eden		Crosby-on-Eden
Edenhall		Edenhall, Langworthy, Great Salkeld
	Lake Dome	
Loweswater		Loweswater, Mockerkin
Ennerdale*		
Kinniside*		
	Pennines	
Ainstable		Ainstable
Kirkland		Kirkland and Blencarn
Kirkoswald*		Kirkoswald, Lazonby#, Penwick, Staffield
Ousby		Ousby
	Coastal Strip	
Whicham		Whicham, Whitbeck
Ponsonby		Ponsonby, Corney#

Notes:
* The Valuation Books for Ennerdale and Kinniside were missing and it was impossible to establish their constituent civil parishes. Only one Valuation Book has survived for Kirkoswald ITP so there is no certainty that all the constituent civil parishes have been identified.
# Lazonby civil parish was part of Kirkoswald ITP, while Corney was part of Bootle ITP. Searle's method of isolating the hereditaments situated in these civil parishes is unknown.
Source: Cumbria CRO (Carlisle), Valuation Books.

Table 10.9 *Cumbria: holdings by unit size c.1910*

Unit size (acres)		No. of units			
		Searle	%	Re-examination	%
	Coalfield Area				
1–200		87	88.8	260	97.0
200–300		3	3.1	5	1.9
300–400		3	3.1	2	0.8
400–800		3	3.1	1	0.4
800–1000		1	1.0		
1000+		1	1.0		
Total		98		268	
	Foothill Belt				
1–200		190	90.5	301	97.1
200–300		8	3.8	4	1.3
300–400		4	1.9	3	1.0
400–800		5	2.4	2	0.7
800–1000		1	0.5		
1000+		2	1.0		
Total		210		310	
	Eden Valley				
1–200		62	83.8	108	87.1
200–300		5	6.8	7	5.7
300–400		2	2.7	5	4.0
400–800		2	2.7	2	1.6
800–1000					
1000+		3	4.1	2	1.6
Total		74		124	
	Pennines				
1–200		182	87.1	238	92.3
200–300		8	3.8	11	4.3
300–400		7	3.3	4	1.6
400–800		5	2.4	3	1.2
800–1000		2	1.0	2	0.8
1000+		5	2.4		
Total		209		258	
	Coastal Strip				
1–200		51	85.0	118	93.7
200–300		2	3.3	4	3.2
300–400		3	5.0	2	1.6
400–800		3	5.0	1	0.8
800–1000					
1000+		1	1.7	1	0.8
Total		60		126	

Notes:
NB Percentages are those for each region.
Source: Cumbria CRO (Carlisle), Valuation Books.

same token it would be expected that the re-examination would also have found fewer small properties than did Searle, whereas in fact there were many more found. This derived from the inconsistent compilation of the documents by the valuation staff. Many of the figures for area – especially for the smaller properties – are not entered into the Field Books, or for that matter into the Valuation Books, and it is necessary to use both in order to achieve a fairly complete coverage. In only referring to the Field Books, Searle got the worst of all worlds by covering too many parishes whilst only having part of the relevant information. The difficulties of using the valuation documents in a straightforward way to deal with temporal comparisons are clearly demonstrated.

Wharfedale: a comparison with the agricultural returns for 1910

The relating of the 1910 documentation to other contemporary or near-contemporary primary sources would seem to be a profitable exercise. In part, the benefits to be gained from such comparisons will vary with the purposes of the research and with the place-specific nature of the other initial data compilation and survival. Some general types of comparable material have been addressed from time to time in this book, at least at a *prima facie* level. Clearly this would include Rate Books (although the information included in the Valuation Books should include relevant information already taken from this source), directories, estate surveys, sale catalogues and other private papers, and Medical Officer of Health reports. There are doubtless many other relevant local and national sources.

The example to be taken here will deal with a comparison of information obtained from the valuation documents on hereditament size and ownership, with that obtained from parish summaries of the 4 June 1910 Agricultural Returns. A case study area was selected from Wharfedale, West Riding of Yorkshire, comprising the parishes of Buckden, Kettlewell with Starbotton, Linton, Grassington, and Conistone with Kilnsey (Fig. 10.2).

Two ITPs were involved. Grassington included the civil parish of that name, together with Conistone with Kilnsey, Hebden and Linton. The other was Buckden which also included the civil parish of Kettlewell with Starbotton. It was necessary therefore to identify the hereditaments in each civil parish by reference to the Valuation Books in the North Yorkshire CRO at Northallerton. At the same time it was noted whether each hereditament was owner-occupied or tenanted. The area was also noted, where possible relying on the figures provided by the valuer as being the more accurate. It is fortunate that the Valuation Books were well compiled, with the valuers' statements of area being present in the majority of cases. Almost all of those not entered in the Valuation Books were later obtained from the Field Books in the PRO. Finally therefore, only twenty of the 282 hereditaments over

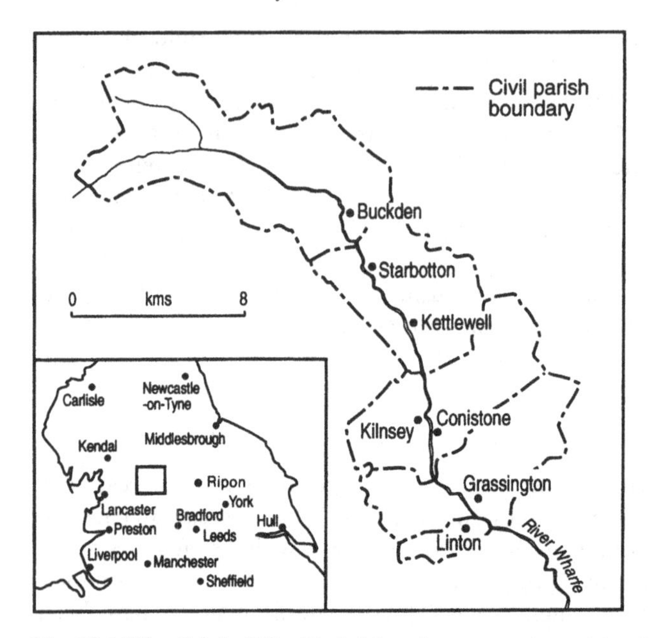

Fig. 10.2 Wharfedale, West Yorkshire: the case study parishes in 1910.

1 acre (7.1 per cent) had to be classified on the basis of the Rate Book estimate for area.[10]

The size-groups used by the compilers of the agricultural returns must necessarily provide the basis for what follows, and the basic information obtained is presented in Table 10.10.

Immediately noticeable is the fact that there were ninety-nine more hereditaments over 1 acre in size identifiable in the Valuation documents than there were agricultural holdings according to the returns. All but eight of these were of 50 acres or less. The 1–5 acre size-group contains over twice as many hereditaments as it does agricultural holdings, while hereditaments in the 6–50 size-group were almost twice as numerous as were agricultural holdings. Kettlewell with Starbotton, and Grassington were particularly discrepant.

These discrepancies are no doubt largely due to the fact that most of these hereditaments were not used for agricultural purposes. The differences were far less in the more agricultural Buckden and Conistone. No note of usage was made during the study and in any event it is not always possible to distinguish agricultural from non-agricultural land in the valuation documents. Moreover – and this applies across the entire size range – hereditaments and holdings are

[10] North Yorkshire CRO Valuation Books for Grassington and Buckden (uncat.). PRO IR 58/43371–3 Field Books; MAF 68/2435 Agricultural Returns 1910, Parish Summaries.

Table 10.10 *Wharfedale, W. Yorkshire: hereditaments and holdings by unit size 1910*

		Size-group (acres)								
		1–5		6–50		51–300		over 300		
Parish		O	T	O	T	O	T	O	T	Total
Buckden	AR	2	1		13	2	20	1	6	45
	VD	1	2	1	11	2	17	1	12	47
Kettlewell/Starbotton	AR		4		13	1	16	1	6	41
	VD	4	10	11	34	3	24		5	91
Grassington	AR		14	1	17	1	12		1	46
	VD	5	23	2	34	1	11		2	78
Conistone with Kilnsey	AR		3	2	13		14		5	37
	VD	2	9	3	13		11		7	45
Linton	AR	1		2	6		7			16
	VD	2	2	2	10		6			22
Totals	AR	3	22	5	62	4	69	2	18	185
	VD	14	46	19	102	6	69	1	26	283

Notes:
O Owner-occupier; T Tenant; AR Agricultural Returns; VD Valuation Documents
Source: N.Yorks. CRO, Valuation Books; PRO, Field Books and MAF 68/2435.

not synonymous. Hereditaments were merged in accordance with normal practice by the valuers, and where this has occurred, they have been regarded here as a single unit. However hereditaments that have not been merged by the valuers may still have formed part of a single holding. In short we have little idea of the relationship between the hereditament and the holding.

Despite the uncertainties, it would appear that agriculturalists in Wharfedale were less likely to be owner-occupiers than were non-agriculturalists, though this is only true for holdings smaller than 50 acres. At a generalised level, it can be noted that the analysis of the valuation documents alone could lead to an exaggerated assessment of the existence of small-scale agriculture in this area, and probably in all areas. It is clear that assessment of agricultural holding structures based on the valuation documents alone is inadequate, although they may well provide a most important complementary and corrective source for the study of the agricultural returns. The question of owner-occupation versus tenanted holdings is brought out very clearly by both sets of data, especially in relation to hereditaments of over 50 acres – the traditional Wharfedale grassland farms.

It may well be that researchers will find it beneficial to use the two categories of material together in the future. J. T. Coppock, among his various discussions of the uses of, and problems of using, the Agricultural Returns, has said that there are no cadastral registers in Great Britain to provide a complete and readily accessible record of landholding.[11] In fact, of course, there are, if only for the short time span prior to the First World War (and also for 1941–3). The valuation documents constitute just such a cadaster for 1909–10. This fact is of vital significance for reaching a solution, for that one year, of a major problem encountered by users of the agricultural returns, and detailed by Coppock.[12]

The problem confronting users of the Returns, and discussed by Coppock, is identical to one facing users of the Valuation material, and indeed other sources dealing with historic farm records. Coppock describes the lack of congruence between the civil parish and what he calls the 'agricultural parish'. The problem arises because agricultural holdings often lay in more than one parish.

Administrative practice varied, but with holdings being returned either in the parish in which most of the land was situated or in that in which the farmstead was located. However, interpretation is further complicated by the flexibility permitted in respect of multiple holdings. What is clear is that there is frequently a discordance between the agricultural parish and the civil parish. In the case of the parish of Nuffield in Oxfordshire in 1941, this practice meant that a third of the agricultural land within the parish was included in the summaries of other parishes, whereas a third of the land returned under Nuffield lay outside that parish. Similar though less dramatic discrepancies were found in nearly all of the 400 parishes studied by Coppock.[13]

This is exactly the same problem encountered with the Valuation documents, although it is unlikely that the 'agricultural parish' will conform with the 'hereditament parish', while both are likely to differ from the civil parish. Of course, the existence of the Record Plans means that the problem of discordance between the 'hereditament' and civil parish is, potentially at least, fairly easily solved.

Coppock himself asserts that a cadaster would greatly enhance the value of the agricultural returns, especially in resolving the problems just discussed. Additional light would be shed on the returns, including owners' and tenants' names, and it would also provide a check on their accuracy. In addition, many hereditaments in some districts have field-by-field land use data entered in the Field Books; those of Isleham ITP, for example, in the Fens contain such

[11] J. T. Coppock, 'Mapping the agricultural returns: a neglected tool of historical geography', in Michael Reed (ed.), *Discovering Past Landscapes* (London 1984), 14–36.

[12] See Chapter 1. The 1941–3 National Farm Survey provides a similar cadaster.

[13] Coppock, 'Mapping the agricultural returns', 32; and see also J. T. Coppock, 'The relationship of farm and parish boundaries – a study in the use of agricultural statistics', *Geographical Studies* 2(1) (1955), 12–26; Coppock,'The parish as a geographical/statistical unit', *Tijdschrift v. Econ. en Sociale Geog.* 51 (1960), 317–26.

material for most holdings, while the Forms 37 also may contain this same information. The Field Books for Godney and Meare in the Somerset Levels similarly contain field-by-field land use data, including remarks on the fields which have become exhausted through peat digging. Analysis combined with the Record Plans would be particularly beneficial here. In many areas the often full descriptions of farm outbuildings could also be analysed in conjunction with returns for livestock etc. while details of tenure are usually available in the valuation documents, but infrequently so in the returns.[14]

In short the relevance and significance of the valuation material for the study of agriculture and land use in conjunction with the Agricultural Returns is enormous. A far more intensive survey should now be conducted than has been possible in this pilot study, and the significance of the material may be considerably enhanced by its incorporation into studies using other contemporary documents in a variety of thematic investigations.

Corsley: a hamlet parish in Wiltshire

In 1905 a student at the London School of Economics received a suggestion from Sidney and Beatrice Webb that she should continue her studies in economic history and social science by pursuing an in-depth study of her parish in which she was living. The result was the fine study of Corsley in Wiltshire by Maud Davies, *Life in an English Village,* which was completed some five years prior to the Valuation and published in 1909. The timing of the study thus provides an interesting example of how previously published material can be re-examined in order to draw on the Valuation material to support or possibly expand on the original – in this case a pioneering sociological study.[15]

The Valuation Books yielded the hereditament numbers within the larger Corsley ITP that related to Corsley civil parish. Since the Wiltshire County Record Office did not accept Form 37, the examination was confined largely to the Field Books. These were mostly completed in detail, although in the case of the larger hereditaments, the descriptions are usually missing, with references being made to the 'file', or to the 'schedule' attached to Form 37. Thus the information for Cley Hill Farm of nearly 450 acres, tenanted by Richard Oxford from Lord Bath, was in 'Note Book 759 p.17'. Descriptive material was commonly attached to Form 37, and once again the detrimental effects of non-acceptance of the latter are demonstrated. One difficulty that arose on close examination of the Field Books was that much of the field survey for the Valuation was actually carried out in 1914, making this a late

[14] For Isleham see Chapter 8, and for Godney and Meare see Chapter 9.

[15] M. F. Davies, *Life in an English Village: An Economic and Historical Survey of the Parish of Corsley in Wiltshire* (London 1909). The influence of Seebohm Rowntree's *Poverty: A Study of Town Life* is apparent, especially in the chapters on poverty in Corsley.

valuation, and leaving a gap of nine years between the two sources.[16] In an internal memorandum in 1912 it was stated of Mr W. T. Howes, District Valuer of the Salisbury District under whose responsibility Corsley lay, that: 'He does not inspire that degree of enthusiasm in his staff . . . and the District is one of the most backward in the Division of Central West.'[17] However, a large proportion of the Provisional Valuations in this area were being objected to, and this was undoubtedly slowing down the operation.

Corsley civil parish is today perhaps overshadowed by its well-known neighbour, Longleat House, on its southern boundary. The parish is on the western border of Salisbury Plain, near the Somerset border, and comprises nine separate hamlets with smaller clusters of buildings, and although Corsley itself is the location for the parish church, most of the population were living in 1905 at Chapmanslade, a village bordering on other parishes also, and along the western and southern borders of the parish (Fig. 10.3).

The Field Books included 349 hereditaments, but three of these (nos.83, 195 and 197) were labelled as 'cancelled' in the Field Books, presumably being incorporated elsewhere; another was the main road; and five were included in adjoining parishes for valuation purposes. Thus 340 hereditaments within the civil parish were actually detailed in the Field Books. In the case of smaller hereditaments the data seem full and detailed, and the comparison of information on cottage property with that supplied by Davies is easily fulfilled.

The area of these hereditaments was 2920 acres 2 roods 2 perches according to the Field Books. This compared with an acreage of 3,056 according to Davies and the 1911 Census. The few small hereditaments without stated areas presumably account for most of the difference, although discrepancies between the boundaries of the civil parish and the 'hereditament' parish should not be discounted.

Landownership was superficially fragmented between forty-seven different owners. However, all but one owned between them a mere 310 acres. The exception was Lord Bath, resident at Longleat, who owned some 2,610 acres in the parish. Table 10.11 shows the distribution of landownership. Several of these 'owners' were in fact copy, life or lease holders of Lord Bath, so that ultimately his potential influence was probably even greater than the figures suggest. Table 10.12 shows that the majority of hereditaments were owned on a freehold basis.

Owner-occupation was unusual. Apart from Lord Bath, who occupied 360 acres of his own land, and the Church, there were only twelve people occupying their own property. The owner of Corsley House, B. W. Davis, occupied some 60 acres of his own land, and A.W. Parish was an owner-occupier of

[16] Wiltshire County Record Office, Domesday Books, Register 126 Corsley and Upton Scudamore ITP (Warminster Division); PRO IR 58/73332–5, Field Books for Corsley ITP, Wiltshire. For Cley Hill Farm see PRO IR 58/73333, Hereditament no. 110.

[17] PRO IR 74/148, 361–2.

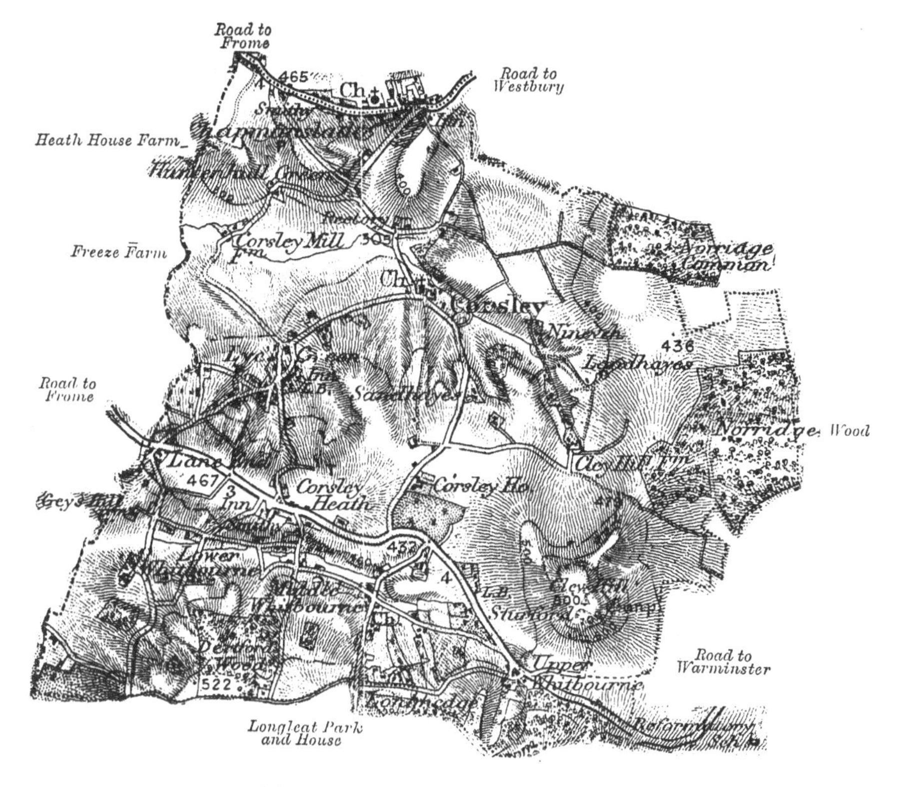

Fig. 10.3 Corsley, Wiltshire *c.* 1905.
Source: Davies, *Life in an English Village*, facing p. 3.

12.5 acres. The remaining owner-occupiers were cottagers with only a few roods of land.

The various properties were let on a variety of tenancy terms. Davies stated that: 'Some of the old cottages are still let on a lease of lives, but ordinary yearly or weekly tenancies have during the last forty years been gradually substituted for this old system as the lives fall in.'[18] But the Field Books showed only four weekly tenancies, with yearly and longer tenancies (leases for lives?) providing the great bulk (Table 10.13).

These figures may conceal extensive sub-letting on a weekly basis, but this seems unlikely as the valuers tended to indicate sub-letting where it occurred. The conclusion must be that Davies overstated the extent of weekly letting, and this is borne out by a study of the length of residence. Table 10.14 shows that many tenants had occupied their premises for a considerable time. This is known because owners, notably Lord Bath, frequently returned the date of

[18] Davies, *Life in an English Village*, 132.

Table 10.11 *Ownership structure in Corsley, Wiltshire*

Area (acres)	No. of owners
<0.25	11
0.25 to 0.9	22
1 to 4.9	6
5 to 24.9	4
>25	5
Total	48

Source: PRO, Field Books IR 58/73332–5.

Table 10.12 *Types of tenure in Corsley, Wiltshire*

Type of Tenure	Hereditaments
Freehold	264
Leasehold	26
Lifehold	10
Copyhold	7
Not stated	33
Total	340

Source: PRO, Field Books IR 58/73332–5.

Table 10.13 *Tenancy terms in Corsley, Wiltshire*

Period	No. of hereditaments
Weekly	4
Fortnightly	1
Quarterly	24
Half-yearly	11
Yearly	127
Leases over one year	51
Not stated	21
Total	339

Source: PRO, Field Books IR 58/73332–5.

Table 10.14 *Length of occupation in Corsley, Wiltshire*

Period (years)	No. of hereditaments
<1	1
1 to 4.9	66
5 to 9.9	42
10 to 24.9	84
>25	9
Total	202

Source: PRO, Field Books IR 58/73332–5.

Table 10.15 *Hereditaments in Corsley, Wiltshire (by unit size)*

Area (acres)	No. of hereditaments
<0.25	79
0.25 to 0.9	97
1 to 4.9	26
5 to 24.9	31
25 to 49.9	10
50 to 99.9	9
100 to 299.9	4
>300	2
Total	258

Source: PRO, Field Books IR 58/73332–5.

commencement of the tenancy on Form 4, and this was transcribed into the Field Books. The base date for the table is 1 October 1910, and only those hereditaments having the relevant information are included.

The overall impression is one of comparative stability of occupancy in the parish, although Davies argued that this was true only for part of the population, and that the rest were constantly moving. The table does not contradict this, since it could easily be the case that those tenants of less than five years' occupancy comprised this 'floating' population.[19]

This hamlet-dominated parish was characterised by large numbers of small and medium holdings. Although Table 10.15 gives no indication of multiple occupation of hereditaments, it is clear that amongst those hereditaments having relevant information, large farms were in a minority. Moreover, the table demonstrates that a substantial number of 'large gardens' of up to 1 acre

[19] *Ibid.*, 135.

Table 10.16 *Rooms in 'cottages' in Corsley, Wiltshire (according to the Field Books and to Davies)*

	No. of cottages		No. of cottages	
No. of rooms	(Field Book)	%	(Davies)	%
1	2	1.5	—	—
2	4	3.0	8	4.9
3	64	47.8	51	30.9
4	33	24.6	45	27.3
5	17	12.7	33	20.0
6	12	9.0	18	10.9
Not stated	2	1.5	—	—
>6	—	—	10	6.1
Total	134		165	

Source: PRO, Field Books IR 58/73332–5; Davies, *Life in an English Village* (London 1909), 133.

were attached to many houses. Some surprising details were sometimes given by the valuer, for example Ellen White's 'good pear tree' on the front wall of her house and the medlar tree on the small lawn in front of the house at Leigh (Lye) Green. The findings on holding size accord fairly well with Davies' figures. She made no attempt to analyse all land holdings, but gave details of the holdings of 'farmers'. Thus she also found only two farms larger than 300 acres, and three between 100 and 300 acres.[20]

Housing conditions in Corsley varied widely. Davies found it impossible to draw a sharp distinction between cottages and houses, and it is equally difficult to do so from the Field Book data. She calculated that there were about 165 'cottages' in the parish, and examined the amount of living accommodation in them.[21] A similar exercise follows to compare her findings with those drawn from the Field Books, for which attention is confined to those inhabited buildings that the valuer described as 'cottages', numbering 134 in total. Table 10.16 shows the results of this analysis.

Davies found that only about one-third of the cottages had three rooms. In the present smaller sample, the figure is almost 50 per cent, and there are thirteen more in absolute terms, although fewer four- and five-roomed cottages were found. The increase in the number of three-roomed cottages may reflect a genuine increase in the years after Davies' study, but may also be due to different methods of counting. Davies counted 'rooms' while the present re-examination counted 'living rooms', and the two clearly may not be identical.

[20] *Ibid.*, 110–11; PRO IR 58/73332, Hereditament no. 65. [21] *Ibid.*, 132–3.

Table 10.17 *Condition of cottage accommodation in Corsley, Wiltshire*

Condition	No. of cottages
Poor	8
Moderate	38
Fair	45
Good	1
Not stated	44
Total	136

Source: PRO, Field Books.

In addition, valuers' comments make it plain that landings were frequently used as bedrooms. Where this was the case, they have been included as a room, but changes in use of the landing with new tenants, or life-cycle changes in household composition, could easily therefore affect the number of 'rooms' identified in a house. Nevertheless, both the present analysis and Davies' survey show that housing was cramped in Corsley before the Great War.[22]

The condition of the cottages, according to the Field Books, accords well with Davies' statement that they were 'on the whole . . . neither specially good nor specially bad . . .' Table 10.17 shows that the valuers described the majority as being of 'moderate' or 'fair' condition. But the fact also that just one cottage could be classed as 'good', out of the ninety-two for which such an adjective is given, is grounds for believing that she was perhaps a little too willing to gloss over some of the deficiencies. She states at one point that 'There is a fair proportion of really good cottages, mainly owned by residents in the parish.'[23] There is little to underpin such a statement from the Field Books. There had certainly been some changes since her survey: Albert Garratt occupied a half-acre property at Leigh Green (Lye Green on the Ordnance Survey map reprinted in Fig.10.3) which had been an 'old cottage in poor repair' and which had been pulled down around 1909. Part only now (October 1914) remained, being covered by corrugated iron and used as a store. Others, such as Hereditament no.80 at Huntenhall Green, were described as 'ruinous' and 'void', or fit only to be pulled down.[24]

Cottages were stated in the Field Books to be almost all built of stone-rubble and brick, with thatch and/or tile roofs being about equally common. The block of cottages at 79–82 Lane End had been completely reroofed with tiles in early 1914, since the old thatch had been rather poor. Water supply was mentioned only infrequently, and was apparently usually obtained from wells, often shared by adjacent cottages. Only one, 25 Whitbourne Hill, had no water

[22] *Ibid.*, 133. [23] *Ibid.*, 132. [24] PRO IR 58/73332, Hereditament nos.3, 48, 80.

nearby, and James Sims, the unfortunate tenant, had to fetch it nearly a half-mile. The cottage was, however, rent-free.[25] On the other hand, a number had taps as part of the 'estate supply'.

Only five cottages were said to have had stables, one a cowpen, and thirteen were said to have had pigsties. Davies noted that 'some' cottagers kept pigs, and it is doubtful whether the stated number of pigsties gives a full picture of the extent of cottage pigkeeping. Nevertheless, it suggests that the 'cottager's pig' was far from ubiquitous in Corsley in 1910. The Field Books also give land use data on a field-by-field basis for a number of hereditaments, and this material confirms Davies' statement that agriculture was predominantly pasture-based.[26]

As well as this work on housing, which could be broadened out by complementing her Chapter XI on 'Houses and Gardens in Corsley', one could also elaborate on Chapter X, for example 'Who the Corsley people are, and how they get their living', by offering details of farms, tradespeople's premises and the conditions of their workshops etc. Thus, her anonymous 454 acre farm, the largest in the parish, was described as: 'Arable and grass. Wheat, oats, beans, sold Frome and Warminster markets. Calves reared. Sheep. Butter-making. Sometimes milk sold. 10 regular [Men and Boys employed, including sons or other relatives].'[27]

This could be none other than the 448 acre Cley Hill Farm. As noted above, the information for this farm was said to be in a separate notebook, but nevertheless it was certainly possible to identify this property, and Richard Oxford, the tenant of Lord Bath. It is also possible to add that the farm ('one of the historical houses of the parish')[28] was held on a yearly tenure and had been so held since Michaelmas 1904, for a rent of £346.

The farm had been sold previously on 25 December 1903 for £8,500, including all the timber. The hereditament was also valued with six others, one being the sporting rights of Lord Bath, and the others including four houses in the hamlet of Chips (or Landhayes), presumably tied cottages, also belonging to Lord Bath.[29] Similarly Davies' anonymous 255 acre holding can be equated with Manor Farm, built for Sir John Thynne in the early seventeenth century. We are told that the farm was: '25 acres arable, the rest grass. Dairy. Milk sold in London. Grow enough wheat to supply own straw, and roots for own supply, 7 [workers].'[30] The hereditament, actually surveyed at 255 acres 1 rood 37 perches, was held from Lord Bath by A. E. James on a yearly tenancy which dated back to Michaelmas 1900, for which he paid £307 10s. Four tied cottages were also included in the same valuation, occupied by Messrs Clements, Snelgrove, Harris and Cowley.[31]

[25] *Ibid.*, Hereditament nos. 40–4 and 13. [26] Davies, *Life in an English Village*, 108.
[27] *Ibid.*, 110. [28] *Ibid.*, 6. [29] PRO IR 58/73333. Hereditament nos. 108, 110–15.
[30] Davies, *Life in an English Village*, 110. [31] PRO IR 58/73333, Hereditament nos. 190–4.

Fig. 10.4 Mr Pearce's workforce at the Corsley Heath wagon works *c.*1905.
Source: Davies, *Life in an English Village*, facing p.126.

Tradesmen's premises have descriptions too. John Pearce was the occupier of several properties from Lord Bath, including Hereditament no. 238 on a thirty-year lease from Michaelmas 1887 at £10, which included a timber shed with a travelling saw, movable steam engine and others carpenters' tools, wheelwright's shop, painting shop, a smithy and various storerooms. Davies mentions painters, carpenters and wheelwrights, a blacksmith and a sawyer. She wrote of Mr Pearce 'wheelwright and builder, living at Corsley Heath' who employed over twenty inhabitants besides another twenty workers not resident in the parish. 'The work carried out by Mr Pearce at his Corsley workshop is mainly that of cartbuilding; he also undertakes house carpentering, painting etc.'[32] A splendid photograph of the workforce was also included (Fig. 10.4). She omitted to say that Mr Pearce had also tenanted a limekiln and quarry at Cley Hill, belonging to Lord Bath, since Michaelmas 1904. Unfortunately women's work is invisible in the Valuation survey, and whilst Davies also writes of the laundry-work, charring, gloving, nursing and midwifery as well as domestic activities, there is nothing to enhance her descriptions.

Other properties can certainly be identified in this way, but other chapters in Davies' account can also be amplified. Thus, Chapter XVI, 'Social life in

[32] PRO IR 58/73334, Hereditament no. 238. The quarry is Hereditament no. 235; Davies, *Life in an English Village*, 122–4.

Corsley', could be supplemented by descriptions of the pubs or of the Reading Room, which was let on a ninety-nine-year lease at 1s per annum from Lord Bath from Michaelmas 1891. It had been built in 1892, but was 'not much frequented' according to Davies.[33] The church and chapels were described in the Valuation Field Books also. The Baptist and Wesleyan chapels were fully described. The Wesleyan chapel at Lane End was a stone, brick and slate building in fair repair by 1914. It had a rectangular chapel room with pitch pine furniture and a small raised dais and preaching desk. There was also a small gallery, and a schoolroom at the rear. Davies described the chapel as having services normally taken by lay preachers from the neighbourhood, and occasionally attended by a travelling preacher. The Baptist chapel at Whitbourne, 'brick, tile and part stucco', had a small organ and an adjoining room for Sunday school and prayer-meetings. It was surrounded by a burial-ground and beyond that was a 'deep tank for use in the rite of baptism'. The Field Book referred to it as a 'circular baptistry in [the] graveyard'.[34]

At this remove in time, one could also now add many names to the anonymous individuals cited. The data obtained from the Field Books therefore allow some comparable examination of a near-contemporary secondary source, and in this case largely confirms that many of Davies' statements held good some five or six years later. Nearly all aspects of her enquiries as published in 1909 can be added to in some way by using the Valuation Field Books.

Conclusion

Benefits and problems must be balanced in any comparison of the Valuation documents with other sources of information. In general the gains from undertaking such work will probably outweigh the difficulties, but this is a matter of judgement for the individual researcher. Certainly the amount of detailed examination of the records which is required should not be underestimated. The great difficulty, as in so many other areas of historical and geographical research, lies in the attempted comparison of records which were compiled for differing motives, and spatial units which might also be more at variance than appears at first sight. But the reconstruction of the past through the integration of its different representations is the very stuff of our interests, and to this effect it is therefore important that the possibilities and problems associated with the valuation material can be appreciated, the more easily to facilitate its use.

[33] PRO IR 58/73335, Hereditament no. 346; Davies, *Life in an English Village*, 276.
[34] PRO IR 58/73333, Hereditament nos. 137 and 138; Davies, *Life in an English Village*, 282–3.

11

The survey in Ireland

The Lloyd George tax legislation has been discussed so far largely in the context of England and Wales. However, the taxation proposals were to apply also to Scotland and to the whole of Ireland, in which places the Westminster procedures were to be modified in various ways. It is certainly very important to remember the manner in which the Irish land wars and Gladstone's involvement with the Irish Land Act 1881, and the Irish attempts to win land reform through a route to peasant proprietorship, also very popular in Scotland, were significant contexts for the Liberal concern with the land in broad terms, and Lloyd George's interest in land reform in particular.

The fact that Henry George's highly influential *Progress and Poverty* was published in 1880, while the Irish Land War (1879–82) was very much in the public eye, gave added spirit to Liberal intentions. George had already published an article on 'The Irish land question' in 1879 which he expanded into a pamphlet, and he joined cause to some extent with Michael Davitt and the Irish Land League soon after the publication of *Progress and Poverty*. For George the solution to Ireland's difficulties was to 'sweep away all private ownership of land and convert all occupants into tenants of the State, by appropriating rent'. State control through taxation was George's aim, and in this he differed markedly from Parnell who was fighting for peasant proprietorship in Ireland. His first visit in late 1881 coincided with the height of the Land League's resistance to the British in Ireland, and his pro-Fenian dispatches for the New York *Irish World* newspaper were guaranteed to intensify opposition within England. Both the Land Nationalisation Society and the English Land Restoration League sympathised with the activities of the Irish Land League and opposed the various land purchase schemes, although their own activities remained focussed on England.[1]

[1] E. P. Lawrence, *Henry George in the British Isles* (E. Lancing, Michigan 1957), 8–39; S. Ward, 'Land reform in England 1880–1914', University of Reading, unpubl. Ph.D. thesis, 1976, 269–70.

In July 1889 Henry George addressed an audience at Toomebridge in Ireland:

That is what we propose by what we call the single tax. We propose to abolish all taxes for revenue. In place of all the taxes which are now levied, to impose one single tax, and that a tax upon the value of the land. Mark me, upon the value of the land alone – not upon the value of the improvements, not upon the value of what the exercise of labour had done to make the land valuable, that belongs to the individual; but upon the value of the land itself, irrespective of the improvements, so that an acre of land that has not been improved will pay as much tax as an acre of land which has been improved.[2]

The combination of severe agricultural distress and political agitation on agrarian and nationalist issues had formed a volatile background to George's visits to Ireland. Despite the internal tensions in the land reform movement, his views gained wide acceptance.[3]

However, to a great extent the Irish part of the land valuation proposals have been ignored in favour of the simultaneous and convulsive political events connected with Home Rule, as indeed, although to a lesser extent, have most of the concerns over land. The political connection between the opposition by the English landed elite to the valuation and the Tory ideological support for the Unionists at this time is an extremely important issue worthy of closer attention than can be given here. The Unionist cause was undoubtedly taken up strongly at this time by the Tories as a deliberate attempt to divert public attention away from the perceived success of the Liberals' Land Campaign, although this is not to deny the deep mutual and intertwined interests of landowning families on both sides of the Irish Sea, as well as the fears of the end to Imperialist ambitions spelled out by Irish autonomy. After all, Lord Lansdowne, Tory leader in the Lords, had once owned over 100,000 acres of Irish land himself, and had spoken against the 'sheer confiscation of property' inherent in Gladstone's 1881 proposals for Irish land reform.[4] State intervention through a series of Land Acts culminating in Wyndham's Land Act 1903 and Birrell's Act 1909 resulted in spectacular transfers of property into the hands of former tenants, and former large landowners such as Lansdowne, Devonshire, Fitzwilliam and Leconfield now sold up their Irish estates. Between 1903 and 1909 alone, 9 million acres were sold, and in the midst of such a land reform movement, it

[2] Lawrence, *Henry George*, 57.

[3] P. Bew, *Land and the National Question in Ireland 1858–82* (Dublin 1978), 99–232; D. Jordan, 'Merchants, "strong farmers" and Fenians: the post-famine political elite and the land war', in C. H. E. Philpin (ed.), *Nationalism and Popular Protest in Ireland* (Cambridge 1987), 320–48; K. T. Hoppeen, 'Landownership and power in nineteenth-century Ireland: the decline of an elite', in R. Gibson and M. Blinkhorn, *Landownership and Power in Modern Europe* (London 1991), 164–80.

[4] D. Spring, 'Land and politics in Edwardian England', *Agricultural History* 58 (1984), 36–7; J. E. Pomfret, *The Struggle for Land in Ireland 1800–1923* (New York 1930), 169.

is perhaps small wonder that the English 1909 Budget proposals seemed less significant.[5]

Irish land surveys before 1910

The Irish Valuation Office began in 1830 with responsibility for a general valuation of Ireland for rating purposes only down to the level of the townland. Field Books and maps created in the making of the Townland Valuation under Acts of 1826–36 thus refer to the townland and its soil and situation. The Books contain the valuers' notes on the soil and the value of the land, based on a set scale of prices for agricultural produce. Each townland was divided into numbered lots based on soil quality.[6]

Meanwhile, the Ordnance Survey had been proceeding with careful mapping of Ireland, from its newly established Dublin headquarters after 1824, with results such as the manuscript boundary sketches of 1826–41, produced under the supervision of the geologist Richard Griffith, head of the Boundary Department, who demarcated county, barony, parish and townland boundaries. The 1:1250 scale was used for this work, and its revision continued thereafter throughout Ireland. Before 1844 the Field Books had contained little information relating to individual holdings, and only houses over a certain value were included. But with the introduction of a new local rating system after the Irish Relief Act 1838 this responsibility was reduced down to the level of the individual tenement. The numbered lots based on soil quality were now usually drawn up to coincide with farm boundaries, and are so shown on the maps. Responsibility for this work was also devolved to Griffith, now appointed as Commissioner of Valuation, whose surveyors now therefore proceeded to map farm boundaries. It is of interest to note that the tenement boundaries did not always coincide with field boundaries on the published sheets.[7]

Following the Tenement Valuation Acts (1846–52) it was resolved that a Tenant Valuation of the whole of Ireland should proceed on a uniform basis for all public purposes, and this became generally known as the Primary or Griffith's Valuation (1852–64). The valuation revision lists contain the name of every occupier of land or buildings, the name of the immediate lessor, a brief description of the tenement (with industrial and public buildings being differentiated from dwelling houses), the area of the tenement, and valuation

[5] D. Cannadine, *The Decline and Fall of the British Aristocracy* (Yale 1990), 104–5.

[6] I am grateful for information kindly supplied by Mr Peter Cassells of the Valuation Office, 6 Ely Place, Dublin. See also J. H. Andrews, *History in the Ordnance Map: An Introduction for Irish Readers* (2nd edn, Kerry 1993). I am also grateful to John Andrews for his help in tracing the 1910 material in Ireland, and also to Frances McGee, Karl McGee and Aideen Ireland of the Irish National Archives, Dublin, and Paul Ferguson, Trinity College, Dublin.

[7] Andrews, *History in the Ordnance Map*, 56; G. Armitage, 'Ordnance Survey land valuation plans', *Sheetlines* 35 (January 1993), 8–9.

figures separately calculated for land and buildings. References to the appropriate sheet of the 1:10560 map and to the individual valuation number accorded to the tenement by the valuers were also given. These numbers also appeared on the Valuation Office's copy of the map as a handwritten annotation in red ink. Some partnership farms held on the old rundale system of intermixed and minute plots had individual parcels bracketed together without separate measurement, and the survey was acknowledged to be less than uniform throughout the country, but otherwise the survey represents an almost complete picture of landownership and tenure in mid-nineteenth-century Ireland.[8]

The survey was finished by 1865 but periodically the information was updated, although the books themselves were never reprinted. The Act of 1852 had provided for an annual revision of the valuation of those tenements liable to frequent alteration and a general revision at periods of greater than fourteen years after the last such general revision. The latter provisions were not enacted, partly because of the costs involved, but the local Government Act 1898 (s.65) did provide for a general revaluation of any county borough on application by the council of that borough. Belfast, Dublin and Waterford did subsequently take advantage of this provision.[9] The maps were updated to keep pace with changes in boundaries and landownership, with several editions of any given 1:10560 sheet now available, such as the 1870–83 edition with tenement details overprinted in orange (Fig. 11.1). Few were to be found outside the Valuation Office, however, since they were printed in very limited numbers. Effectively, each edition of the regular 1:10560 constituted a revision intended for land valuation purposes.[10]

Although the survey continued to use the 1:10560 scale through to the 1890s, town plans were prepared and published by the Ordnance Survey after 1847 and a survey at the 1:2500 scale also began with Dublin, its suburbs and County Dublin in 1864, as a response to government valuers' demands for an accurate map of the faster-growing suburban areas. But the 1:10560 revision project ultimately included most of Ireland by its close in 1913. The social revolution in landownership from the 1880s brought owner-occupation to so many, and it also brought with it the need for accuracy of mapping. In 1887 the 1:2500 survey of Ireland was sanctioned by the Treasury, and work on this lasted from 1888 to 1913. Some 1:2500 sheets were also produced

[8] The historical background to the Griffith Valuation is covered in G. L. Herries Davies and R. C. Mollan (eds.), *Richard Griffith 1784–1878* (Dublin 1980). See also C. G. Eve, 'Systems of land valuation in the United Kingdom', reprinted in *The Country Gentleman's Estate Book* (1914), 101; and J. H. Andrews, *A Paper Landscape: The Ordnance Survey in Nineteenth-Century Ireland* (Oxford 1975).

[9] Unpublished MSS history of the Valuation Department, Dublin, kindly supplied by Mr Peter Cassells.

[10] Andrews, *History in the Ordnance Map*, 57. These added detail to plates engraved in 1841, with lithographic transfer in the 1870s; Armitage, 'Ordnance Survey', 8–9.

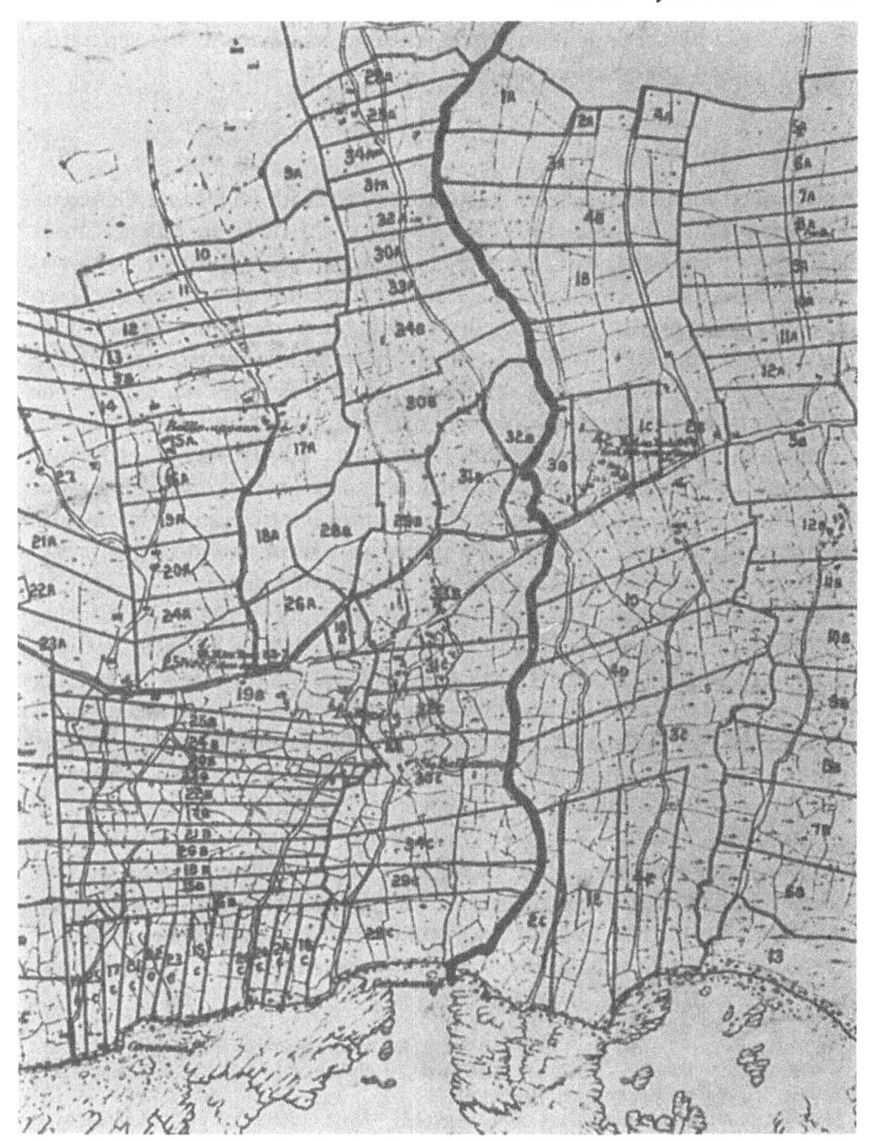

Fig. 11.1 County Galway: 1:10560 map used in the Griffith's Valuation, showing tenement boundaries.
Source: Andrews, *History in the Ordnance Map* (1993), 57.

independently by the Valuation Office for some small towns not separately mapped by the Ordnance Survey.[11]

The 1910 survey in Ireland

The extent to which Ireland was taken into consideration by Lloyd George in the framing of his original 1909 Bill is somewhat uncertain. What is more certain, however, is that Irish members in the Commons felt that Ireland was a distinct case, and many members of the Irish nationalist party felt aggrieved not only in this new attack upon landed wealth, but also in other clauses in the budget which put extra taxes on spirits, tobacco and liquor licences. Indeed, in the words of Stephen Gwynn, 'The case of Ireland is the case of the Colonies which you do not tax.' Huge areas of Ireland had been transferred into peasant proprietorship since 1881, and clearly the 1910 legislation catches the country at a time of great land-related transition. In 1909, for example, £56 million in sales agreements were pending through the Land Commission, and between 1910 and 1915 over 700 estates were dealt with under the activities of the Congested Districts Board in the West of Ireland.[12]

Following the eventual passage of the Bill, the work of the valuation relating to Ireland covered by Part I of the Finance (1909–10) Act 1910 was entrusted by a letter from the Treasury on 18 May 1910 to the Commissioner of Valuation, Richard Griffith's old post, an officer appointed by the Treasury, who acted in this instance as an agent for the Inland Revenue. In a letter of 30 May 1910 it was agreed between the Treasury and the Commissioner that for the present the valuation should be restricted to urban and suburban districts which might be more liable to Undeveloped Land Duty – the duty which levied 1d (later reduced to ½d) in the £.[13] Thus in November 1910 Lloyd George stated in answer to a question in the Commons that those lands were being valued in Ireland, as in Great Britain, that were likely to be productive of revenue, but that because there had been 'two, three or four' valuations in

[11] Andrews, *A Paper Landscape* (1975), 262–3; see also P. Ferguson for the mapping procedures between *c.*1835 and 1865, 'The maps and records of the Valuation Office', University College Dublin, unpubl. BA dissertation, 1977.

[12] Hansard, *Parliamentary Debates* (Commons) Vol.V, VI (May 1909), Col. 1123. The Irish background to Lloyd George's legislation, however, was such a marked feature of politics in the early years of the Liberal administration, with successive measures being passed to allow more and more of the land to be purchased by its former tenants, that it would be difficult to imagine the Chancellor not being cognisant of its Irish connections. For the land reform measures of Wyndham in 1903, the Congested Districts Board and the Liberal Chief Secretary Birrell in 1909, see Pomfret, *The Struggle for Land*, 276–314. For an example of the great transition from landlord domination to early twentieth-century owner-occupation in one parish, Clogheen-Burncourt in south-west Tipperary, see W. J. Smyth, 'Landholding changes, kinship networks and class transformation in rural Ireland: a case study from County Tipperary', *Irish Geography* 16 (1983), 16–35. Holding sizes *c.*1900 were determined from the 1901 Census.

[13] See Chapter 2 for a fuller description of Undeveloped Land Duty.

Ireland it was not thought financially beneficial to mount another complete survey to obtain information which was essentially already available. On the hustings at Deganwy the following month he also noted that 'In Ireland there has been a government valuation and several revisions.'[14]

Technically therefore, the Griffith's Valuation remained in force, and one result of this was that out of an estimated total number of hereditaments ultimately to be valued under the Act in Ireland, estimated by the Commissioner of Valuation at 1,750,000, just 309,000, in urban areas, were valued by 1920. Of these 294,609 had been served with provisional valuations, and just under 200,000 had been settled – a little over 10 per cent of the overall target (for more detail see Table 11.1). Of course, the nature of smallholding in Ireland would have precluded a very large proportion of the tenements from actually being taxed. And in view of the special provisions of section 4(5) of the 1910 Act, conveyances on the sale of registered land to which the Land Purchase (Ireland) Acts applied could be presented to the Registrar of Titles to receive the stamp as normal to show that the necessary formalities had been undergone.[15]

Ireland had its own centres for the presentation of instruments and issue of necessary forms for taxation purposes under the regulations of the Finance Act. Whilst the Commissioner of Valuation undertook the actual work of valuation from Ely Place, Dublin, on behalf of the Commissioners of Inland Revenue, the administration was centred on the Custom House and Four Courts, Dublin, and at the Offices of the Collectors of Customs and Excise at Belfast and Cork (Fig. 11.2). Copies of the necessary forms were periodically sent to Ely Place and to Belfast and Cork from the Storekeeper of Stamps at the Inland Revenue's offices at the Custom House, Dublin.[16] Sir John Barton, the Commissioner of Valuation at Ely Place, held office until 1917 but his railway engineering background may not have been very helpful to his new post which he came to when already in his sixties. In fact he seemed to have an uneasy relationship with his opposite number J. Simpson, the Assistant Comptroller, Stamps and Taxes, Inland Revenue, at the Custom House. During the years following the 1909 Budget formal memoranda flew between them, not least on the subject of Barton's supposed slowness in the valuation.

[14] Select Committee on Land Values, BPP 1920 (Cmd. 556), XIX, Appendix A, 'The valuation in Ireland', 20; 'Inequalities in valuation', *Land Values* (January 1911), 182. For an example of the use of the Field Books and a demonstration of their contents, relating to the First Valuation in the 1830s, see S. A. Royle, 'The socio-spatial structure of Belfast in 1837: evidence from the First Valuation', *Irish Geography* 24 (1991), 1–9. These records are held at the Public Record Office of Northern Ireland (PRONI) as VAL 1B71A–VAL 1B719A.

[15] C. E. Davies, *Land Valuation under the Finance (1909–10) Act 1910* (Estates Gazette, London 1910), 247; in 1882 A. R. Wallace, in making an impassioned plea for land nationalisation, had cited Ireland as having 400,000 holdings under 30 acres, 30,000 under 15 acres, and with 156,000 mud cabins of only one room occupied by 228,000 families – the result of 'pure landlordism' (A. R. Wallace, *Land Nationalisation* (London 1896), 35).

[16] National Archives VO FA1910 Box R6/17/2005, 2011.

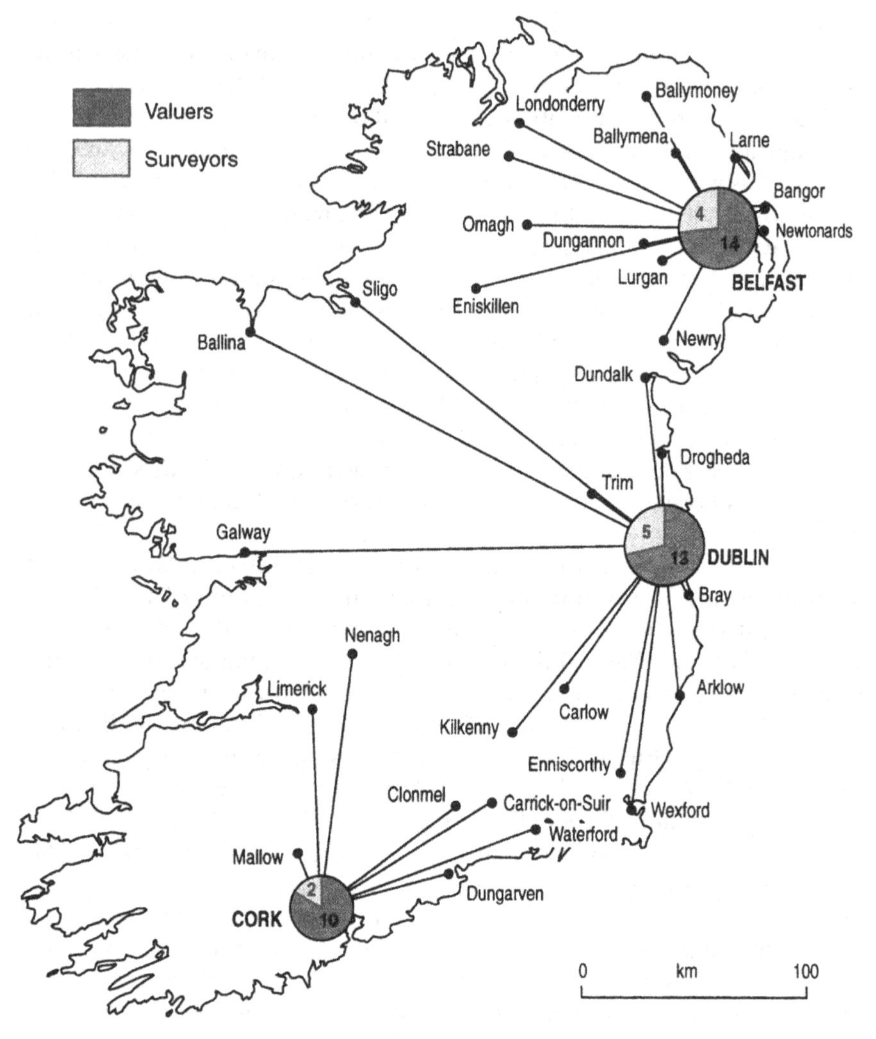

Fig. 11.2 Ireland *c.*1910: the Valuation Districts.
Source: National Archives Office Box R6 (end).

In March 1911 Simpson wrote to Barton, telling him that he (Barton) needed 'a competent man in charge of transferring all the information from the deeds' and that:

I regret the trouble which appears to be experienced by your valuers in many cases in identifying hereditaments. The fault, however, does not appear to lie with the Marking Officers of this Department. Every care is taken to see that abstracts agree with the deeds . . . If however . . . a deed contains an incorrect description, or is deficient in information the importance of which can only become apparent to and be appreciated by a valuer when the work of valuation commences, it does not appear to be within the power of the Marking Officer to obviate the trouble . . .[17]

A tardy approach to dealing with agricultural land was another subject of tension between the two men. Barton wrote to Simpson on 7 April 1911:

My duties . . . are at present confined to the Cities, Urban Districts and Towns, and the land immediately surrounding them which may be liable to Undeveloped Land Duty. The work in these areas will keep my staff fully-employed for the next 2 or 3 years at least.[18]

In a letter of 2 October 1911 Simpson wrote in connection with an enquiry about one estate:

With regard to the valuation required for IVD purposes I was certainly under the impression that it had been definitely arranged that all 'occasion' cases whether relating to rural or urban properties should be dealt with specially and in advance of the general valuation to be made under Section 26 of the Act and instructions to this effect were sent to me by the Board of Inland Revenue. If, as I gather from your letter of the 25th Ultimo, it is not intended for the present at least to take up the valuation of property in rural areas where a transfer on sale or lease has taken place or the property has passed on death I apprehend that much difficulty will be experienced in administering the Act. Moreover such a course will cause inconvenience to the responsible parties who are entitled to know within a reasonable time whether any IVD is payable and until the valuation is made the Department is not of course in a position to give this information.[19]

Barton replied fully, and noted that:

The land is mainly agricultural and is therefore, I conclude, exempt from IVD (Section 7 of Finance Act) . . . As I have already stated, my duties are at present confined to making the original valuations in the Towns and Urban Districts. Outside these areas there will be practically no Increment or Undeveloped Land Duty payable under the Finance Act. If I were to take up the Occasional cases arising in these rural districts I would have to stop most of the work in the Towns: I would also have to get the consent of the Chancellor of the Exchequer with whom I arranged that I should take the course now being followed.[20]

By far the bulk of the administration of the 1910 survey, normally referred to as the General Valuation of Ireland, was centred on Dublin, and the progress of the valuation was made the slower as a result. In their 54th Report for the Year ending March 1911, the Commissioners of Inland Revenue referred somewhat bitterly to 'the unavoidable centralisation of the work in

[17] *Ibid.*, Box R6 (end)/1004. [18] *Ibid.*, Box R6 (end)/1065. [19] *Ibid.*, Box R6/17/2278.

[20] *Ibid.* On 22 March 1912 Barton was again writing in the same terms to Simpson to inform him that he would not take the 'Occasion' cases but hold them over 'until I get some place to store them'. He would only take estate duty cases in the Rural Districts since they were nearly totally exempt from Increment Value Duty and Undeveloped Land Duty (National Archives VO FA1910 Box R6/18/4168).

the Dublin office'.[21] Inspectors from Dublin were seldom welcome anywhere in Ireland, and cultural and political difficulties certainly existed. Innumerable queries were also dealt with and staffing and office space problems caused delays. Owners faced with Form 4 asked for more time to complete it because they lived in the United States or England. The solicitors acting for the Loretto Convent at St Stephen's Green, Dublin asked for more time to sort over 800 deeds on three separate sets of properties. The Secretary of the Great Northern Railway Company (Ireland) wrote from the Amiens Street terminus in Dublin on 2 October 1911 to seek more time for completing Form 4 'as we have been exceedingly pressed with special and urgent business, including present labour troubles'.[22] Others asked for refunds to defray the expense of having their solicitors undertake the completion, whilst land agents resented the extra work being undertaken on the government's behalf. One case of forced entry by a valuer to Sandford Terrace, Dublin, had to be sorted out. Furthermore, following the Dyson case in October 1911, many wrote in to enquire whether they still had to complete Form 4. One anonymous letter of 23 October ran: 'I would refer you to the judgement in Kings Bench Division London today in reference to your Form IV and would advise you to send out no more of them.'[23] A circular letter, declaring Form 4 to be legal in Ireland, was sent out in December 1911. The offending paragraph affecting legality in England was not contained in the Irish version.

Clearly concerned at the rate of progress, in February 1911 Sir John Barton wrote to his Valuers in Belfast:

Dear Sir,
I have received the Return of your work from 6th January to 6th February which is most disappointing. In England the valuers are averaging 20 cases a day and these are complete valuations while yours are only 2/3 done.

I see no reason why an Irish valuer should not be able to do as much as an English one – your average is a little over 2 per day, and in England they have not nearly as much information as you have.

Yours faithfully,
John Barton
Commissioner of Valuation

In reply the valuers all stated that they were astounded at the rate at which the English valuers were working. Thomas McConnell, responsible for the Castlereagh, Hillsborough and Newtonards areas, replied:

Sir,
I am in receipt of your letter of the 8th Inst. which contains a very serious imputation, and in addition states that English valuers are doing 20 cases per day of Undeveloped

[21] 54th Report of Commissioners of HM Inland Revenue, Year ended 31 March 1911, Chapter VIII. [22] National Archives VO FA1910 Box R6/17/2274.
[23] National Archives VO FA1910 Box R6/17/ 2475.

land similar to that at which I am at present engaged, and further that I am only able to do about 1/7 of the work which is being done across the water.

I take the liberty of expressing a doubt if any person be he English or Irish, could possibly do 20 cases per day similar to that work I have been doing . . . would require to inspect 60 cases each on 2 days per week . . .

A slight effort of imagination will show how impossible it is to inspect 60 cases in one day . . . I would have to do 9 of such cases per hour.

I can assure you that no time is being lost in endeavouring to carry out this work, but such as it is at present I regret that the rate of progress cannot be exceeded.

I am Sir,
Your obedient servant
Thomas E. M. McConnell[24]

It is of interest to note that in September 1911 a firm of chartered accountants and rent agents from Belfast wrote directly to Lloyd George:

We would be obliged by your assisting us to fill in Form 4 by further information than is contained in instructions issued with the Form. If members of the local branch of the Valuation Office were competent to instruct us we would not trouble you, but they like ourselves seem undecided as to the procedures to be pursued in regard to filling up the Form.

There then followed details of the properties in question. The letter was sent internally to Dublin and six days later a letter came to Barton from the Accountants stating that their difficulties had now been 'thoroughly explained' by the Belfast Office.[25]

In April 1917 the Acting Commissioner of Valuation at Ely Place wrote to Somerset House to explain the slow progress in Ireland. In part he frankly blamed inadequacies in his staff, but by 1917 the enlisting of men had also taken its toll on progress whilst work in Ireland was spatially very scattered: across the country the work was done by five 'travelling valuers' who were based at Dublin, Belfast and Cork. Inexperience in dealing with rural valuations was also manifest. Mr Ward, chief valuer at Belfast, was uncertain in September 1918 (very late in the day, one might think) as to whether peat bog workings were to be classed as agricultural, and thereby excluded from the taxation. He was curtly informed from Dublin that they were not agricultural and therefore not exempt. The travelling valuers were therefore required to report on the valuation of peat bogs since 'information on the subject is much needed', arising out of queries relating to such land on the Verner Estate in Armagh, covering several townlands with large areas of raised cutover peat

[24] *Ibid.*, Box R6 (end)/942.

[25] *Ibid.*, Box R6/17/2221. Doubtless the tension produced between Dublin and Belfast would have been just one more issue which threatened the relations between the two cities: 'The city [Belfast] threatened on at least half-a-dozen issues between 1911 and 1914 to go its own way' (S. Gribbon, 'An Irish city: Belfast 1911', in D. Harkness and M. O'Dowd (eds.), *The Town in Ireland* (Historical Studies 13, Dublin 1981), 206).

bog south of Lough Neagh in Lurgan and Armagh as well as Tyrone. Sales of peat from here were taking place since 1907 but had gone unreported. The particulars were collected by November 1918, and valuers were initiated into the different qualities of peat affecting its price per acre and its sale price in Portadown.[26]

Some complications arose because of the different situation in Ireland into which the legislation was translated. The Irish diaspora certainly meant that some landowners were now living abroad, and even if traceable, letters could certainly go astray. Difficulties arose when provisional valuations were being dealt with for property that was in process of being sold through the Land Courts under Section 1 of the Irish Land Act 1903. Doubts as to who the legal owner actually was at this time resulted in the provisional valuation being served on one Land Judge, Mr Justice Ross, who promptly repudiated ownership. From the High Court of Justice came a letter to the Chief Clerk at the Valuation Office cautioning him against any repetition of his action (14 January 1913).[27] Neither would the Registrar of the Court agree to accept the valuation, and ultimately Simpson's suggestion that it should rest with the Receiver-Examiner was accepted. Unfortunately the error was repeated five years later after the war, when a Belfast office mistake sent the valuation to the Judge once more. Simpson informed Barton of the error, and Barton, on the edge of retirement, had to apologise.

The Irish situation certainly caused problems in personnel terms. By 1920, at the very end of the valuation procedures, T. H. Burns MP wrote to Stanley Baldwin at the Treasury to allege that of the six men in the Belfast valuation office, five of whom were ex-soldiers, only two got increased wages and both were Catholics. Baldwin investigated and replied that in fact one man was a Catholic but the other was a Presbyterian, and both were given rises solely on merit.[28] By September 1920 letters were being sent to temporary clerks terminating their employment on cessation of the taxation. Also at this time there was a serious security matter to be dealt with. In a confidential letter of 17 February 1920 from the Chief Secretary's office at Dublin Castle, John Joseph Moran, who worked in the Valuation Office at Dublin, was noted as being the secretary of the Irish Transport and General Workers Union (ITGWU) committee acting for the new Cooperative stores at Swords, Dublin, where he lived. The committee was said to be 'composed solely of Sinn Feiners', and his house was supposedly used for Sinn Fein meetings, where his brothers, also Sinn Feiners, also lived. Carroll, successor to Barton, replied that he had seen Moran and told him to cease contact with the Coop, and further that he had pointed out the consequences of 'his constant association with professed republicans, and how inconsistent it is with his Oath of

[26] National Archives VO FA1910 Box R6/24/14160; 15340.
[27] *Ibid.*, Box R6/17/18187; R6/18/4211. [28] *Ibid.*, Box R6/17/14958.

Table 11.1 *Progress of the valuation in Ireland by June 1919*

1 Total no. of Identification numbers in County Boroughs, Urban Districts and Towns		308,974
2 Total no. of other Identification numbers in Ireland		1,460,451
3 Total no. of Provisional Valuations not issued in no.1 above		14,365
Made up of (a) Agricultural land	9,030	
(b) Statutory companies	3,459	
(c) Miscellaneous	1,876	
4 Total number of Provisional Valuations served		294,609
5 Total no. of objections to no.4 above		15,200
6 No. of Provisional Valuations served since 3 June 1914 (still open to objection)		92,796

Source: National Archives VO FA1910 Box R6/24.

Alliegance'. Moran told Carroll that everyone he knew in Swords was a Sinn Feiner, but, Carroll disingenuously wrote, 'he has given me his word of honour that in future he will not associate with Sinn Feiners beyond bidding them the time of day, and that he will not frequent houses of an objectionable character'. This was not the end of the story however, for on 5 December 1920 Moran was arrested by Military Police and imprisoned in Collinstown camp. He sent a letter to Carroll explaining his absence from the office, together with the keys to his desk and some correspondence he was working on. He was released without charge on 23 December and returned to work on 29 December, bringing more wrath down on his head for an absence which he later explained as having been due to a cold, but for which he received no pay.[29]

With such difficulties as these to contend with, the numbers of Irish transactions consequently lagged behind those of Scotland until 1917, and before the onset of war in 1914 IVD in Ireland yielded just £765, compared with over £8,000 in Scotland, for example (Tables 11.2 and 11.3). Undeveloped Land Duty, the declared main initial thrust of the taxation in Ireland, yielded nothing until 1913, and only came close to matching the income from Scotland in 1913 and 1914 (Table 11.4).[30] Simpson, probing as ever, wrote in June 1919 for statistics on progress in Ireland, and statistics were provided in a letter of 12 June 1919 (Table 11.1). The policy of restricting valuation to Cities, Towns and Urban Districts (126 'urban' areas) was clearly still in place and had been confirmed by a Treasury letter of 10 February 1915. The 1.4 million properties not valued were mainly agricultural land. Of the

[29] *Ibid.*, Box R6/17/15272.

[30] 64th Report of Commissioners for HM Inland Revenue, Year ended 31 March 1921 (BPP 1921 (Cmd. 1436), XIV, 156).

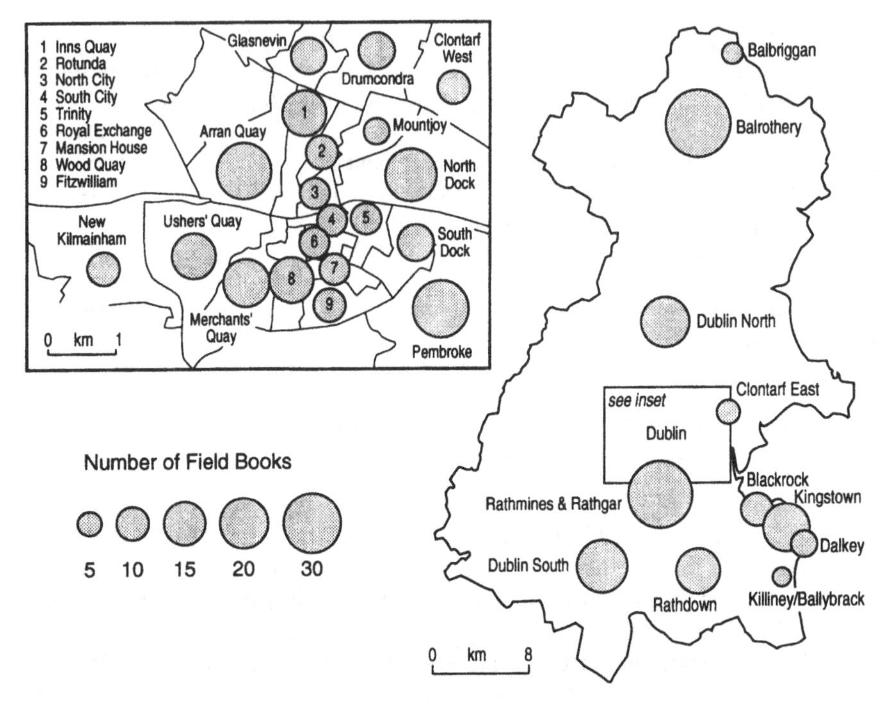

Fig. 11.3 County Dublin: distribution of Field Books by wards and districts. *Source:* National Archives, Dublin, Valuation Office Finance Act Boxes 16–52.

provisional valuations served in urban areas, there were still over 15,000 to which objections had been raised, mostly in Dublin and the northern cities. Even in the more rural south and west, the Cork office had concentrated very largely on urban areas, and by 1921 the office here was left with 391 Field Books, of which just thirty-four touched on rural districts.[31]

The 1910 documents in Ireland

The numbers of Field Books completed as a result of the survey in Ireland is consequential upon the decision to concentrate on the urban and suburban districts first. Fig. 11.3 shows the numbers of Books for the different areas within the City and County of Dublin and Fig. 11.4 indicates the geographical spread of the Field Books by county. The primacy afforded to the large towns is certainly indicated by the larger numbers of Field Books available for Dublin and Cork compared with the largely rural western counties. Thus Dublin has 448 books and Cork 148, compared with Roscommon's nine or

[31] National Archives VO FA1910 Box R6/24.

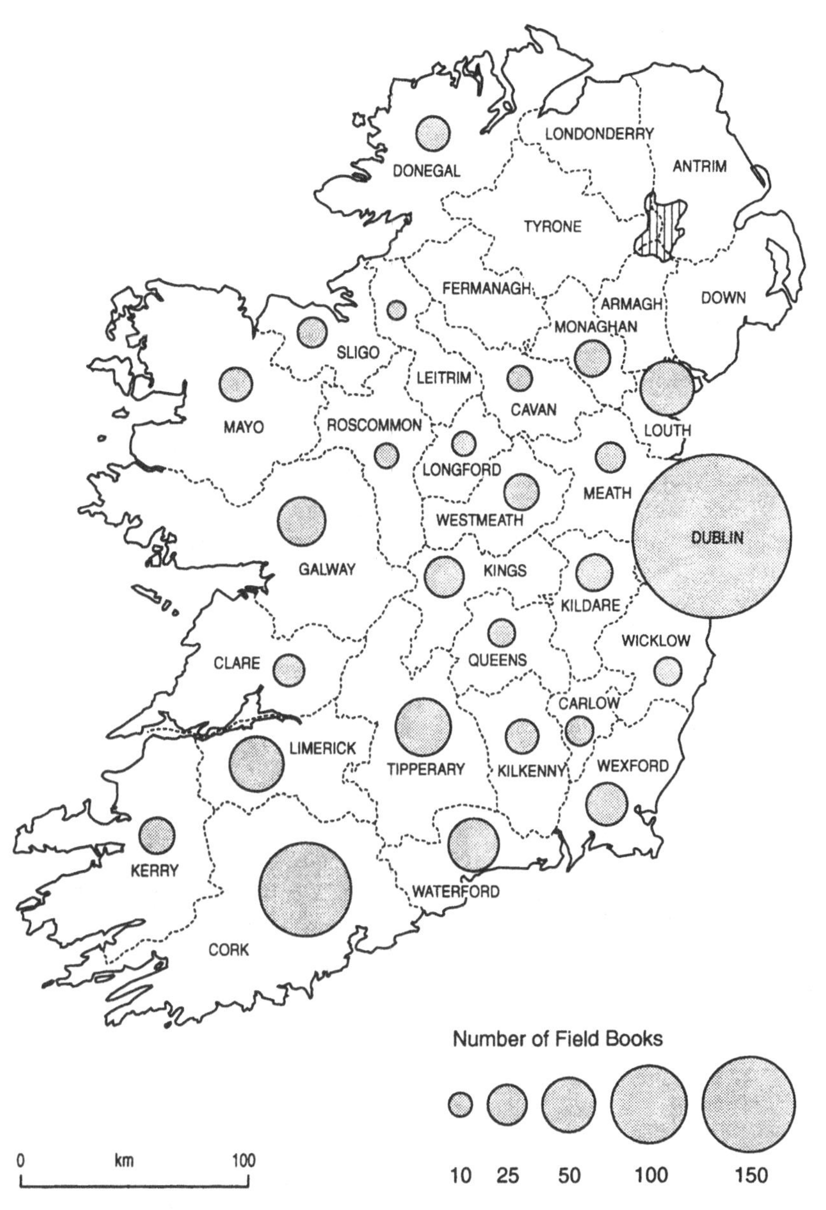

Fig. 11.4 The distribution of Field Books relating to the Irish Republic (by county). *Source:* National Archives, Dublin.

Leitrim's five. The distribution of Field Books for Dublin itself reflects in part the processes of active development and urban expansion in the later nineteenth century. The inner city areas, characterised by the traditional high density tenement dwellings – about 5,000 by 1914 giving Dublin the position of the most congested city in the British Isles – had relatively high numbers of Field Books in consequence at such wards as Arran Quay, Inns Quay, Ushers Quay and Wood Quay. In contrast were the larger properties at the more wealthy southern suburbs such as Rathmines and Rathgar. Kingstown (Dun Laoghaire) was also developing inland around its older port nucleus. The superior Pembroke estate, built up in the 1870s near Ballsbridge, where the Royal Dublin Society grounds hosted the annual Dublin Horse Show, as a 'new town' of Dublin, has a large number of Field Books which would repay study for the insight they offer into upper-middle-class housing standards. The estate also contained properties in what the Agent, Fane Vernon, referred to as 'the slummy portion of the city' as well as more salubrious areas to the south.[32] Both Rathmines and Pembroke fiercely fought off attempts by Dublin to extend its boundaries to include them in the early twentieth century.

Because of the prior work of the Valuation Department in Ireland, much of the preliminary work of valuation was able to be effected before the owner was called upon to complete a Form 4. Identification of hereditaments was made, at least in part, through the maps kept in the Land Registry Office, and the valuers were requested to consult relevant maps there. The issue of Form 4 was made at a later stage in the process than in Great Britain, to allow the owner an opportunity to claim deductions or give an estimate of value. By 31 March 1911, for example, just 6,000 forms had been issued, and 190,030 by 31 March 1912, compared with the great and vilified August 1910 distribution in Great Britain.[33] During the summer months of 1911 temporary valuers, clerks and 'computers' were being recruited. Many were disappointed. G. L. Banford, a valuer for the past fifteen years and an Assistant Land Commissioner 1899–1907, was recommended by Sir Horace Plunkett, and the Master of the Rolls and others. From Kells, Co. Meath he sent in a bound volume of printed testimonials, only to be told that 'when the time comes for dealing with the Rural Districts your application will be duly considered'.

[32] In September 1911 Fane Vernon, agent for the Pembroke Estate, wrote to Sir John Barton ('My Dear Barton') asking for 200 blank copies of Form 4 for his own use, as well as asking for a form on which to object to the provisional valuations received. He assumed that the sixty days allowed for objections 'would not be rigidly adhered to by your department', but was informed by Barton that there would have to be very special circumstances which would be referred to the Commissioners in London (National Archives VO FA1910 R6/17/2158, 2164). Until 1903 Lord Pembroke's agent had been automatically a Pembroke councillor, and Vernon would have been a well-known personage in Dublin (M. Daly, 'Late nineteenth and early twentieth century Dublin', in Harkness and O'Dowd, *The Town in Ireland*, 232).

[33] 'The progress of the valuation', *Land Values* (November 1912), 270; 54th Report of Commissioners of HM Inland Revenue, Year ended 31 March 1911, BPP 1911 (Cd. 5833), XXIX, 167; 55th Report, Year ended 31 March 1912, BPP 1912–13 (Cd. 6344), XXIX, 155.

Appointments were only sought from men who knew the Dublin, Cork and Belfast areas.[34]

The records themselves largely replicate the pattern for Great Britain. The Field Books are however set out in a different format (Fig. 11.5a). The different questions being asked are numbered, there is a standard entry for the dimensions of the buildings (frontage, depth, height, 'cube contents'), with space for up to ten buildings per hereditament, or tenement as referred to in this case. The plan is positioned in the middle of p.1 of the Book and there are only two pages per hereditament. Frequently the plans were separately drawn on tracing paper and pinned into the relevant section of the Field Book. There is provision for a description of the hereditament, as in the English case ('Remarks and Valuation Details'), but there is also simply a small space left (Q.8) for 'Condition of Buildings'. Among the new questions posed to Irish owners are one on the separate values attributed to machinery, timber and shrubs (Q.17), a separate question asking for information on capital expenditure separately on land and on buildings since the purchase by the owner (Q.12) and an obvious question about the address, left off the English form.

The valuation was designed, despite Lloyd George's November 1910 remarks, to ascertain the four standard values as in England. The Field Book's second page is therefore devoted to these values, beginning with the Gross Value. It is interesting that the Gross Value is defined in the Irish Field Books as the 'Market value of fee simple free from incumbrances, burdens, charges or restrictions other than rates or taxes'. There is no comparable definition in the English Field Books, although the essential meaning was the same. The Full Site Value, Total Value and Assessable Site Value were then calculated in the usual way.[35]

Each Field Book contains three pages of instructions at the front, explaining the procedures for completing the Book. The lack of space compared with the English versions seems to have generated a less tidy completion by the valuers, and the space reserved for a plan of the hereditament is frequently taken up with calculations. The space for an index at the back of each Field Book was by no means always used, although those for North City Ward in Dublin, for example, are indexed by street. Many of the latter Field Books contain the phrase – 'For details see Valn Note Book No. [blank]'. These were the office notebooks containing the full information from the previous surveys

[34] National Archives VO FA1910 Box R6/12/1–62 and R6 (end)/1074, 1078. The documents consist of valuable detailed applications, biographical details and supporting references from aspirant temporary valuers; results of examinations and final decisions are also recorded.

[35] See Chapter 4 for a full description of the values, and the debates surrounding their calculation and imposition. In the case of appeals, the Irish Reference Committee consisted of the Lord Chief Justice of Ireland, the Master of the Rolls in Ireland, and the President of the Surveyors' Institution. The latter might, however, appoint a member of his Institute to the Committee who had special knowledge of the Irish situation (Finance (1909–10) Act 1910, S.33 (5)).

mentioned above. Many, as with the Roscommon Field Books, were stamped to indicate that Form 37 had been filled in, or that an agreement had been reached – many of them in 1915 – a far later date than most English Books and possibly indicative of a greater degree of continuity through the first part of the War. Thus the Union Workhouse Fever Hospital and its land at Boyle was valued in September 1915.[36]

Other documents connected with the valuation in Ireland were adaptations of the basic models as already seen in England and Wales. Forms 36 and 37–Land, for example, provide all the same information, except for minor changes in wording to allow for the different land legislation and fixed charges arising in Ireland. Crown Rent is included, but not the deductions associated with the redemption of Land Tax or the cost of enfranchisement of copyholds.

Sales and presentation of Irish Ordnance maps reached new heights just prior to the passing of the 1910 legislation. Most sales were accounted for by the 1:10560 series, peaking in 1908, and the 1:63360 series which rose considerably in the first six or seven years of the century, presumably as a response to the passing of the Wyndham and Birrell Acts. But substantially increased sales of the 1:2500 peaked in 1909–10 with the need for accurate urban and suburban valuations following the Finance Act 1910 which effected the new legislation, and the town plans also show a small rise for the same reason in 1910.[37] An example of the 1:1050 series is shown in Fig. 11.5b, although this is a 1915 map (and in poor condition, hence the crack marks shown). An example of the same scale in use in 1911 for the Finance Act valuations is given for the town of Elphin, Co. Roscommon, from a Working Sheet copy from the Valuation Office (Fig. 11.6).

In 1922 a separate Dublin government survey office marked the creation of the Irish Free State following the 1921 Anglo-Irish Treaty, and the 1920 repeal of the Finance (1909–10) 1910 Act with its obligation to a valuation survey. By June 1921 memoranda were being circulated in Ireland detailing all forms which were to be destroyed, including copies of Form 4, for example, and those to be retained.[38] The numbers of cases dealt with for each of the

[36] National Archives VO FA1910 Box 78 Roscommon. There were nine books covering Roscommon, three dealing with Roscommon itself, four with Boyle and two with rural districts near the towns. The Fever Hospital is Boyle Hereditament no. 414. The war did however interrupt valuation proceedings, and many young men left their civil duties for the war. By 1919 some were asking for instructions as to resuming their posts (National Archives VO FA1910 Box R6 24/14744).

[37] Andrews, *A Paper Landscape*, Fig. 12 p. 271. Andrews makes no mention of the increases in the 1:2500 series and the Town Plans being related to the Lloyd George legislation.

[38] National Archives VO FA1910 Box R6/17/15391. This bundle of papers also includes specimen copies of all Irish valuation forms then in use (both 'Land Values' and 'VO' forms being detailed separately). Some forms were to be returned to be adapted to new and separate uses within Northern and Southern Ireland. The Belfast Office now transferred Field Books, maps, Forms 4 and Increment Value documents for Donegal and Monaghan to Dublin (National Archives VO FA1910 Box R6/24/15497). The Cork office also sent a memorandum to indicate the amount of material left over from the valuation procedures in June 1921, which included amongst a mass of Field Books and maps, approximately 55,323 provisional valuation files!

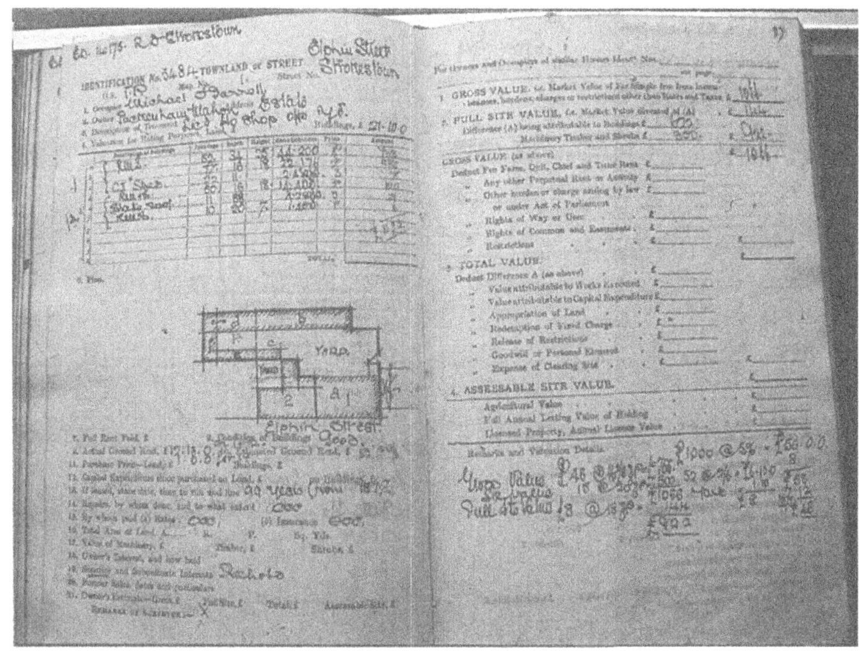
IDENTIFICATION No. 34 8/4 TOWNLAND or STREET Elphin Street Strokestown

1. Occupier Michael O'Farrell
2. Owner Packenham Mahon Estate

Elphin Street

1. GROSS VALUE
2. FULL SITE VALUE
3. TOTAL VALUE
4. ASSESSABLE SITE VALUE

Remarks and Valuation Details

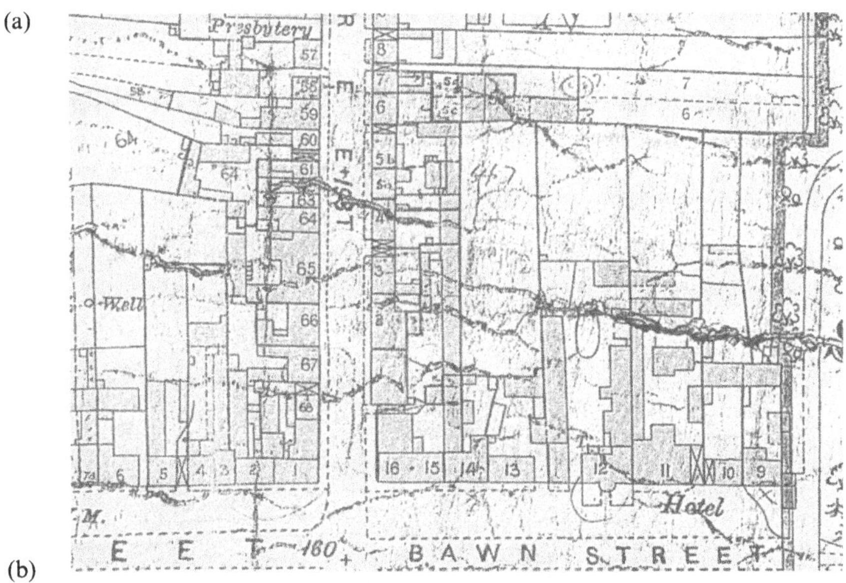

(a)

(b)

Fig. 11.5 (a) The Field Book entry for Michael O'Farrell's hereditament (Packenham-Mahon estate), Elphin Street, Strokestown, Co. Roscommon. (b) The 1:1050 series Ordnance Survey (Phoenix Park, Dublin) sheet showing Michael O'Farrell's hereditament at No.1 Elphin Street, Strokestown, Co. Roscommon. *Source:* National Archives VO FA1910 Box 78, Roscommon 401–600; Valuation Office, Ely Place, Dublin.

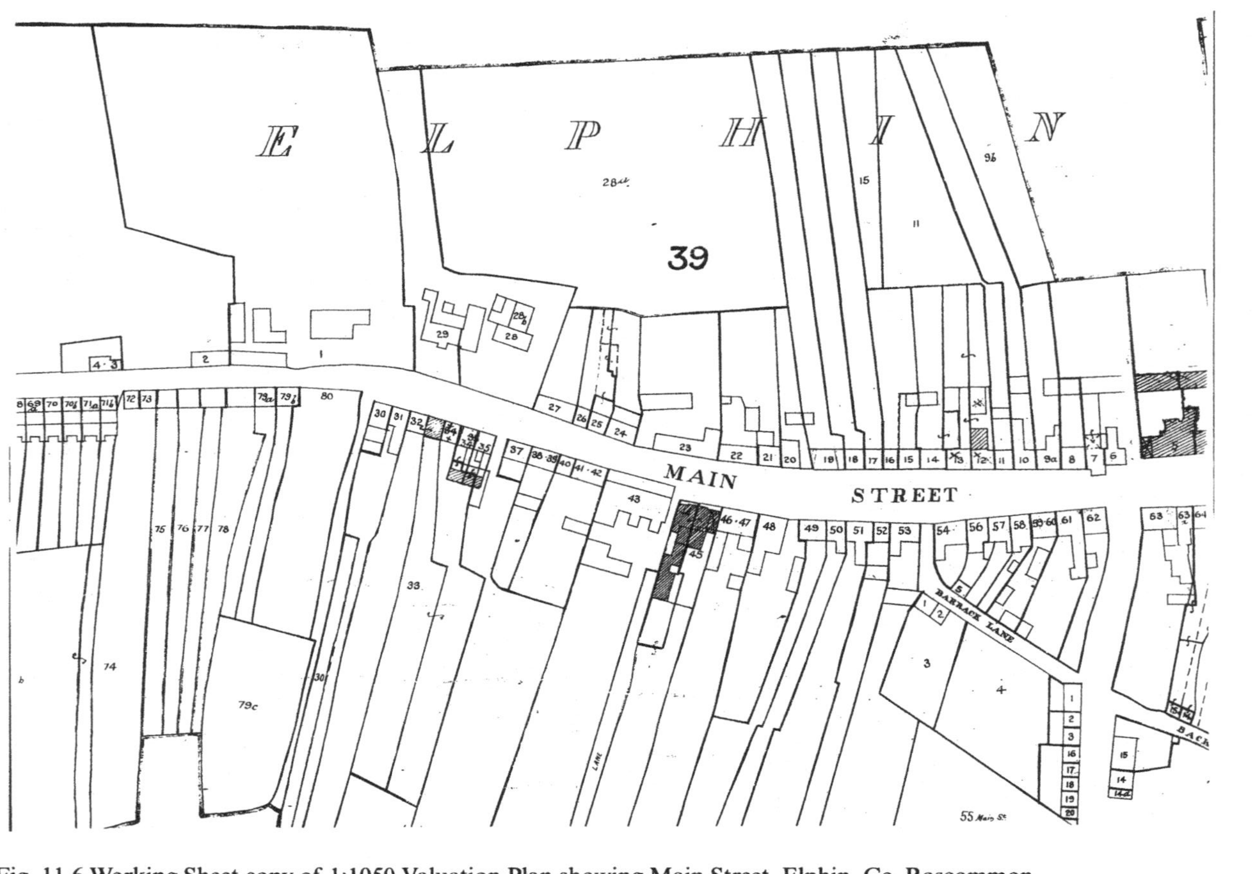

Fig. 11.6 Working Sheet copy of 1:1050 Valuation Plan showing Main Street, Elphin, Co. Roscommon.
Source: Valuation Office, Dublin.

Table 11.2 *Increment Value Duty in the United Kingdom: number of transactions*

Year ending 31 March	England	Scotland	Ireland	United Kingdom
1911	123,799	7,192	4,486	135,477
1912	172,540	7,893	6,138	186,571
1913	177,572	7,566	6,720	191,858
1914	188,290	8,083	6,302	202,675
1915	151,807	7,330	5,589	164,726
1916	121,326	6,052	5,772	133,150
1917	108,729	4,126	5,900	118,755
1918	132,227	6,952	7,567	146,746
1919	207,881	7,495	9,020	224,396
1920	491,332	17,370	14,187	522,889
1921	540,094	14,859	15,377	570,330

Notes:
Transactions: particulars being presented under Section 4(2) of the 1910 Act, or received from Registrars of Lands or Sasines under Section 4 (5).
Source: 64th Report of Commissioners for HM Inland Revenue, Year ended 31 March 1921 (BPP 1921 (Cmd. 1436), XIV, 156).

different duties in England, Scotland and Ireland during the life of the legislation are detailed in Tables 11.2 to 11.5.

The administration of the records in Ireland

The Irish Public Records Office accessioned records from the Valuation Office on four separate occasions: in 1947, 1949, 1977 and 1979. These accessions included the records of the Boundary Surveys 1825–1923; copies of books used in making the Townland Valuation 1830–46; records connected with the 1910 valuation; and Valuation Office correspondence etc. through to the 1940s. These records are available for consultation at the National Archives in Bishop Street, Dublin. The Field Books are bound in parish volumes and arranged in baronies and counties but there is no specific catalogue or finding aid available at present. The maps are still retained at the Valuation Office in Ely Place.

The situation in Northern Ireland is more uncertain. There is no knowledge at the Public Record Office Northern Ireland (PRONI) of any material specifically relating to the Lloyd George Finance (1909–10) Act 1910. Valuation material is available here by way of valuation books and maps for 1830; manuscript or printed and bound editions with maps for the Griffith's Valuation

Table 11.3 *Increment Value Duty in the United Kingdom: number of assessments made and net receipt (£) of duty (excluding duty from minerals)*

Year ending 31 March	England		Scotland		Ireland		United Kingdom	
	No.	Rec't	No.	Rec't	No.	Rec't	No.	Rec't
1911	10	102	1	25	—	—	11	127
1912	777	5,698	63	392	19	34	859	6,124
1913	813	8,356	148	7,729	86	158	1,047	16,243
1914	2,871	23,941	209	7,711	142	573	3,222	32,225
1915	4,849	36,209	196	5,440	182	1,102	5,227	42,751
1916	2,377	34,843	102	3,223	168	2,432	2,647	40,498
1917	3,780	54,946	190	3,929	55	985	4,025	59,860
1918	6,655	84,274	206	4,891	61	3,061	6,922	92,226
1919	8,769	153,163	220	13,871	106	3,324	9,095	170,358
1920	8,437	120,076	85	6,461	158	2,273	8,680	128,810
1921	183	−359,131	—	−35,505	—	−8,656	183	−403,292

Notes:
Negative sums in 1921 due to repayment of IVD following repeal of the legislation.
Source: 64th Report of Commissioners for HM Inland Revenue, Year ended 31 March 1921 (BPP 1921 (Cmd. 1436), XIV, 157).

1848–64; and the volumes and maps constituting the annual revaluations thereafter until the First General Revision 1935.[39] The years following the 1909–10 legislation are not singled out for any special treatment. Documents relating to these years are thus differently formulated from those 1910 documents of other parts of the United Kingdom, and relate to the ongoing annual revaluation. There is, for example, no attempt to calculate any of the four values associated with the Act. Neither is there any evidence that the activities of the Belfast valuation office between 1910 and 1920 have been preserved in Dublin, either at the National Archives or at the Valuation Office. This is not, of course, to deny the value to the researcher of the extant valuation documentation which actually does offer information which is otherwise unavailable for England and Wales (Table 11.6).

About two weeks after the passing of the Irish Free State (Agreement) Act on 14 April 1921 the Anti-treaty force under Rory O'Connor seized the Four Courts in Dublin. In the attack by the Provisional government forces, the Four

[39] Public Record Office Northern Ireland (PRONI): 1830 Valuation VAL 1B and maps VAL 1A; Griffith's Valuation VAL 2B and maps VAL 2A; Annual Revision List volumes VAL 12B; First General Revision 1935 VAL 3B and maps VAL 3A. I am grateful for the help of Michele Neill (PRONI) for assistance with this information.

Table 11.4 *Reversion Duty: number of assessments made and net receipt (£) of duty*

Year ending 31 March	England		Scotland		Ireland		United Kingdom	
	No.	Rec't	No.	Rec't	No.	Rec't	No.	Rec't
1911	47	249	1	8	—	—	48	257
1912	419	22,482	12	16	12	123	443	22,621
1913	1,318	44,349	89	594	7	3,031	1,414	47,974
1914	1,646	79,386	30	401	11	648	1,687	80,435
1915	186	19,099	5	21	8	193	199	19,313
1916	301	11,546	1	13	30	237	332	11,796
1917	434	17,779	1	3	4	227	439	18,009
1918	611	22,064	1	9	16	1,091	628	23,164
1919	562	38,752	—	—	3	469	565	39,221
1920	163	15,480	—	—	11	257	174	15,737
1921	1	−228,748	—	−552	—	−4,547	1	−233,847

Notes:
Reversion Duty payable on the determination of any lease after 29 April 1910.
Negative sums in 1921 due to repayment of Reversion Duty following repeal of the legislation.
Source: 64th Report of Commissioners for HM Inland Revenue, Year ended 31 March 1921 (BPP 1921 (Cmd. 1436), XIV, 158).

Courts, including the treasury of the Public Record Office, was destroyed, along with large amounts of Irish official records covering the period of the Union down to the late 1830s.[40] To compensate for this, one extremely important and significant concession has been made, and one which renders the 1910 material in Ireland extremely interesting – the 1911 Census enumerators' schedules covering both Northern Ireland and the Republic (together with those for 1901) have been released in their entirety for use by the public. The matching of the 1910 material and the 1911 Census now gives a platform for local studies in Ireland which cannot be matched anywhere in the British Isles. The amount of census information available for 1911 is anyway quite extraordinary by English census standards, and Table 11.7 gives details of what may be found therein. All households were required to complete Form A and, where applicable, other forms which dealt with various topics, among them house building and quality (Form B1), out-offices and farmsteading (B2), sick

[40] H. Wood, 'The tragedy of the Irish public records', *Irish Genealogist* 1 (1937), 67–71. The Census enumerators' books of 1813 and 1821–51 were largely destroyed in 1922 while the returns for 1861–91 were never preserved (S. A. Royle, 'Irish manuscript census records: a neglected source of information', *Irish Geography* 11 (1978), 110–25).

Table 11.5 *Undeveloped Land Duty: net receipts (£)*

Year ending 31 March	England	Scotland	Ireland	United Kingdom
1911	2,351	—	—	2,351
1912	28,130	817	—	28,947
1913	94,378	1,887	1,587	97,852
1914	255,435	9,649	9,832	274,916+
1915	3,090	5,540	22	8,652$
1916	−795*	168	−11*	−638*
1917	−187	—	−9*	−196*
1918	76	1	−3*	74
1919	59	—	3	62
1920	52	—	—	52
1921	−138,757*	−10,169*	−2,744*	−151,670*

Notes:
Negative sums due to repayments of Reversion Duty in excess of receipts are highlighted as minus figures and with an asterisk.
+This figure largely represents arrears from the preceding years.
$ Judicial decisions from 28 February 1914 entailed the suspension of assessment and collection.
Source: 64th Report of Commissioners for HM Inland Revenue, Year ended 31 March 1921 (BPP 1921 (Cmd. 1436), XIV, 158); Select Committee on Land Values, BPP 1920 (Cmd. 556), XIX, 41, Evidence of Mr C. J. Howell Thomas, Appendix G.

persons (C), lunatics and idiots not in institutions (D), paupers in workhouses (E), details of agricultural holdings (M1, with abstracted details on M2), and amalgamation of agricultural holdings (M3). Not all these were preserved, however, and of those given as examples above, only B1, B2 and E are extant, and for a typical urban street only B1 is normally available. But the information on this is directly comparable with the 1910 material (Table 11.7).

The existence of such germane material to a study of the 1910 data in Ireland highlights the need for a synthesis of Census Enumerators' Sheets, Field Books and Record Sheet Plans. Returning to the example given of Michael O'Farrell's property in Fig. 11.5 above, the Census schedules inform us that on the night of 2 April 1911 O'Farrell, a merchant aged 45 and a bachelor, was living with four relatives, a servant and a shop assistant. The property was returned as a 'first class' house, having eight windows at the front and eleven rooms. There were also out-offices and farmsteadings such as stables, a coach house, piggery, turf house, shed and three stores.[41]

Unfortunately the retention of the Record Sheet Plans by the Valuation

[41] National Archives Census 1911 Roscommon 111/10.

Table 11.6 *Information in the annual revaluation volumes for Northern Ireland*

A. URBAN AREAS

(e.g. North Belfast Working Men's Club, Danube Street, Shankill Ward, Belfast, PRONI VAL 12a/7N/12)

Page 1.
Street, ward, OS reference number, Street no., Map no.

1. Occupier
2. Immediate Lessor
3. Description of Tenement
4. Approximate Age
5. Frontage
6. Description of Walls and Roof
7. No. of Storeys, excluding Basement
8. No. of Sitting-rooms and Bed-rooms, excluding Kitchen
9. If Offices, Warerooms etc – No. of rooms in each letting
10. Whether fitted with Gas and Bath-room
11. Full Rent paid
12. Ditto if let in tenements
13. Actual or Estimated Ground Rent or Head Rent
14. Actual or Estimated Cost of Construction
15. Reputed Price if Purchased
16. Expenditure by Occupier on purchased or leased premises
17. If Lease, state date, time to run, and fine, if any
18. Repairs – by whom done, and to what extent
19. Insurance – by whom paid
20. Rates paid by Landlord, if any
21. Area of land, if one rood and upwards
22. General Remarks
23. Valuation and how arrived at

Page 2.
Annotated sketch plan of premises

B. RURAL DISTRICTS

(e.g. Bleach works, Dunadry, Antrim Rural District, PRONI VAL 12A/1/15)

Page 1.
Electoral Division, Ward, Town, Street, Townland, reference to map
(Questions are not numbered)

1. Occupier
2. Immediate Lessor
3. Description of Property
4. Actual (or estimated) Rent
5. Ground Rent
6. Cost of construction
7. Reputed Price if purchased

Table 11.6 (*cont.*)

B. RURAL DISTRICTS
8. How held
9. If Lease, length of term and fine, if any
10. Expenditure by Occupier
11. Repairs and Improvements, by whom done
12. Taxes, by whom paid
Page 2.
Sketch plan of buildings (index letters for each block)

Table 11.7 *The 1911 Census returns for Ireland: contents*

Each household member was required to furnish details on FORM A as to:

1. Names
2. Relation to household head
3. Religion
4. Education (mainly regarding literacy, i.e. whether can read and/or write)
5. Age
6. Sex
7. Occupation
8. Marital status
9. Duration of marriage
10. Number of children born to the existing marriage
11. Number of children still living
12. Where born
13. Language spoken (Irish, Irish and English, English only space to be left blank)
14. Whether deaf, dumb, blind, idiot

For each house Form B1 gives the following information:

1. A number (not always the house number)
2. Whether it was built or building
3. Type of tenure
4. Whether inhabited or not
5. Whether wall and roof materials were perishable or not
6. Numbers of rooms and windows
7. Its 'class' (1st class to 4th class, depending on a formula based on permanence of materials, numbers of rooms and windows etc.)
   In addition there were details collected about houses in multiple occupation

Source: S. Royle, 'Irish manuscript census records', *Irish Geography* 11 (1978), 121–2.

Office in Ely Place, Dublin, does not make for easy consultation, and some of the earlier maps associated with the survey cannot easily be located. The maps here should perhaps best be regarded as equivalents to the Working Sheet series in England, since they have been allowed to deteriorate greatly over the years with continuous use.

To some extent, users of the 1911 Census will be able to substitute the Field Books for missing census forms B1, but to date the use of the 1910 material remains extremely limited. Such additions as the 1910 material can make to household studies in Ireland might therefore be applied to the tenements and corporation dwellings in working-class Dublin as studied by Daly and Aalen, since the material also includes schedules of the dwellings of the Dublin Artisans' Dwelling Company Ltd which give details of dwellings, streets, widths, frontages etc. It is perhaps a pity that the absence for most rural areas renders its further use rather spatially selective.[42] It is, of course, also the case that paradoxically the existence of the 1911 Census and the annual revaluation material renders the use of the 1910 material less important for historical studies in Ireland. Clearly the 1910 material has to be seen as a further addition to the large amount of extant early twentieth-century documentation available, rather than as a vital text in an otherwise poorly documented area, as in the rest of the British Isles.

[42] M. Daly, 'Social structure of the Dublin working class 1871–1911', *Irish Historical Studies* 23 (1982), 121–33; Daly, 'Late nineteenth and early twentieth century Dublin', 243; F. H. Aalen, 'Approaches to the working class housing problem in late Victorian Dublin: the Dublin Artisans' Dwelling Company and the Guinness (later Iveagh) Trust', in R. J. Bender (ed.), *Neuere Forschungen zur Sozialgeographie von Irland,* Mannheimer Geographische Arbeiten 17 (Mannheim 1984), 161–84; and 'Health and housing in Dublin *c.*1850 to 1921', in F. H. Aalen and K. Whelan (eds.), *Dublin – City and County: From Prehistory to Present* (Geography Publications, Dublin 1992), 279–304. See National Archives Box R6/24/14290, 2834. Studies such as that by Fitzpatrick on rural communities in Mayo, Clare, Cork, Monaghan and Wexford which relate the 1911 Census to the Revision Books of the General Valuation of Ireland are particularly instructive here (D. Fitzpatrick, 'Irish farming families before the First World War', *Comparative Studies in Society and History* 25 (1983), 339–74).

12

The survey in Scotland

Landed wealth in the form of the ownership of vast acreages and estate-centred power was displayed more fully in Scotland than anywhere else in the United Kingdom, and the names of the dukes of Sutherland, Buccleuch and Richmond were synonymous with such wealth. In Scotland *c.*1880 1,758 owners held 17.5 million acres in estates of over 1,000 acres, compared with the 4,376 such owners in England holding 12.8 million acres.[1]

The revolutionary changes in landscape and farm layout which had largely occurred by *c.*1850 testified to this power and ensured that the 1910 survey mirrors a relatively young, but frequently highly capitalised rural landscape. In the crofting counties the socio-economic basis was, of course, sharply different, yielding small-scale farming and dual occupations. However, in Scotland as in the rest of the country, landowning at the end of the nineteenth century no longer offered the rewards it once held. During the last twenty years of the century there were sales of portions of estates, such as those of the Duke of Fife who had originally owned nearly 250,000 acres in Aberdeen, Banff and Elgin, as well as sales by Lord Aberdeen, the Marquess of Tweeddale or the Marquess of Queensberry.[2]

Irish agrarian disturbances during the second half of the nineteenth century were to some extent also echoed in the Highlands and Islands of Scotland, where rent strikes, police battles, raids on deer forests and other activities during the 1880s kept crofters and the authorities in a state of some tension. The crofters' cause was also widely espoused by urban and lowland Scots through such media as the Inverness newspaper *The Highlander*, and the newly formed Highland Land Law Reform Association in 1883 (renamed the Highland Land League in 1887).

By April 1882 Henry George was reporting on the clashes between the

[1] D. Cannadine, *The Decline and Fall of the British Aristocracy* (Yale 1990), 9.

[2] For more detail on the landscape changes and attendant processes at this time see D. Turnock, *The Historical Geography of Scotland since 1707: Geographical Aspects of Modernisation* (Cambridge 1982); Cannadine, *Decline and Fall*, 108.

Scottish crofters and police from Glasgow at the 'Battle of the Braes' on Skye, following the trail of Michael Davitt and the Land League's agents from Ireland to the crofting counties. In dispatches from June to August 1882 George expounded on his land nationalisation proposals, and gained fame through being arrested in Ireland as a suspicious character, and having the arrests reported in *The Times*. During the 1880s his views were debated on strenuous lecture tours (as in 1884 and 1885) within the wide firmament of agitation over landownership within Scotland, Ireland and England. Urban sympathisers among the middle classes of Edinburgh and Glasgow were not slow to respond. Henry George included such audiences in his lecture tours, and he was heard with great interest. Glasgow Liberals soon turned his ideas also to urban issues, pressing for urban land reform during the 1880s and 1890s. English and Scottish Land Restoration Leagues were founded on Georgite principles at this time, and there was now a perceptible widening of opinion between the disciples of George and the followers of Wallace's Land Nationalisation Society or the more radical Social Democratic Federation. By showing that their views were non-socialist, George's followers now proceeded to lay the ground for closer links with the Liberal Party, primarily through George's influence upon the radical Liberal Joseph Chamberlain, who preached Georgite views on land reform in Glasgow and the Scottish Highlands, thereby furthering the cause of Scottish radicalism. During the 1890s both Liberal and radical politicians took up the single-tax cause.[3]

The manifesto of the Scottish Land Restoration League stated that what was required was: 'such a full and complete Restoration of the Land of Scotland to the Scottish people as will secure to the humblest and weakest of our number his just share in the land which the Lord our God has given us.'[4] Landed property belonged to the people, since its value was socially created, and during the later 1880s George made it clear that land was to be restored to the people via the single tax, an important point which, although first remarked upon by Scottish newpapers such as the *Edinburgh Evening News* or *Dundee Courier and Argus* as early as 1884, had not generally been appreciated.

The idea of using the taxation of land values as a means of raising a local tax was first raised in the Glasgow City Council in 1889, but little was achieved before 1895 when a motion to petition parliament in favour of a local land tax was accepted. Within two years the movement snowballed, and in March 1897 the 'Glasgow Bill' was passed by the Glasgow Council and introduced in 1899 and then again in 1905. Although there was to be little success with a Conservative government, the movement essentially spear-

[3] E. P. Lawrence, *Henry George in the British Isles* (E. Lancing, Michigan 1957), 89–128.

[4] *Ibid.*, 51.

headed the campaign for the taxation of land values for some years, and by rendering it in a municipal forum, made it the more important for Liberals to support.

The Highlands continued to demonstrate unrest at the beginning of the twentieth century as overcrowded crofting communities looked for more land from the advancing deer forests and graziers' lands. Communities such as those on Skye and Barra and in other parts of the Hebrides were in a state of great unrest through to the end of the Conservative administration. At the 1906 Liberal election victory, the seventy-two Scottish constituencies returned sixty Liberals and two Labour candidates, all of whom had supported land value taxation to some degree, and who numbered among themselves such leading members of the new government as Asquith and Campbell-Bannerman. The latter had supported the Dunfermline Council and other Scottish assessing bodies in 1896 with their campaign to include land value taxation within local taxation. Sweeping land reforms in the Highlands of Scotland were included within the Liberal election pledges, and the crofting constituencies, which had all voted for Liberal candidates, now looked forward to change.[5]

The new Liberal Scottish secretary, John Sinclair (Baron Pentland from 1909), was certainly determined to put pledges into action. In the first year of Sir Henry Campbell-Bannerman's administration the Select Committee on the Scottish Land Values Bill, presided over by Alexander Ure, a member of the United Committee for the Taxation of Land Values and later Lord Advocate for Scotland, made various recommendations, including:

> That a measure be introduced making provision for a valuation being made of land in the burghs and counties of Scotland apart from the buildings and improvements upon it, and that no assessment be determined upon until the amount of that valuation is known and considered.[6]

In its first session, the Liberal government therefore proposed Sinclair's Taxation of Land Values (Scotland) Bill to pave the way for future land taxation. This was essentially the 'Glasgow Bill' once more. It provided for a valuation in Scotland, a rate not to exceed 2s in the £ on site values, and the taxation of undeveloped as well as developed land. The Bill also provided for a sweeping land settlement programme in Scotland which would be presided over by a new body, to be called the Board of Agriculture for Scotland. Passed by an enormous Commons majority, the Bill was rejected in the Lords in 1907, giving a hint of the worse troubles to come. A second

[5] J. Brown, 'Scottish and English land legislation, 1905–1911', *Scottish Historical Review* 47 (1968), 72–85.

[6] Report of the Select Committee on the Land Values Taxation, etc. (Scotland) Bill, 13 December 1906 (reprinted in *Land Values* (January 1907)).

government Bill was again passed through the Commons, but was then subjected to wrecking amendments in the Lords. Thereafter Sinclair was lauded by many of the crofters for his efforts and the 'selfish, cruel' Lords were reviled.[7]

The valuation of land and its attendant taxation was, of course, only one aspect of Lloyd George's attack on the landowning classes. He also turned his attention to the game laws, sporting culture and shooting land.

> Take the Scottish deer forests. There you had scores of thousands of industrious, hard-working, thrifty, happy crofting families – who have produced some of the most gallant defenders of the Empire, all swept away with the disastrous brush of landlordism – swept clean as if they were dust – clean from the board. What for? Purely in order to provide a few weeks' pleasure every year for a few rich plutocrats.[8]

In Scotland deer-stalking withdrew vast areas of moorland from grazing uses, as well as closing them to the growing numbers of walkers, and Lloyd George drew attention to such matters in his Bedford speech of October 1913 and again in his Glasgow speech of February 1914. The report of the Scottish Land Enquiry Commission was published in 1914 just before the outbreak of war, drawing attention to such issues. But the Liberals, as interested as anyone in the prestige and pleasures of landownership and 'compromised by their complicity in the culture of the rural ascendancy', could make little headway. And as in England, a landowners' defence association brought strong political pressure to bear in the shape of the Scottish Land and Property Association, founded in 1906, primarily as a reaction to the Small Landholders (Scotland) Bill of that year which had threatened to extend the Crofters Act 1886 to the rest of Scotland. Local associations, such as the Dumfriesshire Land and Property Association, or the Ayrshire Agricultural Defence Association, were merged into the national body, which attempted to ameliorate the attacks on the landed classes.[9]

Nevertheless, general proposals were incorporated into the Land Campaign, and smallholdings, land settlement, rent reductions and the cancellation of rent arrears were encouraged in Scotland by the Small

[7] R. Douglas, 'God gave the land to the people', in A. S. A. Morris (ed.), *Edwardian Radicalism 1900–1914: some Aspects of British Radicalism* (London 1974), 148–61; Lawrence, *Henry George*, 120; For further analysis of the situation in south-west Scotland, see R. H. Campbell, *Owners and Occupiers: Changes in Rural Society in South-West Scotland before 1914* (Aberdeen 1991), 97–159. Campbell's analysis would certainly have benefited from use of the Valuation documents in the Scottish Record Office.

[8] D. Lloyd George, *The People's Will* (London 1910), 94.

[9] A. Offer, *Property and Politics 1870–1914: Landownership, Law, Ideology and Urban Development in England* (Cambridge 1981), 369–71; B. B. Gilbert, 'David Lloyd George: the reform of British landholding and the Budget of 1914', *Historical Journal* 21(1) (1978), 117–41; Douglas, 'God gave the land', 156–7. For a response to Lloyd George on the deer-forest question, see 'The arch-depopulator of the Highlands', *The Nineteenth Century* (June 1914), 1413. See also Chapter 2.

Landholders (Scotland) Act of 1911. A Board of Agriculture for Scotland was instituted in 1912 to implement the Act, although a thicket of obstructions from landowners and sheep farmers awaited its attempts to deal with land settlement on behalf of the crofters. Land raids by crofters continued, and a guerrilla warfare on the land lasted well into the 1920s, beyond the scope of this volume.[10]

Administration of the 1910 valuation

Following a visit to Edinburgh by Sir Robert Thompson and Mr E. Harrison, members of the committee appointed to draw up regulations for dealing with the duties on Land Values, it became clear that Scotland would need to administer its own procedures from Edinburgh. Scotland was accordingly treated as a separate Division for the administration of the valuation, controlled by a Chief Valuer as from August 1911, originally J. W. Fyfe, assisted by an Assistant Chief Valuer (equivalent in rank to an English Superintending Valuer), who answered to the Board of Inland Revenue through the secretariat located in Scotland. Tours of inspection of the Scottish offices by Chief Valuers from London were made from time to time. The Scottish Division was divided into nine Districts at first, then ten Districts by 1912 and twelve Districts by 1914. Distances could be vast in the Districts: the Inverness district included the counties of Inverness, Ross and Cromarty, Sutherland, Caithness, Orkney and the Shetlands![11]

The Scottish equivalent of the Land Valuation Officers were the Assessors under the Scottish Lands Valuation Acts, who had been responsible for making up the local Valuation Rolls for counties and burghs, and fifty-eight such officials were in post by September 1911. The Rolls established a uniform valuation of land throughout Scotland subsequent to the Lands Valuation (Scotland) Act 1854, and listed every property annually with its description and situation, owner, tenant and occupier, and value. Since 1857 the assessors had also acted as the Crown Surveyors of Taxes of the Inland Revenue, although some were practising solicitors. By 1914, thirty out of the thirty-four counties had appointed Crown officials. These were assisted by thirty Temporary Valuation Assistants, eight Temporary Draughtsmen, and forty Temporary Clerks, as well as sixteen Valuers on the permanent staff. The presence of the assessors meant that in practice the Scottish procedures should

[10] An excellent account of the general crofting problem at this time is given in J. Hunter, *The Making of the Crofting Community* (Edinburgh 1976), Chapter 11 'Land raids and land settlement, 1897–1930', 184–206.

[11] Chief Valuer's Archive, Box 3/5 includes a handwritten letter from Edgar Harper to C. J. Howell-Thomas detailing the itinerary of an inspection by him of the Scottish offices. Valuation Department, Inland Revenue, *List of Valuation Districts with the Income Tax Divisions in each District and of the District Valuers in Great Britain* (HMSO, London, October 1910), 14.

have been far easier to set in train than those for England and Wales where no such comparable local machinery existed.[12]

Procedures for the presentation of instruments on the occasion of a transfer of land on sale, interest in land or the grant of any lease for a term of more than fourteen years, or any feu of land or the creation of any ground annual (see below) differed from those in England since presentation at the General Register of Sasines, the Burgh or other local register sufficed to provide the general particulars required to initiate valuation proceedings.[13] By 1915 over 19 million acres in Scotland, comprising 396,000 provisional valuations from 1.3 million hereditaments, had been tackled. The large size of the estates here gave an average size nearly three times that of the English hereditaments, but by far the lowest total value per acre for any of the Divisions. Thus Scotland at £26 average total value per acre compared unfavourably with the English average of £130 (and certainly with the North London figure of £5,197!). Average total value per hereditament was also the lowest for any Division at £375 compared with the average for England and Wales of £515.[14]

Differences in Scottish land law entailed some corresponding differences in treatment of the 1910 legislation in Scotland.[15] Sites for building were acquired, for example, through the payment of Feu Duty, an annual payment proportional to the capital value of the land involved, normally at 5 per cent. Builders did not therefore have to lay out large sums of money in acquiring land for development, and retained correspondingly more capital for the development itself. In addition, many builders did not at first enter into complete contracts, but signed preliminary agreements, only obtaining the complete feu contract with its title to the land after the buildings were erected. In this system it was often possible to arrange that a larger proportion of the Feu Duty was allocated to the built-up land than the actual pro rata rate on the land, with part of the land being relieved of Feu Duty. When buildings were erected on an area cleared in this way of Feu Duty he might then create a ground annual rent of equal amount to the Feu Duty. This rent can in turn be sold at a profit for the builder, usually at twenty years' purchase. In his evidence to the Select Committee on Land Values in 1919 the Chief Valuer (Scotland), Mr Alexander Blair, noted how this method of development actually created Minus Site Values, since the difference between the Gross Value and Full Site Value could actually be greater than the Total Value, yielding a

[12] Inland Revenue Valuation Department, *Instructions to Valuers* Part I (HMSO, London 1910), 34 ('Special instructions to valuers in Scotland'). Separate annual listings were prepared for each burgh and county (C. Sinclair, *Tracing Scottish Local History: A Guide to Local History Research in the Scottish Record Office* (HMSO, Edinburgh 1994), 25–6).

[13] C. E. Davies, *Land Valuation under the Finance (1909–10) Act 1910* (London 1910), 246–7.

[14] Select Committee on Land Values, BPP 1920 (Cmd. 556), XIX, 5. Evidence of Mr Alexander Blair.

[15] Finance (1909–10) Act 1910, S.42 (1–4). The English terminology had to be translated into the approximate Scottish equivalents for teinds, servitudes, fiars etc.

negative figure for the Assessable Site Value. One case tried before the Court of Session in Edinburgh in March 1912 held that there could not be a Minus Assessable Site Value, and that therefore the Assessable Site Value had to be taken as nil. In addition, the ground annual rent was a means of avoiding Increment Value Duty, by burdening the property concerned with a ground annual rent which could then be offset against the value at an 'occasion' when IVD might otherwise be expected.[16]

The arguments over the adverse or otherwise effects of the Scottish feuing system were fought over in the months following the 1910 Budget.[17] Clearly, the existence of the Feu Duty and ground annual created great difficulties in applying the valuation procedure in a meaningful way over much of Scotland. In addition, the Site Value of many flats raised complex problems since there was an intricate mixture of private and communal interests which could result in the owners of upper-floor flats escaping liability to Increment Value Duty, at the expense of the whole burden being borne by the ground-floor owners.

The records

Because of the differences in the existing mechanisms of valuation in Scotland, valuers copied details of ownership and occupation etc. from the Register of Sasines into the Valuation Books ('Domesday Books') and thence into the Field Books. But the Scottish records also contain an additional item of great interest and importance for one locality, namely the Draft Notebooks. These give some idea as to how surveyors from a particular office undertook the process of valuation, and contain illustrations such as scratch elevations or building plans which do not appear in the Field Books themselves. In the latter the reader is enjoined merely to see the relevant Sketch Book (the phrase constantly used by the surveyors themselves, rather than the probably more accurate 'draft notebooks' as used by the Scottish Record Office) for more details of the property (Figs. 12.1 and 12.2). However, there are some instances where although the Field Book has the relevant referral to the Sketch Book, the latter in fact has very little information and no plan. Thus in an example

[16] Select Committees on Land Values, BPP 1920 (Cmd. 556), XIX, 5. Evidence of Mr C. J. Howell Thomas, 13. It should be noted that the Conservative press in Scotland was put into a 'state of wonder and conjecture' by the proposals – see, for example, Robert Guy in the *Glasgow Herald* 11 May 1910 on 'Scottish feuing and Increment Duty', and a reply on 13 May by Alexander Mackendrick, President of the Scottish League for the Taxation of Land Values. For Minus Assessable Site Values and their resolution to a nil status see Herbert v. Commissioners of Inland Revenue Minus Site Values, details of which appeared in the *Land Union Journal* March, April and May 1912. In the case of appeals against valuation which got as far as the Reference Committee, the Scottish Committee consisted of the Lord President of the Court of Session, the Lord Justice Clerk, and the Chairman of the Scottish Committee of the Surveyors' Institution (Finance (1909–10) Act 1910 S.33 (5)).

[17] For a summary of some of the Scottish correspondence in the pages of the *Glasgow Herald*, *The Scotsman* and the *Morning Post*, see *Land Values* (July 1910), 6.

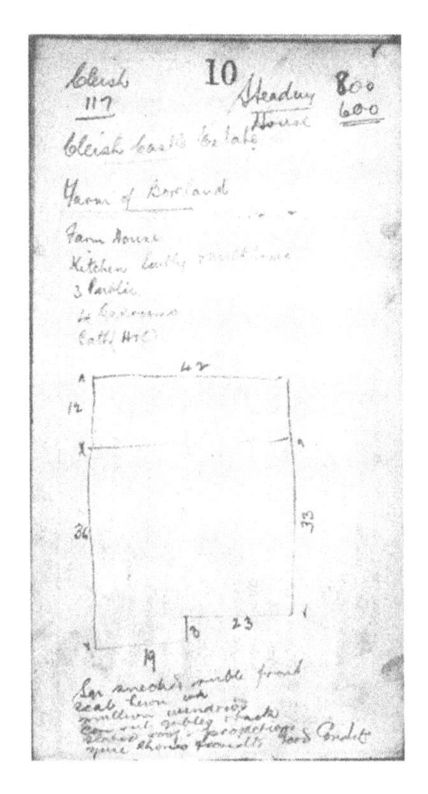

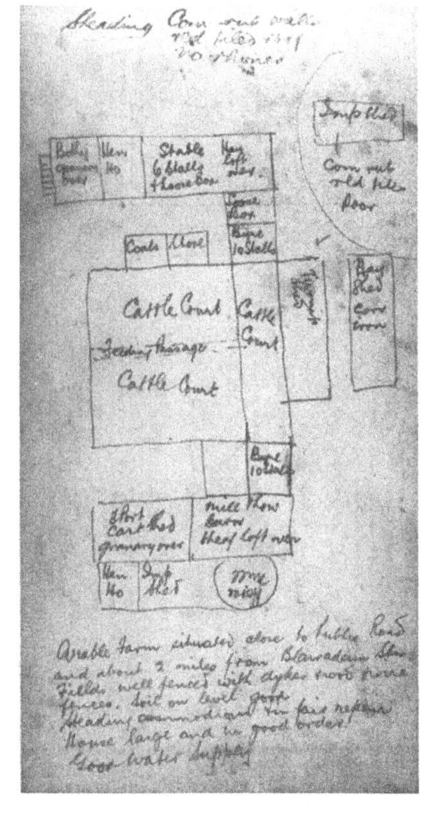

Fig. 12.1 Draft Notebook entry for Boreland Farm, Cleish Castle estate, Kinross. *Source:* SRO IRS 88/8.

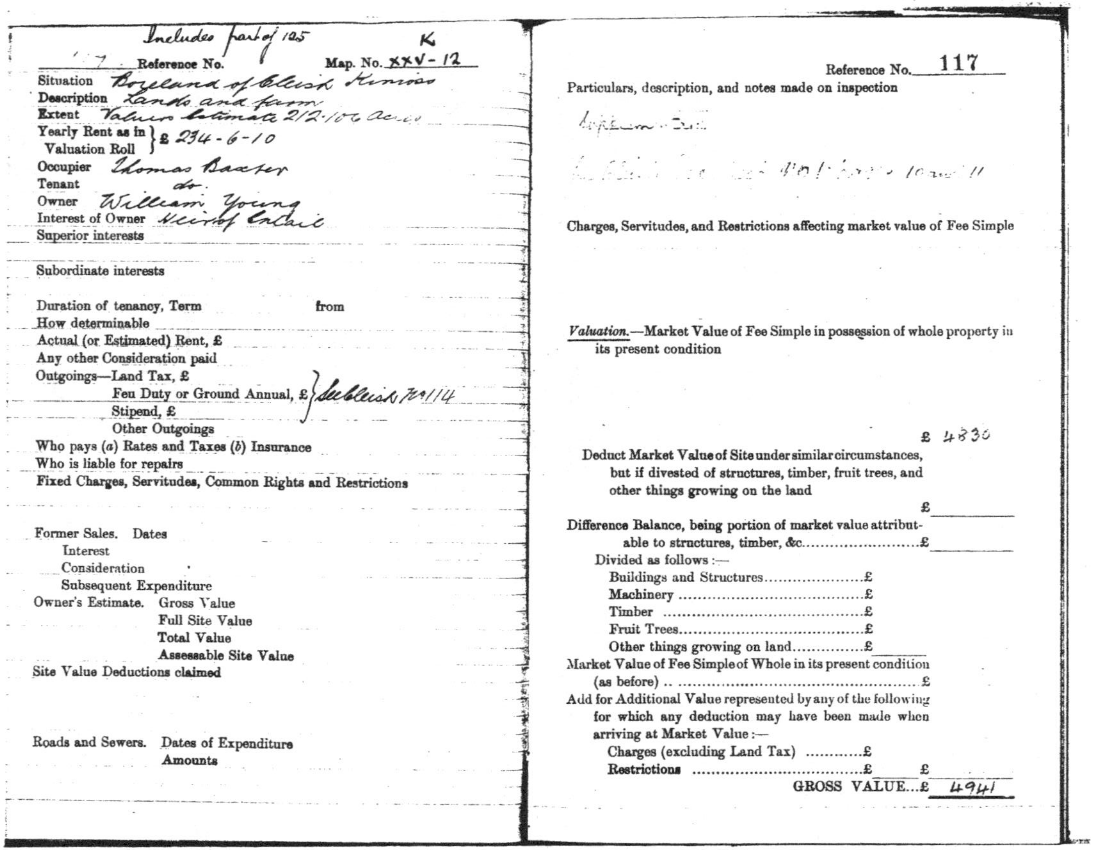

Includes part of 125

117 Reference No. Map. No. XXV-12 K

Situation Boreland of Cleish Kinross
Description Lands and farm
Extent Valuers estimate 212·106 acres
Yearly Rent as in Valuation Roll } £ 234-6-10
Occupier Thomas Baxter
Tenant do.
Owner William Young
Interest of Owner Heir of Entail
Superior interests

Subordinate interests

Duration of tenancy, Term from
How determinable
Actual (or Estimated) Rent, £
Any other Consideration paid
Outgoings—Land Tax, £
Feu Duty or Ground Annual, £ } See Cleish No 114
Stipend, £
Other Outgoings
Who pays (a) Rates and Taxes (b) Insurance
Who is liable for repairs
Fixed Charges, Servitudes, Common Rights and Restrictions

Former Sales. Dates
Interest
Consideration
Subsequent Expenditure
Owner's Estimate. Gross Value
Full Site Value
Total Value
Assessable Site Value
Site Value Deductions claimed

Roads and Sewers. Dates of Expenditure
Amounts

Reference No. 117

Particulars, description, and notes made on inspection

Charges, Servitudes, and Restrictions affecting market value of Fee Simple

Valuation.—Market Value of Fee Simple in possession of whole property in its present condition £ 4830

Deduct Market Value of Site under similar circumstances, but if divested of structures, timber, fruit trees, and other things growing on the land £

Difference Balance, being portion of market value attributable to structures, timber, &c. £
Divided as follows:—
Buildings and Structures £
Machinery £
Timber £
Fruit Trees £
Other things growing on land £
Market Value of Fee Simple of Whole in its present condition (as before) £
Add for Additional Value represented by any of the following for which any deduction may have been made when arriving at Market Value:—
Charges (excluding Land Tax) £
Restrictions £ £
GROSS VALUE... £ 4941

Fig. 12.2 Field Book entry for Boreland Farm, Cleish Castle estate, Kinross.
Source: SRO IRS 70/02.

Table 12.1 *Surviving Draft Note Books in the Scottish Record Office*

Angus (various)	53 books	(IRS 88/1–3)
Dundee City	29 books	(IRS 88/4–5)
Fife (various)	28 books	(IRS 88/6–7)
Kinross-shire	22 books	(IRS 88/8)
Perthshire	180 books	(IRS 88/9–17)
Total	312 books	

from Kinross, we find Cleish Hereditament no.84, Bogside, with a Note Book only containing the hereditament number, value (£17 10s), name of the property (Land Bogside Hill) and the tenant's name (Wallace Bell) – the rest is blank. The tangle can also worsen where occupation is of a dispersed property, where the reader can be referred backwards and forwards between Field Book and Note Books, and between parcels, with none of them actually having the information.

The surveyors used their small Note Books on site, and those for the area covered by the Perth and Dundee District Valuation Offices survived and have been retained by the Scottish Record Office as samples of the record in seventeen boxes.[18] At Perth the District Valuer was James Glen, and we find his valuers such as John Leonard or A. McNaughton keeping rough notes in the official issue notebooks with 'Blackfriars House, Perth' handwritten on the front cover. The former dated his Note Book as March 1915. There are altogether 312 of these small volumes covering such areas as the City of Dundee, the county of Kinross-shire – with such parishes as Cleish, Fossoway, Kinross, Orwell and Portmoak – and much of Perthshire and various parishes in Angus and Fife. Some Note Books are indexed at the front, but this was clearly an option for individual valuers and often a luxury that time did not permit.[19] The number of Note Books for these areas is given in Table 12.1 above.

The Note Books are frequently very rough, and show how the surveyors went around the buildings measuring and estimating as they went, often turning the Note Book upside down in the process. In many cases the entries have been crossed through or ticked, presumably as they were later entered up into the Field Books. In Kinross there are separate books dealing with estates, such as that at Dalhousie, with hotels such as the Palace Hotel at Abergeldy, and with collieries.[20] In Kinross Burgh Note Book no.3 records the Kinross

[18] Scottish Record Office (SRO) Survey Books (Drafts) IRS 88. I am very grateful for the kind assistance of Martin Tyson, Historical Interview Room, Scottish Record Office, in connection with the following paragraphs.

[19] Examples of indexed books are those relating to Kinross-shire, Orwell parish no. 4; Kinross parish no. 2; Fossoway parish no. 4 (SRO IRS 88/8).

[20] SRO IRS 88/8 'Estates' and 'Collieries'.

gasworks, with a detailed diagram included. At Orwell, Note Book no.4 has information on the county council's small Infectious Diseases Hospital at Milnathort with its sixteen beds.[21] From time to time there are cross-references to the Valuation Roll or to the Sasine extract for details of a recent sale. At Portmoak the valuer working on the Kinneston estate was clearly told that the estate was sold 'about 4 years ago' for £17,000. He made a note in his book to 'see if this is correct' from the Sasine records.[22]

The use of maps, Field Books and Note Books in combination can throw an enormous amount of light on properties in areas lying on either side of the Firth of Tay, and this area would repay close attention in the future. Some illustrative examples of detail from the Blairadam estate, Cleish, Kinross, are given here.[23] The Blairadam estate belonged at this time to Sir Charles Elphinstone Adam, Lord Lieutenant of Kinross 1909–11, a descendant of the celebrated architect Robert Adam and who was created a baronet in 1882 in recognition of his father's public services abroad and on behalf of Gladstone at home as a whip and MP for Clackmannan and Kinross-shire. The estate, 2,869 acres in the late nineteenth century, bordered Loch Leven on the Kelty Water and was centred on the mansion amidst a large and finely wooded park.[24]

The estate comprised many large, predominantly arable farms and several collieries at this time. There was the 379 acres of Lochran, occupied by Robert Meiklam, just 200 yards from Blairadam station (on the Perth–Edinburgh main line) and 'hard on the main public road' but with poorly drained fields and a poor water supply. Sunnyside comprised 248 acres at a mile from the station, 'backlying. Land very boggy and not drained. Stony. Sub-soil with clay – fencing good.' Water for the houses might come from wells or nearby burns. Flockhouse Farm (so-named in the Note Book but not named in the Field Book) was subject, as were many of the properties, to Sir Charles Adam's shooting rights. It possessed a 'Fair road of access though steep' and it was 'all laid out in grass. Fields well fenced with drystone dykes. Poorly drained. Good water supply.' The tenant had erected a drier about 100 yards long at a cost of about £100, something also undertaken by the tenant of nearby Blairfordle Farm. Boreland Farm (Figs. 12.1–12.3) demonstrates the extent to which a somewhat uninformative Field Book entry can be illuminated by the Note Books and the use of the Record Sheet plan: the 212 acre hereditament being described in terms of its farmhouse ('large and in good order') and steading ('commodious and in fair repair'), with sketches. This, too, was an arable farm, situated close to the public road and about 2 miles

[21] SRO IRS 88/8. [22] SRO IRS 88/8 'Portmoak'.
[23] SRO IRS 88/8 Cleish 10 (Draft Notebooks); 70/02 (Field Books).
[24] Details from Burke's *Landed Gentry* vol. I (London 1939), 3; J. M. Wilson, *Imperial Gazetteer of Scotland* vol. I (nd), 172. The latter records that in the Park were the Keiry Crags, 'a romantic spot' favoured by Sir Walter Scott.

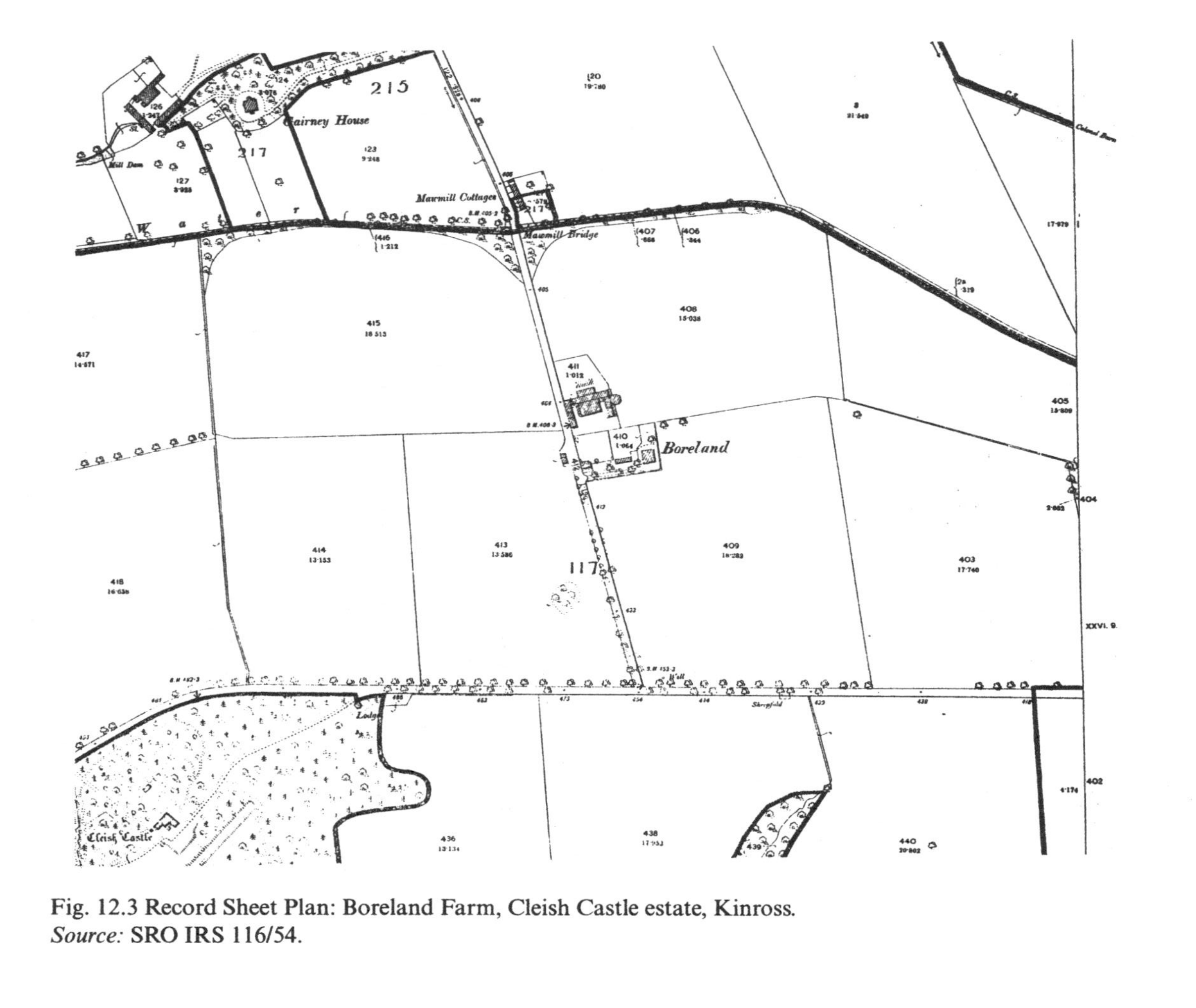

Fig. 12.3 Record Sheet Plan: Boreland Farm, Cleish Castle estate, Kinross.
Source: SRO IRS 116/54.

from Blairadam station. The fields were well fenced, and the soil on the level ground was good. There was a separate ploughman's cottage, of just two rooms and in poor condition. Many such sketches show typical hexagonal, single-storeyed horse engine houses, the horse mills, although not all appear to be operating at the time of the survey, and indeed most had supposedly gone out of use by about 1840.[25] The sketch plans provide fascinating material for a comparative study of the farmtouns of this eastern part of the Central Lowlands. The layout of the steading (often on three sides around a central cattle court), its houses, sheds, byres, hen houses, cart and gig barns, horse mill, coal store and hay lofts can all be traced.

There is much detail on the mansion house and offices of Sir Charles Adam himself, described as having a 'very unpretentious elevation but well furnished inside', and with its attendant gardener's house, coach houses, gardens with their southern aspect and vinery, glasshouse, fruit room, mushroom house and tomato house. There is also a great deal of information on James Ferris the factor's house at Dullomair. The parish church at Cleish, built in 1832 with a chancel and tower added in 1897, accommodated about 400 people, and there was good detail given of its exterior square coursed walls of local freestone and slated roof, and of its interior furnishings, much also renovated in 1897. Similarly Cleish School, built in1835 of the same materials, was stated to have two class rooms taking sixty-six and thirty pupils respectively. Inside the school was wood-lined to about 4′ 6″ and each room had a stove. The 'Preaching Hall' is also noted.

All cottages were included in the Note Books, with sketches no matter how small the accommodation. Hereditament no.16 at Kelty Bridge was included with Nos. 16–31 as a group, with none being separately distinguished in the Field Book. There was a cross-reference here to the Valuation Roll, but it included a small sketch and information to the effect that it 'contains 2 apartments, water in street, com[mon] rub[ble] walls, red tiled roof, stone chimney heads in fair condition'. However 'the gable of one house is falling away from the gable of the other owing to pit workings underneath' although it was otherwise said to be in fair condition. Presumably the cracked walls noted at the Post Office and Shop of Robert Shorthouse might similarly have been caused by the underground workings.

The Note Books have full details on the estate collieries. Indeed, there are two Note Books, written in two separate hands, which give these details and although they are broadly similar, the details are not quite identical. Indeed, the spellings of key place-names sometimes differ, suggesting non-local surveyors, as in the colliery name of 'Blairfordle' or 'Blairfordel'. This colliery,

[25] IRS 88/8, 10–11; I. H. Adams, *Agrarian Landscape Terms: A Glossary for Historical Geography* (Institute of British Geographers Special Publication No.9, London 1976), 55. See also David K. Cameron, *The Ballad and the Plough: A Portrait of Life in the Old Scottish Farmtouns* (London 1978), 28.

and that at Blairenbathie (alternatively spelled 'Blair & Bathy'), have details given on their machinery, buildings and output (Table 12.2). Of the former it was said that 'the output of coal per day when pit was working 150/160 tons – 100 from pit and 50 to 60 from mine.' Of Blairenbathie, leased to the Fife Coal Co. Ltd, it was written that the average daily output was 450 tons but 'at the date of inspection 13/11/14 the output was only 330 tons'. A rough sketch of the colliery shows the location of its winding engines, pumping engine and gear, two shafts, boilers, loading bank, workshops and offices.[26] Further away there were also collieries at Kelty as the estate extended towards Dunfermline and Cowdenbeath. The Kelty colliery was also leased by the Fife Coal Co. Ltd who themselves leased houses from the estate. Also at Beath, James Ferris owned property – two houses and gardens at Church Row – but these appear not to have been very salubrious since the 'servant says house is very unhealthy – the rooms being much too small'.

The Field Books also differ from those used in England or Ireland since they needed to record those outgoings peculiar to Scotland. They recorded the 'Yearly Rent as in Valuation Roll', the tenant as well as the occupier, Feu Duty and ground annual noted above, paid by a vassal (who had complete security in the property) to a person with a superior interest, and who might well retain certain rights to it. It did not record tithe, or teind, as such but did instead record the equivalent 'stipends'. The fourth page, concerned with the calculations of values, therefore also took account of such deductions as Feu Duty, ground annual, servitudes. The overall layout is otherwise the same as for the English books and as such does not resemble those for Ireland (Figs. 12.2 and 12.4).

Coverage was as variable as in England. At one extreme there were detailed descriptions of tenements such as those in Glasgow or listings of individual field acreages by Ordnance Survey parcel size as at Newton of Strathenry, near Leslie, Fife, and on the other hand there were the injunctions to see the information kept elsewhere. The Strathenry valuers were particularly detailed in their coverage of properties. The Field Book entry for Strathenry House, owned by Robert Tullis, also listed the acreages and rents of the properties making up the estate – fourteen in all totalling 2,016 acres and worth £1,690 in rents. Outgoings included land-tax for the entire estate, stipend, heritors' assessments and the 'Leslie Poor burden'. The estate, which included an old castle, had been purchased in 1905 for £40,000 and the owner had subsequently spent £6,163 on improvements, mostly on the house itself, which was described as a 'comfortable country mansion' (Fig. 12.4).[27]

Urban properties are often well illustrated and described. Part of Nos. 3–15 Ewing Place, Glasgow (Figs. 12.5 and 12.6), for example, consisted of

[26] SRO IRS 88/8: (Blairadam Colliery) 1–4; (Cleish no. 2) 24–8.
[27] SRO IRS 66/431, Hereditament no. 42.

Table 12.2 *Blairfordel and Blairenbathie collieries, Kinross-shire, as detailed in the Draft Note Books*

Blairfordel

Machinery

1 pair winding engines complete (Tod & Sons, Dunfermline)
16″ diameter cylinders 3′ stroke, 7″ diameter drum
put in 1902 £100

Steam engine air compressor (old, poor condition) (Ingersoll Rand Co.)
Steam cylinder 22″ diameter with 24″ stroke
Air cylinder 24¼″ diameter×24″ stroke
1000 c.ft of free air per minute
£90 value

1 pair winding (haulage) engines and drum, new in 1883, 14″ bore×2′ 6″ stroke
Engine in fair order, winding gear in poor order (done)
£50

Motor haulage gear with motor installed in 1912. Good condition

BP motor and generator new in 1912. Good condition

Screening plant & bunkers & engine; scraper, table, riddles etc. 1909 £100 fair condition

Two ‘Lancashire’ steam boilers with Galloway tubes working from 60 to 75 lbs per sq inch. Insurance test for pressure 75 lbs
£100 each. Fair condition

Receiver for air compressor £4. Obsolete nowadays
2 pulley wheels £5
Fan and fan engine. Engine 9″ diameter 24″ stroke £7.10s

800 galls per min by combined pumps

Erections and structures

Chimney stack 6′ square at bottom outside and about 36′ high. Poor condition £10
Shed for winding engines. Corrugated iron £50
Concrete seat to winding engine £10
Concrete stand for boilers; fireclay brick packing and casing £20
£250 for all buildings including Pithead frame

Smithy – brick walls, tiled roof £15
Fan house – brick c[orrugated] i[ron] roof £10

Output of coal per day when pit was working 150/160 tons –
100 from pit, 50 to 60 from mine

Blairenbathie

Machinery

2 pair coupled winch engines 21″ diameter 4′ stroke cylinders. 200 IHP. 10 years old at 1909, value £250 each at 1909. Good condition.

Table 12.2 (*cont.*)

1 pumping engine HP cylinder 18″×5′ LP cylinder 3′×5′ about 80 HP. Value £250. Good condition

Screening plant & housing: wood frame, corrugated iron roof and walls. Value £900. Engine for driving same £50. 18″×4′ 6″ cylinders, about 30 HP. Poor condition.

4 'Lancashire' boilers 28′ by 7′ 6″. No Galloway Tubes. 80 lbs to 100 lbs per sq. inch pressure. Value £220 each

2 small donkey pumps £5 each. steam engine 8″ diameter by 12″ stroke 10 HP. Dynamo by T. Parker Ltd. Electric lighting set £10. Good condition

Fan 12′ diameter £60 including engine 16″×2′ cylinder about 20 HP. Good condition

Crab engine £70 8″×20″ cylinder about 15 HP

Donkey pumps 5″ square 7″ diameter cylinder 10″ stroke about 10 HP. Poor condition

1 Rugler pump at pit bottom. 7″×14″, 3′ stroke. Value £320

2 pit head frames (timber) £60 each. 4 pulleys at £5 each

3 electric motors & haulages since 1909

Structures and buildings
Smithy, office etc.
built of 9″ brickwork; arched windows; slated roof; ci rhones and condts on front; 87′×21′×16′ high. In good repair

Casing for boilers. Fireclay brick

Chimney stack: built of brick with moulded cornice at top; 14′ diameter at bottom and about 100′ above ground. Built on concrete foundation. Good condition.

No. 1 engine house (winding engines) 'only fair'
14″ brick walls; arched windows; elevated roof; ci rhones & condts; in good order; 33′×24′×18′ high
No.2 ditto 33′×33′×18′

Pump engine house
9″ brick, slated roof, ci rhones; 11′×9′×14′ high. Fair

Fan house and fan engine ho.: 9″ brick walls, slated roof, ci rhones; good repair; 19′×12′×12′ and 21′×26′×8′

Concrete seat to pumps 33′×18′×9″ thick

Small house for crab engine, corrugated iron in 1909 say £8

Average output per day 450 tons. At date of inspection 13/11/14 the output was only 330 tons

Source: SRO IRS 88/8: Note Books Cleish No.2 and Blairadam Colliery.

42 Reference No. Map. No. XXVII. 2. E.

Situation Strathenry, near Leslie.
Description House, Offices, Gardens & Shootings of Strathenry.
Extent 15·30 Acres: 17·267

	Ord. No.	Ac. dec.
	352	1·886
	337	14·573
Part of	336	0·218
"	355	0·200
"	298	0·352
	317	0·688
		17·267

Yearly Rent as in Valuation Roll } £130.
Occupier Robert Tullis
Tenant Do
Owner Robert Tullis Esq. of Strathenry, Leslie, Fife.
Interest of Owner Proprietor.
Superior interests Total Rents & Area of Strathenry Estate viz.
Subordinate interests

No.	Rents £ s d	Area Ac. dec.
42	130·0·0	17·267
43	337·10·9	302·668
44	60·0·0	210·652
45	100·18·6	143·892
46	171·7·0	114·571
47	226·3·2	207·730
48	161·4·3	206·469
49	150·7·11	359·993
50	296·12·6	243·908
51	44·0·0	174·795
52	20·0·0	8·765
53, 54	14·10·0	0·279
55	12·0·0	1·242
56	36·0·0	24·174
Total	1690·14·1	2016·405

Duration of tenancy, Term from
How determinable
Actual (or Estimated) Rent, £312.
Any other Consideration paid
Outgoings—Land Tax, £16·10·5 for all Strathenry Estate
Feu Duty or Ground Annual, £
Stipend, £74·18·4 for all Strathenry Estate
Other Outgoings £45·14·10 Heritors' Assessments } for all Strathenry Estate; £20·0·0 for Leslie Poor
Who pays (a) Rates and Taxes (b) Insurance (a) Tenant (b) Owner.
Who is liable for repairs Owner.
Fixed Charges, Servitudes, Common Rights and Restrictions

Former Sales. Dates 12 May 1905 — for the whole Estate —
Interest Do
Consideration £40,000
Subsequent Expenditure £3181·6·8 on House, Stables, &c. £1111·15·0 on Gamekeeper's House & Kennels.
Owner's Estimate. Gross Value £4293·1·8.
Full Site Value
Total Value
Assessable Site Value
Site Value Deductions claimed Yes.

Total "Subsequent Expenditure" for Strathenry Estate:—

No.	£ s d
42	4293·1·8
43	710·0·0
44	0·0·0
45	18·0·0
46	153·0·0
47	6·0·0
48	170·8·9
49	0·0·0
50	325·0·0
51	217·0·0
52	14·0·0
53	0·0·0
54	81·0·0
55	175·18·3
56	0·0·0
	£6163·8·8

Roads and Sewers. Dates of Expenditure
Amounts

Reference No. 42.

Particulars, description, and notes made on inspection
This is a comfortable Country Mansion, gardens and grounds. See full particulars on three separate sheets.

Charges, Servitudes, and Restrictions affecting market value of Fee Simple
*Proportion of Stipend £11·18·9 @ 25 years' purchase £298·8·9 say £298.

Valuation.—Market Value of Fee Simple in possession of whole property in its present condition
Estimated Rental £312·0·0
Less *Land Tax 2·12·8
*Heritors' Assessment 7·5·9
*"Leslie Poor" burden 3·3·9 .. 13·2·2 £298·17·10 @ 18 years' purchase £5381·0·0 say £5380
Add *Shootings 7/2 @ 15 years' purchase £5·7·6 say 5.
£5385
Deduct Charges (as above) 298
* See Apportionments on separate sheet. £5087

Deduct Market Value of Site under similar circumstances, but if divested of structures, timber, fruit trees, and other things growing on the land
Ac. dec. 5·000 @ £160 per ac. £800·0·0; 12·267 @ 20 " " 245·6·10 } = £1045·6·10 say £1045.
Less Charges (as above) 298 £747

Difference Balance, being portion of market value attributable to structures, timber, &c. £4340
Divided as follows:—
Buildings and Structures £4340.
Machinery £
Timber £
Fruit Trees £
Other things growing on land £
Market Value of Fee Simple of Whole in its present condition (as before) £5087
Add for Additional Value represented by any of the following for which any deduction may have been made when arriving at Market Value:—
Charges (excluding Land Tax) £298
Restrictions £ £298
GROSS VALUE £5385

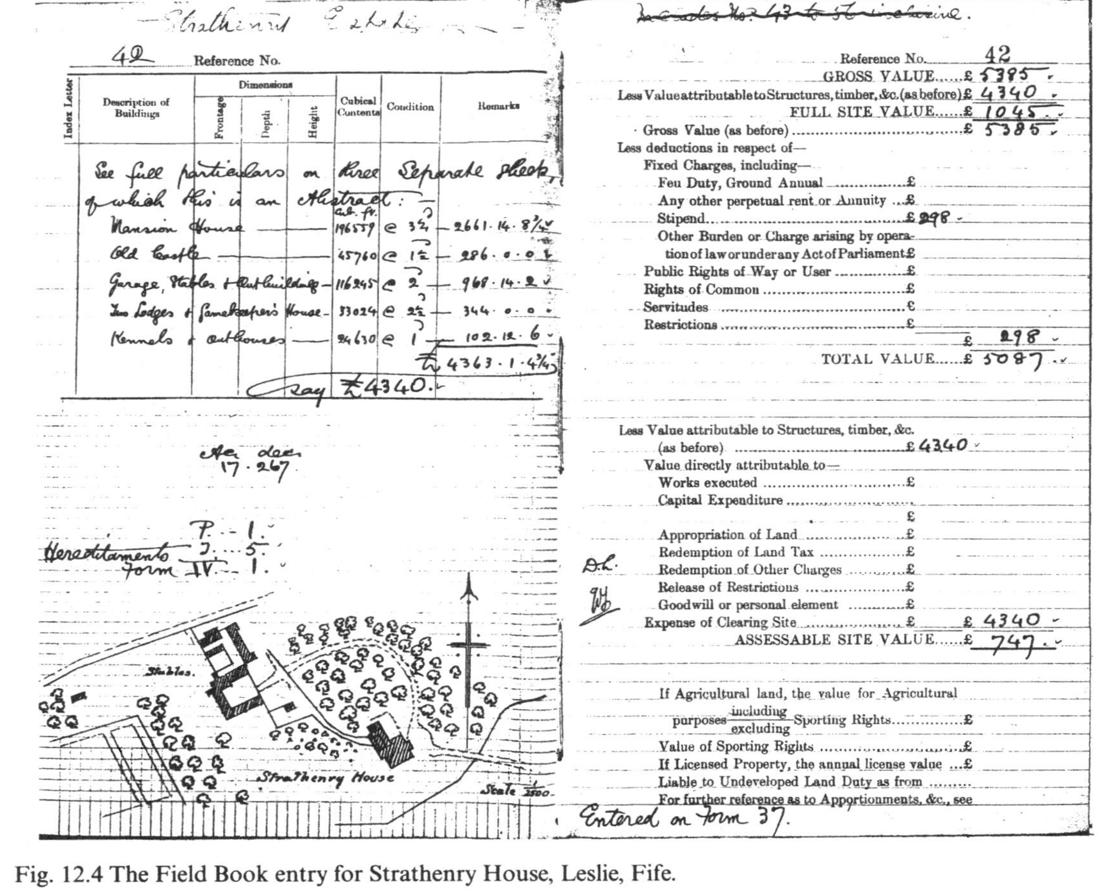

Strathenry Estate

42 Reference No.

Index Letter	Description of Buildings	Dimensions: Frontage	Depth	Height	Cubical Contents	Condition	Remarks
	See full particulars on three separate sheets, of which this is an Abstract:—						
	Mansion House				cub. ft. 196559	@ 3¼	2661. 14. 8¾
	Old Castle				45760	@ 1½	286. 0. 0
	Garage, Stables & Outbuildings				116245	@ 2	968. 14. 2
	Two Lodges & Gamekeeper's House				33024	@ 2½	344. 0. 0
	Kennels & outhouses				24630	@ 1	102. 12. 6
							£4363. 1. 4¾
					say £4340.		

Ac. dec. 17.267

Hereditaments Form P. 1. J. 5. IV. 1.

Reference No.	42
GROSS VALUE	£ 5385
Less Value attributable to Structures, timber, &c. (as before)	£ 4340
FULL SITE VALUE	£ 1045
Gross Value (as before)	£ 5385
Less deductions in respect of—	
Fixed Charges, including—	
Feu Duty, Ground Annual	£
Any other perpetual rent or Annuity	£
Stipend	£ 298
Other Burden or Charge arising by operation of law or under any Act of Parliament	£
Public Rights of Way or User	£
Rights of Common	£
Servitudes	£
Restrictions	£
	£ 298
TOTAL VALUE	£ 5087
Less Value attributable to Structures, timber, &c. (as before)	£ 4340
Value directly attributable to—	
Works executed	£
Capital Expenditure	£
Appropriation of Land	£
Redemption of Land Tax	£
Redemption of Other Charges	£
Release of Restrictions	£
Goodwill or personal element	£
Expense of Clearing Site	£ 4340
ASSESSABLE SITE VALUE	£ 747
If Agricultural land, the value for Agricultural purposes including/excluding Sporting Rights	£
Value of Sporting Rights	£
If Licensed Property, the annual license value	£
Liable to Undeveloped Land Duty as from	
For further reference as to Apportionments, &c., see	

D.L.

Entered on Form 37.

Fig. 12.4 The Field Book entry for Strathenry House, Leslie, Fife.
Source: SRO IRS 66/431 Hereditament no.42.

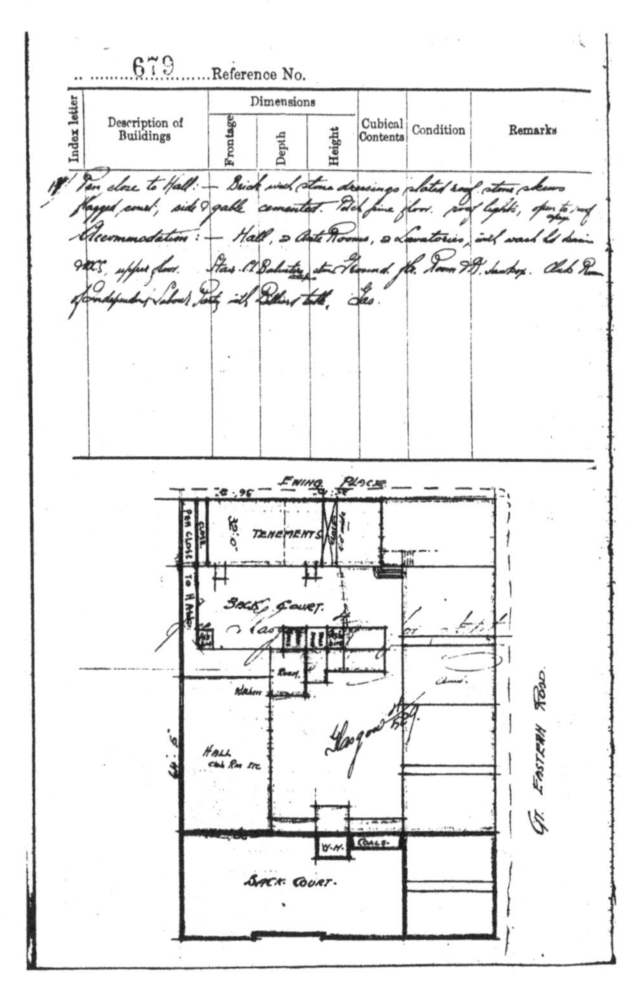

Fig. 12.5 Glasgow, Nos. 3–15 Ewing Place in the Scottish Field Book (p.3).
Source: SRO IRS 67/67 Hereditament no.679.

Hall. 2 Ante Rooms, 2 Lavatories, with wash hd basin & WCs, upper floor. Stair, C[ast] I[ron] Balusters, stone Ground flr. Room & K[itchen]. Jawbox. Club Room of Independent Labour Party with Billiard Table. Gas.[28]

The mapping scales and procedures used appear not to differ greatly from those used in the rest of the United Kingdom. The standard 1:2500 was employed in most areas, but for densely populated urban areas the scale might be 1:500, as for example in the inner areas of Glasgow (Fig. 12.6).[29]

[28] SRO IRS 67/67, Hereditament no. 679.

[29] I am grateful for the help of Mr Hugh Hagan, West Register House, Scottish Record Office, for the retrieval of the maps used in Figs. 12.3 and 12.6.

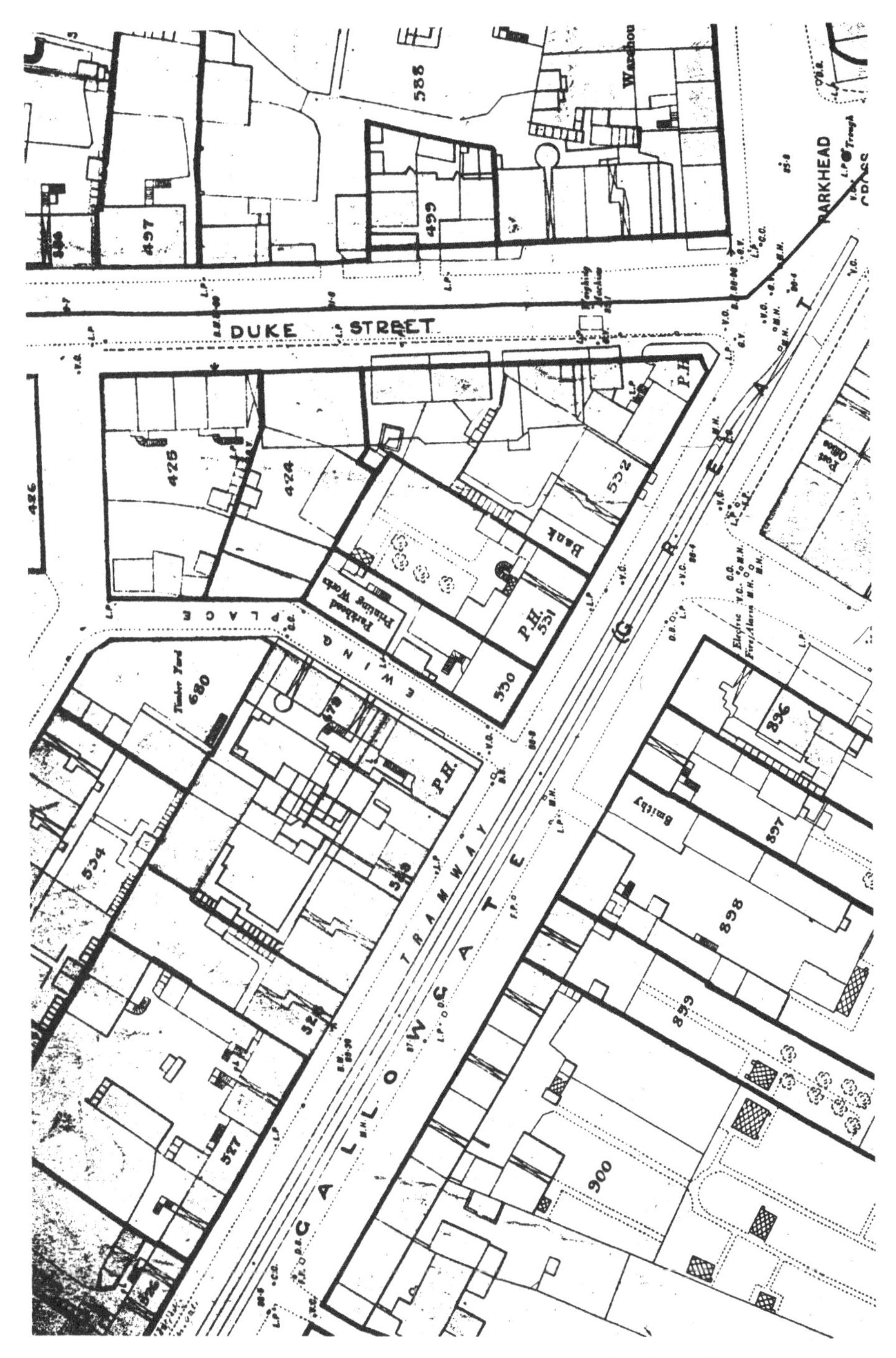

Fig. 12.6 Glasgow, the Ewing Place area as shown on the Valuation Office 1894 second edition Ordnance Survey 1:500 sheet.
Source: SRO IRS 118/936.

The administration of the records in Scotland

The involvement of the Scottish Record Office (SRO) with the 1910 material dates from a false start in 1961 when a proposal was mooted to transfer the Valuation Books from the Inland Revenue to Aberdeen University. However, the proposal fell through primarily because it was decided that the Valuation Books should be destroyed, since it was felt that the information was available in more detail in the Field Books.[30] Accession of the maps and Field Books to SRO began with the arrival of the bulk of the maps in 1985 and the Field Books after 1990. Differences in the functions of the Inland Revenue in Scotland compared with England were largely responsible for the later arrival of the documents in the public domain in Scotland. The Inland Revenue here did not carry out the task of rating valuation unlike their English counterparts, since it was performed in Scotland by the local authorities. Thus, whereas two sets of files were available to the English Inland Revenue, one of which could thus be disposed of, such a situation did not prevail in Scotland. At the time in the mid-1970s when the English Inland Revenue were negotiating to hand over their 1910 material to the PRO, the Scottish Inland Revenue argued that it could not afford to part with the material, especially for the administration of rural properties. Subsequently, it has really been considerations of shortage of storage space that have forced a reconsideration, which has taken place since 1983, with the results as set out above. There are otherwise no Forms of Return held by SRO, although copies of Form 4 may yet be found amongst estate or solicitors' records.

The SRO now has 7,764 Field Books (there were twelve from West Lothian which were not transferred from the Valuation Office), ranging from 630 for Ayrshire down to just twenty-six for Nairn or forty-one for Bute, including Arran. Sutherland, with its huge properties, has only sixty-nine Books. The Books are arranged by counties and cities, within counties by parishes and burghs, and within cities by wards. There appear to be no Forms of Return otherwise retained for Scotland, and there is no knowledge of any other sets of relevant material since the county repository system on the English model does not apply here and all such records would be centralised in Edinburgh. The Record Sheet Plans are also stored there, and indeed are recommended as the easiest way to access the required property in the Field Books. They are arranged by scale and by the numerical order of the Ordnance Survey sheet references as originally held by the District Valuation Offices, a system retained on their transfer to the Scottish Record Office.[31]

[30] Personal communication from Ian Hill, Head of West Search Section, Scottish Record Office, 22 June 1995. I am grateful to Ian Hill for elucidation on the recent archival history of the records.

[31] Scottish Record Office, *Valuation Office Records: Records created under Finance (1909–10) Act* (SR509183.053) (Edinburgh nd).

The records have been available for public inspection in Scotland only since 1990, and the repertory has only recently been compiled. It is also unfortunate that the records are at present stored in a suburban repository in Edinburgh, Thomas Thompson House, with a 24-hour delay after ordering them at West Register House. It is to be hoped that wider knowledge and use of the records in Scotland might then exert pressure on the Scottish Record Office to relocate the records more conveniently. One reason for the delay in the use of these records may be that Scottish genealogists, unlike their English counterparts, have adequate alternative sources in the Registers of Sasines, Valuation Rolls and latterly the Land Register. The main users, primarily local historians, legal consultants and railway enthusiasts who are anxious to obtain accurate material for their reconstructions, clearly do not amount to a sufficient pressure group for easier public access in Scotland.

Conclusion

The particular social and political environments into which the 1910 material was introduced in Scotland necessitated some variations in procedures, documentation and degrees of success. The Scottish valuation system was more akin to that being administered in England than that of Ireland, and the variations in documentation and procedures reflected only the differing history of land tenure and land law operating north of the border. Nevertheless, a great difference did exist in that once again an ongoing system of land valuation was already in place in Scotland before 1909, with the result that the annual Valuation Rolls were available for inspection by the valuers. The administration of valuation by the local authorities in Scotland has also brought differences in the subsequent history of the 1910 records in Scotland, and a reluctance on the part of the Scottish Inland Revenue to part with the records at all – hence a very recent deposition at the Scottish Record Office, and accordingly little knowledge of the source amongst Scottish historians.

Apart from the empirical information yielded by the Scottish material, this chapter has revealed how important the extant documentation remains in terms of gaining insight into the procedures of the survey itself. The main value of the Scottish material in understanding the processes of valuation lies in the (to date) unique collection of Note Books for the Perth and Dundee Valuation Offices which were used to collect information in the field, before its final transfer to the Field Books. This information is the most detailed of any so far found within the entire survey documents, and, when used in conjunction with Field Books and maps, presents the reader with a wealth of information for that area.

Although the material has not yet found its way into the larger academic or public consciousness in Scotland, it is to be hoped that regional and local historians, historical geographers and genealogists will now take up the challenge, not only of examining the documents, but of exerting pressure on the Scottish Record Office in Edinburgh to accord the material the importance it most certainly deserves. Finding aids need to be written or amplified, and access needs to be much easier than at present.

13

Conclusion: a discourse of state power 1909–1914

The Lloyd George Chancellorship of the Exchequer was a stormy, eventful and potentially epoch-making period. Class differences were highlighted by his office in a way which has seldom before or since been so overtly wrought, and the ways in which his 'Domesday' legislation was interlinked with the wider issues of Commons vs. Lords or even 'People vs. Lords' has been explored in the preceding chapters.

In this volume attention has been focussed on the evaluation of a highly significant source for the investigation of the land of the British Isles in the Edwardian period. Throughout the chapters concerned with the 'excavation' of the records there have been many contextual currents running which could not be explored for fear of subverting the intention of the text. However, great care has been taken here to ensure that the 1910 material is seen not just as inert documentation covering metres of shelving but as a means of communication between people in the period 1909–20. The messages that were being communicated were sometimes superficially related to the problems of taxation and valuation and sometimes at depth – so far beneath the surface indeed that they touched on the assumed, taken-for-granted world of society at that time. Not only excavation then, but a 'thick description' is required, to elucidate not only what the documents 'say' on the surface, but what they do not say – what was conventional, and what was being subtly (or not so subtly) attacked through the entire mechanism of the valuation survey.

Thus one very important wider issue that reappears within the whole process of valuation is the extent to which the onward march of urban-industrial capitalism faced the rearguard action of older rural systems of production, culture and social control. Another interconnected issue is that of the increasingly visible involvement of the state in the lives of its individual members, replacing Victorian discreet and indirect taxation, surveillance and policing methods. Connected with this came the growth of bureaucracy: in

1851 there had been 39,100 civil servants, rising to 116,400 by 1901 and 172,000 by 1911.[1]

This would be too crudely portrayed as a political issue of progressive and radical Liberal vs. landowning and patrician Tory, since the complexities of ever-shifting political allegiance and pragmatism always had to be constantly renegotiated within the ideologies of the polar extremes of both groupings. The Liberal Party itself was by 1909 a rich mixture of old-fashioned Liberal dogma, 'aristocratic whiggery' and twentieth-century *fin de siècle* modernism. One commentator declared that 'To reduce the Liberal Party to a definition would be like attempting to reduce the glandular contours of a circus Fat Lady by simply talking her thin'![2] One of the internal Cabinet battles in the so-called 'Cabinet of all the talents' that Lloyd George had to fight, in order to move the legislation forward, was against the landowning Liberals who were as committed as were their Tory counterparts to the survival of a landed hegemony. Although there had been a dramatic reduction in the presence of landowners in the Commons from 1906, the Liberal Cabinet still maintained a grandee presence in the likes of Ripon and Elgin, Crewe and Carrington, Grey, Herbert Gladstone and Lord Tweedmouth. Their stubborn persistence in what Lord Robert Cecil referred to as a 'party of cardsharpers' was however quickly overshadowed by the more dynamic middle-class presence of Asquith, McKenna, Morley, Haldane, Runciman, Samuel, and Lloyd George himself. Nevertheless, as late as 1908 the Liberal Cabinet had sufficient numbers of the Old Guard effectively to block any concerted land legislation and meaningful social reform. Only with the resignation of Campbell-Bannerman in the spring of that year, did Lloyd George and Churchill obtain sufficient power to press more radical policies into effect. And even then the Cabinet was very uneasy about the projected legislation, deliberating as John Burns, the Labour and Liberal MP declared 'like nineteen rag-pickers round a 'eap of muck',[3] since many felt that it was actually unworkable. Hobhouse, then Financial Secretary to the Treasury, felt admiration for Lloyd George's audacity but contempt for the actual formulation of the Bill. The subsequent 1909 legislation offended many traditional Liberals, such as Lord Rosebery, who saw it as 'inquisitional, tyrannical and socialistic' and who left the Party. Indeed, a short-lived group of Liberal MPs, known as the 'Troglodytes', was actually formed to fight the progress of the taxes upon landed wealth.[4]

However, what the 1906 administration advanced through its final recogni-

[1] P. Thane, 'Government and society in England and Wales, 1750–1914', in F. M. L. Thompson (ed.), *The Cambridge Social History of Britain 1750–1950* vol. III (Cambridge 1990), 57.

[2] G. Dangerfield, *The Strange Death of Liberal England* (1935; Pedigree Edn, London 1980),72.

[3] *Ibid.*, 20.

[4] For an example of such splits in the context of Liberal policy in Scotland see J. Brown, 'Scottish and English land legislation 1905–11', *Scottish Historical Review* 47 (1968), 75 fn. 6; D. Cannadine, *The Decline and Fall of the British Aristocracy* (Yale 1990), 184–210; E. David (ed.), *Inside Asquith's Cabinet: From the Diaries of Charles Hobhouse* (London 1977), 7.

tion of the essential necessity of an attack upon landlordism right across the British Isles, from Sussex to Skye and from Lincolnshire to Limerick, was a fascinating complex of individual and institutional utterances, signs, symbols and practices making up a discourse on the very nature of land and landownership. As belief systems were challenged by the elected government of the day, in ways that had previously mainly been the field of anti-landlord and pro-landlord pressure groups in the later nineteenth century, a fundamental rethink was in process here. And it is this rethink that is reflected in the 1910 documents and by the bureaucracy surrounding them. Therefore we find government publications putting not only a 'national' and 'official' viewpoint, but also a rich variety of ideologies proffered by way of counter-discourses of both a radical and reactionary variety, by land tax societies and defence groups such as the Land Union which campaigned inside parliament and outside through a host of books, pamphlets, newspaper articles and speeches. This was, after all, what one Conservative observer referred to as a new cultural age to rival that of the Stone, Bronze and Iron, 'the League Age. For new leagues and associations rise on every side with startling rapidity.'[5] The literary output on both sides was prodigious and consisted of the whole gamut of literature from academic and professional texts, through to political tracts, to 'commonsense' articles, to satirical books such as the wonderful *Alice in Plunderland*,[6] to doggerel verse.

Highly talented individual actors move across this structural struggle for power – Lloyd George the political genius, Henry George the influential American presence, Captain Ernest Pretyman the masterly exponent of the details of the legislation, Wedgwood the enthusiastic 'single-taxer' who later joined the Labour Party – and so on. Their dialogues and those of their supporters inside and outside Whitehall indeed constitute a complex discourse about the nature of Edwardian state intervention and the locus of power.[7] But less visible presences were also there: and the 'multiple voices that animated daily life' have also been explored to some extent here, such as the personal history of the Valuation Officer from Dorset or the inter-office intrigues and personnel problems in Dublin. [8]

Here then is the very stuff of Foucault's 'General History' – 'living, fragile,

[5] *Conservative Agents' Journal* (October 1910), 105, cited in F. Coetzee, *For Party or Country: Nationalism and the Dilemmas of Popular Conservatism in Edwardian England* (Oxford 1990), 3. For a fuller analysis of these groupings see R. J. Morris, 'Clubs, societies and associations', in Thompson, *The Cambridge Social History of Britain 1750–1950* vol. III, 395–443.

[6] Loris Carllew, *Alice in Plunderland* (London 1910).

[7] For further material on the subject of discourse analysis in historical geography see C. Philo, 'Thoughts, words and "creative locational acts"', in F. W. Boal and D. N. Livingstone (eds.), *The Behavioural Environment* (London 1989), 205–34; P. Cloke, C. Philo and D. Sadler, *Approaching Human Geography* (London 1991), 195. For the contrast between 'Total History' and 'General History' see M. Foucault, *The Archaeology of Knowledge* (London 1972), 10.

[8] The quotation is from D. Gregory, *Geographical Imaginations* (Oxford 1994), 256.

NOBLE ARTIST (re-entering dressing-room): "NO GOOD. THEY WON'T LISTEN TO THE 'RUINED LANDOWNER.'"

MR. BALFOUR: "QUICK, JUMP INTO THIS JOAN OF ARC ARMOUR AND TRY THEM WITH 'THE SAVIOUR OF MY COUNTRY.'"

(By special permission of the Proprietors of the "Manchester Evening News."

Fig. 13.1 'The unpopular turn.'
Source: Lloyd George, *The People's Will*, 33.

pulsating history' which demonstrates the shifts, confusions, schisms and uncertainties of everyday dealing with the land in Edwardian England. The concept that landed power bowed the knee and handed over the reins of government at this period can thus be demonstrated, through this study of the land valuation legislation, to be another one of the historian's shibboleths ripe to be dismantled.[9]

Some spirit of the discourse can be seen in Figs. 13.1 and 13.2, both from

[9] See W. D. Rubinstein, 'New men of wealth and the purchase of land in nineteenth-century England', *Past and Present* 92 (1981), 125–47; J. V. Beckett, *The Aristocracy in England 1660–1914* (Oxford 1986); F. M. L. Thompson, *English Landed Society in the Nineteenth Century* (London 1963); Cannadine, *Decline and Fall*, *passim*.

Fig. 13.2 'Alice and the Cabinet Minister in the garden at Jumble Hall.'
Source: Carllew, *Alice in Plunderland*, 45.

1910 when the struggle was at its most intense. Fig. 13.1 is taken from the *Manchester Evening News* and illustrates a collected book of speeches by Lloyd George. Very topically at this time, it portrays a Lord as an unpopular music-hall act, being assisted by the leader of the Tory opposition, to try a different image. The 'ruined landowner' act had not worked, it was time to try the 'saviour of my country' tactic. Such ridicule was countered in many ways. One cartoon alluded to Lloyd George as demagogic as he sits in the garden of Jumble Hall, which was portrayed as the seat of:

> the Tootlers, no further away than Sussex. Part of the Tootler's fortune, which was founded, pretty securely, on a motor-business and some lucky speculations in rubber and oil, had been employed in transmogrifying the rambling, cosy, old red-brick Jumble Hall into a defiant, stone-fronted 'Castellated Mansion.'[10]

Here we find the contemporary iconography of a gothicised home counties, *nouveau riche* landscape and its means of income being employed against Lloyd George in an attempt to demonstrate a background which was not only unstable, but equally importantly was not that of an acceptable English rural landowning lineage. Lloyd George was clearly of another age: as Grey was to explain in 1921, 'As to politics, I am not the sort of person that is wanted now . . . Lloyd George is the modern type, suited to an age of telephones and moving pictures and modern journalism.'[11] We might add also the motor car and holidays on the Riviera as two other symbols of the age which he embraced wholeheartedly.

The shifting alliances and political posturings of the period 1906–14 obviously revolved around many issues other than the land. But in their dealings with land-related issues the Liberal government revealed a class-based legislation which was interventionist, partial and redistributory. It attacked one class – the owners of land – and avoided attacks on smallholders, small farmers, cottagers and those without land. And after setbacks in the courts, it sought to redefine its objectives so as to exclude farmland as well. Its targets most particularly and immediately were those who were deemed to be holding up the process of urban expansion and redevelopment of poor housing quality by refusing to sell land and waiting for speculative currents to push up land prices.

Land indeed was seen as an icon around which hinged a wide variety of social, economic and political issues, involving social justice and economic efficiency.[12] The taxation measures should also be seen as a component in a complex of measures between 1880 and 1910 which also included a series of Allotments and Smallholdings Acts, Land Development schemes and concepts for developing 'land colonies' such as those of the Land Settlement Association.

[10] Carllew, *Alice in Plunderland*, 44.
[11] Cannadine, *Decline and Fall*, 215.
[12] S. Ward, 'Land reform in England 1800–1940: a summary', *Agartorteneti Szemle: Historia Rerum Rusticarum* 18 (1976), 37–44.

But always in the background there was also the fear that the Finance Act (1909–10) 1910 was really about something else: that its insistence upon the need for a valuation of all property was a first step towards some dramatic state nationalisation of the land. The real motives of Lloyd George have been explored earlier, but there were certainly many who did indeed wish to pursue this objective, either through the single tax with no compensation in the Henry George philosophy, or through outright nationalisation with compensation as expounded by A. R. Wallace and the Land Nationalisation group. Knowledge, after all, was power. And in finally acquiring an unprecedented amount of information about the boundaries, contents and worth of the landed estates the government was extending the 'authoritative resources' available for its surveillance of its people, and thereby its power over them. The forms and maps engendered by the survey constitute a source of knowledge which was not available to the general public, and the surveyors were enjoined to ensure that their maps were fully completed with no 'silences' upon them.[13]

So here is a historical geography that must deal not only with an evaluation of a documentary source of great complexity, but also, in order to understand and utilise it the better, must deal with the intentions and fears of all classes of people, in all parts of the British Isles. The 1910 material is essentially geographical. The bureaucracy set up to administer it was regionally defined from the very beginning, with fourteen Divisions in England and Wales divided in turn into 118 Districts composed of the pre-existing Income Tax Parishes (themselves normally groupings of civil parishes), and monitoring of progress was established on a basis which fundamentally revolved around the comparison of progress rates between the regions.[14] Scotland and Ireland demonstrated distinctive administrative frameworks to deal with their different prior administrative spatial structures and land-related cultures. The pre-existing Scottish feuing system, for example, had to be taken fully into account, whilst the existence of an efficient Valuation Office, which had functioned in Ireland for many years, was a boon to the Liberals in London. The documents themselves have a historical geography, in the context of their genesis, progress and current archival status. Reportedly uniform, the process and documents of the valuation demonstrated regional variations and even quite local differences, as bureaucratic practices at the local level were superimposed on nationally conceived strategies. Furthermore Lloyd George's own ideas have a strong spatial content. If smallholders and small farmers were to be excluded from the tax, and urban landowners hit severely, then there were clearly spatial implications. The small farmer of Wales or the peasant proprietor of Ireland might escape altogether but not so the owners of the 'Broad Acres' of the English lowlands or the metropolitan and suburban land hoarder.

[13] J. B. Harley, 'Maps, knowledge and power', in D. Cosgrove and S. Daniels (eds), *The Iconography of Landscape* (Cambridge 1988), 277–312. [14] See Chapter 3, Table 3.6.

The outcomes were also spatially variable, with the values being challenged and negotiated at greater length in some Districts than in others, to the chagrin of senior Inland Revenue officials who were themselves answerable to the Chancellor. And the documents have received markedly differing treatments across the United Kingdom since their original inception and especially since their release into the public domain because archivists have themselves responded differently depending on degrees of knowledge and pressure, and power in acquiring the means and space to store the documents adequately.

This aspect of *difference* in the conception, establishment, process of valuation and subsequent documentary history of the 1910 material underlies the details entered into in the foregoing chapters of this book. The documents themselves show both uniformity and specificity, both conformity and non-conformity, elements which lend uncertainty and, perhaps, charm to this source.

In one sense this book deals with an addition to the lengthy list of fiscal surveys and taxation records that have been studied by historical geographers in the past. All the caveats and difficulties inherent in dealing with records that were not set up by geographers for geographers, but by fiscal and bureaucratic minds, must be remembered. It is to be hoped that the potential and the problems, as well as the full context of the records, have been presented in the foregoing chapters. This is also therefore an exercise in the reconstruction of an ideological debate as well as in the evaluation of a source.

And in an important sense, this taxation legislation was also different. Grigg was able to map the Income Tax, Schedule A Returns, which exist for all parishes in 1818, 1844 and 1860. The Land Tax yields can be mapped, even if their precise relation to property acreages remains debatable. Poor Rates could similarly be mapped, as have the 1334 Lay Subsidies, 1377 Poll Tax or the 1524–5 Returns.[15] But in the case of the 1910 material, we cannot map the yield. We are presented with an enormous amount of information in the records, all of which was supposedly aimed at achieving the calculation of values of various kinds. But the actual amounts of income to the Treasury from the taxation on land values remained very limited. Indeed, the main classes of document that we handle, the Field Books, Valuation Books, Maps and Forms of Return, tell us what the valuer thought was the worth of the property, but that worth would not be realised until an 'occasion' took place to trigger the tax, such as a death which resulted in a transfer of the property, and which might then yield some Increment Value Duty. But by definition, such 'occasions' could

[15] See, for example, D. Grigg, 'The changing agricultural geography of England: a commentary on the sources available for the reconstruction of the agricultural geography of England, 1770–1850', *Transactions of the Institute of British Geographers* 41 (1967), 73–96; M. E. Turner and D. Mills (eds.). *Land and Property: The English Land Tax 1692–1832* (London 1986); H. C. Darby, R. E. Glasscock, J. Sheail and G. R. Versey, 'The changing geographical distribution of wealth in England 1086–1334–1525', *Journal of Historical Geography* 5 (1979), 249–56.

not be in the documents that we handle. We therefore have the situation where we might, for example, map Total Value in a parish or town. This might give interesting spatial insights into such localities, but they do not represent 'real' taxes paid. They represent negotiated values which were often fiercely debated at the local level, and which were the basis of tax calculations, not their eventual outcome. In some senses, although the records are taxation documents, their importance lies more in the large and detailed amounts of ancillary information that was collected in order to establish the basis for taxation, than in any information about the tax yield itself.

There are many directions that future research can take which will build upon this work. A full history of the Valuation Office should be undertaken, which could then be compared with that of Ireland, or indeed other European countries. The number of local studies that could be undertaken would be innumerable, and comparative studies using other data from other periods or other localities raise many intriguing possibilities. The Record Sheet Maps and Working Sheets, so diffusely arranged across the country, should also receive their fair share of attention, and at least one study based at the Public Record Office has pointed the way.[16] The early twentieth century was also a time of very advanced political comment and journalism, and a fascinating study awaits the researcher into the ways in which the media handled the whole affair. The study has by no means exhausted all the ramifications of the material itself. Little has been written here, for example, about the Mineral Rights Duty, the only part of the legislation which escaped repeal in 1920, and which is now ripe for a thoroughgoing analysis. Much more could also be written about the incidence of the Undeveloped Land Duties and the Reversion Duties which were introduced at the same time as the Increment Value Duty but which were overshadowed by the latter.

Technically, this book can now also be supplemented by a comprehensive survey of landownership *c.* 1910 based on the Valuation material. Using computer-aided analyses, we could approach an overview which might serve to add to the debates about landownership in England and Wales on the eve of the Great War. And the advent of Geographical Information Systems gives a spatial component to such studies which will also serve to deepen our understanding of the spatial variations in landownership at many different scales of enquiry. In 1982 Brian Harley warned that 'Amidst a search for new philosophies, methodologies and techniques to vitalize the practice of historical geography, the independent study of historical evidence has been neglected.'[17] This need not be the case. Philosophy, methodology and sources constantly

[16] W. Foot, *Maps for the Family and Local Historian* (Public Record Office 1994).

[17] J. B. Harley, 'Historical geography and its evidence: reflections on modelling sources', in Alan R. H. Baker and Mark Billinge, *Period and Place: Research Methods in Historical Geography* (Cambridge 1982), 261.

speak to each other in ways which enrich our understanding of the past, and as our technical capacity to deal with large volumes of data improves, so our confidence in interpreting the patterns increases.

It is indeed to be hoped that such analyses will be undertaken, and that their objectives will indeed serve to enrich the ideas put forward in this book. It is also to be hoped that archivists and researchers will cooperate to bring the documents forward from what is in some cases virtual isolation in out-repositories, and that the lessons learned from the difficulties encountered in archiving this material will be absorbed. It is also to be hoped that the warnings and dangers about the complexity of the material will also be heeded, and given this, that we can move forward confidently dealing with the largest data bank in British history.

Form 4—Land.

DUTIES ON LAND VALUES.

(*Finance (1909-10) Act, 1910.*)

REFERENCE: To be quoted in all communications.

RETURN TO BE MADE BY AN OWNER OF LAND OR BY ANY PERSON RECEIVING RENT IN RESPECT OF LAND.

(*Penalty for failure to make a due Return, not exceeding £50.*)

Reference to Instructions (Form 2—Land). [*See p. 259*].			This space is not for the use of the person making the Return.
See Instruction 2.	Particulars extracted from the Rate Books.	Parish	
		Number of Poor Rate . ..	
		Name of Occupier	
		Description of Property	
		Situation of Property	
		Estimated extent..	Acres Roods
		Gross Estimated Rental (or Gross Value in Valuation List*) ..	£
		Rateable Value ..	£
	(*Applicable to the Metropolis only.)		

IMPORTANT.—As the Land is to be valued as on 30th April, 1909, the particulars should be furnished, so far as possible, with reference to the circumstances existing on that date.

See Instruction 3.	**I. Particulars required by the Commissioners, which must be furnished so far as it is in the power of the person making the Return to give them.**	
See Instruction 4.	(*a*) Parish or Parishes in which the Land is situated.	
	(*b*) Name of Occupier ..	
See Instructions 1 and 3.	(*c*) Christian Name and Surname and full postal address of the person making the Return.	
	(*d*) Nature of Interest of the person making the Return in the Land:—	
	(1) Whether Freehold, Copyhold or Leasehold.	1
	(2) If Copyhold, name of the Manor.	2
See Instruction 9.	(3) If Leasehold, (i.) term of lease and date of commencement (including, where the lease contains a covenant for renewal, the period for which the lease may be renewed), and (ii.) name and address of lessor or his successor in title.	3 (i.) 3 (ii.)

FORM 4—DUTIES ON LAND VALUES—*continued*.

<table>
<tr><td></td><td>(e) Name and precise situation of the Land.</td><td colspan="4"></td></tr>
<tr><td>See Instruction 2.</td><td>(f) Description of the Land, with particulars of the buildings and other structures (if any) thereon, and the purposes for which the property is used.
(House, Stable, Shop, Farm, &c.</td><td colspan="4"></td></tr>
<tr><td></td><td>(g) Extent of the Land, if known.</td><td>Acres.</td><td>Roods.</td><td>Perches.</td><td>Yards.</td></tr>
<tr><td></td><td>(h) If the Land is let by the person making the Return, state :—
(i.) Whether let under Lease or Agreement, or
(ii.) If there is no Lease or written Agreement, whether let by the Year, Quarter, Month, or Week.
(iii.) If let under Lease or Agreement—
(a) Term for which granted.
(b) Date of commencement of term.
(c) Whether granted for any consideration in money, paid or to be paid by the Tenant, in addition to the Rent reserved,* or
(d) Upon any condition as to the Tenant laying out money in Building, Rebuilding, or Improvements.*
(iv.) Amount of Yearly Rent receivable.
(*If so, give full particulars.)</td><td colspan="4">(i.)
(ii.)
(iii.) (a)
(b)
(c)
(d)
(iv.) £</td></tr>
<tr><td></td><td>(i) If the person making the Return is also the Occupier, state the Annual Value; i.e., the Sum for which the Property is worth to be Let to a Yearly Tenant, the Owner keeping it in repair.</td><td colspan="4">Annual Value £</td></tr>
</table>

FORM 4—DUTIES ON LAND VALUES—*continued.*

	(*k*) Amount of Land Tax (if any) and by whom borne.	£ borne by
	(*l*) Amount of Tithe Rent-charge, or of any payment in lieu of Tithes, for the year 1909, and by whom borne.	£ borne by
	(*m*) Amount of Drainage, or Improvement Rate, or of any similar charge, and by whom borne.	£ borne by
	(*n*) Whether all usual tenants' Rates and Taxes are borne by the Occupier, and, if not, by whom.	
	(*o*) By whom is the cost of repairs, Insurance, and other expenses necessary to maintain the Property borne?	
See Instructions, page 1, foot-note †.	(*p*) Whether the Land is subject to any :— (i.) Fixed Charges (exclusive of Tithe Rentcharge entered in space (*l*)), and, if so, the Annual Amount thereof. (ii.) Public Rights of Way.. (iii.) Public Rights of User.. (iv.) Right of Common .. (v.) Easements affecting the Land. (vi.) Covenant or Agreement restricting the use of the Land, and, if so, the date when such Covenant or Agreement was entered into or made. (Full particulars should be given in each case.)	Annual Amount £ Date when made
	(*q*) Particulars of the last sale (if any) of the Land within 20 years before 30 April, 1909, and of Expenditure since the date thereof :— (i.) Date of Sale .. .. (ii.) Amount of Purchase-money and other consideration (if any). (iii.) Capital expenditure upon the Land since date of Sale.	(i.) (ii.) (iii.)

FORM 4—DUTIES ON LAND VALUES—*continued.*

See Instruction 5.	(*r*) Observations, with description, extent, and precise situation of any part of the Land which the Owner requires to be separately valued.	
	(*s*) If the person making the Return desires that communications should be sent to an Agent or Solicitor on his behalf, the name and full postal address of such Agent or Solicitor.	
See Instruction 6	*(*t*) (i.) Does the person making the Return own the minerals comprised in the Land?	(i.)
	(ii.) If so, state:—	
	(*a*) Whether the minerals were, on 30 April, 1909, comprised in a mining lease or being worked by the proprietor.	(ii.) (*a*)
	(*b*) Whether the minerals are now comprised in a mining lease or being worked by the proprietor.	(*b*)
	(iii.) If not, state the name and address of the proprietor of the minerals.	(iii.)
	(* Minerals not comprised in a mining lease or being worked, are to be treated as having no value as minerals, unless the proprietor of the minerals fills up space (*w*) below.)	

I hereby declare that the foregoing particulars are in every respect fully and truly stated to the best of my judgment and belief.

Dated this day of 191 .

____________________ Signature of person making the Return.

____________________ Rank, Title, or Description.

FORM 4—DUTIES ON LAND VALUES—*continued.*

	II. Additional particulars which may be given, if desired.	
See Instructions 7, 8, and 9.	(*u*) Value of the Land as defined in Instruction 7, and estimated by the Owner, with particulars how arrived at:— (i.) Gross Value .. .. (ii.) Full Site Value .. .. (iii.) Total Value .. .. (iv.) Assessable Site-Value .. (v.) Particulars how Values arrived at* (* May be given on separate sheet of paper, if desired.)	(i.) £ (ii.) £ (iii.) £ (iv.) £ (v.)
See Instructions 7, 8, and 9.	(*v*) If the Owner does not desire to furnish his estimate of of the Value of the Land, but intends to claim a Site-Value deduction under Instruction 7 (iv.), (*a*), (*b*), (*c*), or (*d*), or under Instruction 9 (i.), (*a*), the intention should be stated. A form will then be sent in due course for particulars of the claim to be given.	
See Instructions 6 and 10.	(*w*) Nature, and estimate of the Capital Value of any minerals not comprised in a mining lease and not being worked, which have a value as minerals.	Nature Capital Value £

______________________ Signature.

______________________ Date.

Appendix 2 Select list of forms used in the Valuation

At least 183 forms were promulgated by the valuation department during the period that the land clauses operated. Most were simply 'internal' forms, such as office memoranda, stores requisitions and expenses claims, while many others were standard letters of one sort or another. There were two main categories – 'Land' forms and 'Valuation Office (VO)' forms. The list that follows includes only those most likely to be encountered by researchers, does not include material specific to Ireland, and are all 'Land' forms.

No.	*Name of Form*
1	Notice to make Returns
2	Instructions for making Returns on Form 4
3	Notice to make Returns (statutory companies)
4	Form of Return for owners of land
5	Notice to make Returns (Mineral Rights Duty)
6	Notice to make Returns (Unworked Minerals)
6A	Notice to make Returns (Unworked Minerals – Revised version)
7	Claim for Site Value Deductions
8	Notice to furnish information
8A	Notice to furnish information (Mineral Rights Duty)
9	City of London: Notice to owners etc. of blocks of offices, flats etc. let in separate tenements
10	Demand for Returns (1st reminder)
11	Demand for Returns (2nd reminder)
16	Notice to owners etc. of blocks of offices, flats etc. let in separate tenements
18	Undeveloped Land Duty (1911–12): Notice that assessment is to be made of stated property
20	Valuation Book: title page
21	Valuation Book: 2nd page
22	Valuation Book: main pages
27	Valuation Book: (Mineral Rights Duty) title page

No.	*Name of Form*
28	Valuation Book: (Mineral Rights Duty) main pages
29	Valuation Book: (Undeveloped Land Duty) title page
30	Valuation Book: (Undeveloped Land Duty) main pages
35	Standard letter to accompany Form 36
36	Provisional Valuation
37	Provisional Valuation (office copy)
38	Amended Provisional Valuation
39	Amended Provisional Valuation (office copy)
40	Book of Original Valuations
40A	Report on Original Valuations
40B	Site Value on the occasion
46	Letter to accompany Form 48
47	Provisional Valuation: Minerals (office copy)
48	Provisional Valuation: Minerals
49	Letter to accompany Form 51
50	Amended Provisional Valuation: Minerals (office copy)
51	Amended Provisional Valuation: Minerals
53	Letter to accompany Original Site value
54	Original Site Value
55	Original Site Value (office copy)
56	Letter to accompany Form 57
57	Amended Original Site Value
58	Amended Original Site Value (office copy)

Appendix 3 Valuation documents held in local repositories

The list is based on returns from local repositories to the 1985/6 questionnaire. The existence of holdings does not imply any degree of completeness. This was impossible to assess, since many offices had not (and indeed still had not by 1995) yet listed the material, and even where listings do exist, they are not always fully reliable. In any event the task was beyond the resources at our disposal in the pilot study.

It should be noted that many repositories contain material relating to areas outside of their current jurisdiction. This is due partly to the fact that the Valuation Department of the Inland Revenue was not always organised on a strictly county basis, and also that county boundaries have altered since, most notably as a result of local government reorganisation in 1973. Further reorganisation in the 1990s will again affect the holdings.

Repository documentation

Key: Y Yes; N No; NK Not Known; NI No information;
*Information obtained on personal visit.
+Form 37 retained only if no Valuation Books held. Otherwise destroyed.
$The respondent at Sheffield Library claimed that maps were not accepted. The South Yorkshire CRO claimed that the Library destroyed them.

	Valuation Books	Forms 37	Working Maps	Misc.
ENGLAND				
Bath City Library	Y	N	Y	Y
Bedfordshire CRO	Y Sample only	Y	Y	Y
Berkshire CRO	Y	Y	Y	NK
Bristol City RO	Y	Y	Y	NK
Buckinghamshire CRO	Y	Y	Y	NK

	Valuation Books	Forms 37	Working Maps	Misc.
Cambridgeshire CRO (Cambridge)	Y	Y	Y	Y
Cambridgeshire CRO (Huntingdon)	Y	Y	Y	Y
Cheshire CRO	No Reply			
Cleveland CRO	Y	Y	N	NK
Cornwall CRO	Y	Y	Y	Y
Coventry Library	Y	Y	Y	NK
Cumbria CRO (Carlisle)	Y	Y	Y	Y
Cumbria CRO (Kendal)	Y	N	Y	NK
Cumbria CRO (Barrow)	N	Y	N	Y
Derbyshire CRO	Y	Y+	Y	NK
Devon CRO (Exeter)	Y	N	Y	NK
Devon CRO (Plymouth)	Y	N	N	NK
Dorset CRO	Y	N	N	NK
Dudley Library	Y	Y	N	NK
Durham CRO	Y	N	Y	NK
Essex CRO (Chelmsford)	Y	N	Y	NK
Essex CRO (Southend)	N	N	N	NK
Gloucestershire CRO	Y	Y	Y	Y
Hampshire CRO	Y	Y	Y	Y
Herefordshire CRO	Y	Y	Y	NK
Hertfordshire CRO	Y	Y	Y	Y
Hull City RO	Y	Y	N	NK
Humberside CRO	No Reply			
Isle of Wight CRO	N	N	N	NK
Kent AO (Maidstone)	Y	Y	Y	NK
Kent AO (Folkestone)	N	N	N	NK
Lancashire CRO	Y	N	Y	Y
Leicestershire CRO	Y	Y	Y	NK
Lincolnshire CRO	Y	Y	Y	NK

LONDON

Bexley Library	Y	N	N	NK
Brent (Grange Museum)	Y	N	Y(?)	NK
Camden Library	No Reply			
Ealing Library	Y	N	N	NK
Enfield Library	Y	N	N	NK
Greenwich Library	Y	N	N	NK
Hackney Library	Y	N	N	NK
Hammersmith Library	Y	N	N	NK

	Valuation Books	Forms 37	Working Maps	Misc.
Haringey (Bruce Castle Museum)	Y	N	N	NK
Hillingdon Library	Y	N	N	NK
Hounslow Library	Y	N	N	NK
Islington Library	Y	N	N	NK
Kensington and Chelsea Library	Y	N	N	NK
Lambeth (Minet Library)	Y	N	N	NK
Lewisham Library	Y	N	N	NK
Newham (Passmore Edwards Museum)	Y	N	N	NK
Southwark Library	Y	N	N	NK
Tower Hamlets Library	Y	N	N	NK
Waltham Forest (Vestry House Museum)	Y	Y	Y	NK
Manchester Library	N	N	Y	NK
Merseyside CRO	Y	Y	N	NK
Norfolk CRO	Y* No Reply			
Northamptonshire CRO	Y	Y	N	Y
Northumberland CRO	Y	N	Y	Y
Nottinghamshire CRO	Y	Y	Y	NK
Oxfordshire CRO	Y	Y	Y	Y
Sheffield City Library	N	N$	N	Y
Shropshire CRO	Y	Y	Y	Y
Somerset CRO	Y	N	Y	NK
Southampton City RO	Y	N	N	NK
South Humberside CRO	Y	Y	Y	Y
Staffordshire CRO	Y	N	Y	NK
Suffolk CRO (Ipswich)	Y	N	N	NK
Suffolk CRO (Bury)	Y	N	Y	NK
Surrey CRO (Kingston)	Y	Y	Y	NK
Surrey CRO (Guildford)	N	N	N	Y
Sussex (East) CRO	Y	Y+	Y	Y
Sussex (West) CRO	Y	Y+	Y	Y
Tyne and Wear CRO (Newcastle)	No Reply			
Tyne and Wear CRO (North Shields)	No Reply			
Walsall RO	N	N	N	Y
Warwickshire CRO	Y	Y	Y	NK

	Valuation Books	Forms 37	Working Maps	Misc.
Wiltshire CRO	Y	Y	N	NK
Worcestershire CRO	Y	Y	Y	NK
Worthing Library	N	N	Y*	NK
Yorkshire (North) CRO	Y	NI	NI	NI
Yorkshire (South) CRO	Y	Y	N	NK
Yorkshire (West) CRO (Wakefield)	Y	Y	Y	NK
Yorkshire (West) CRO (Leeds)	N	N	Y	Y
WALES				
Clwyd CRO	Y	Y	Y	NK
Dyfed CRO (Aberystwyth)	Y	Y	N	NK
Dyfed CRO (Carmarthen)	Y	Y	Y	NK
Dyfed CRO (Haverfordwest)	Y	Y	Y	Y
Glamorgan CRO	Y	Y	N	NK
Gwent CRO	Y	Y	N	NK
Gwynedd CRO (Caernarfon)	Y	N	Y	NK
Gwynedd CRO (Llangehui)	Y	N	N	NK
Gwynedd CRO (Dolgellau)	Y	N	N	NK
Powys Library	Y	Y	N	Y
National Library of Wales	No Reply			

Appendix 4 Questionnaire sent to archivists

LANDOWNERSHIP AND SOCIETY IN EDWARDIAN ENGLAND
THE FINANCE (1909–10) ACT 1910 RECORDS

We thank you for your cooperation, which will be acknowledged in any subsequent publication.

WE WOULD BE VERY GRATEFUL FOR COPIES OF THE RELEVANT FINDING AIDS DEALING WITH INLAND REVENUE RECORDS AND OTHER MATERIAL, WHICH MAY OBVIATE THE NEED OF ANSWERING MANY OF THE QUESTIONS BELOW.

NAME OF REPOSITORY ....

AREA OF RESPONSIBILITY ....

1. THE 'DOMESDAY BOOKS'

1a. Did you accept the 'Domesday Books'? (YES/NO) (Please delete as appropriate)

1b. Do you have a complete coverage (i.e. every parish) for the area for which you are responsible? (YES/NO)

1c. If not, please tell us how many parishes are missing.

....

1d. Does there appear to be any pattern in the missing parishes? Are they in a particular geographical area? Are they areas of large estates?

....

....

....

(please continue on a separate sheet if necessary)

1e. Do you have any coverage outside your area of responsibility? (Y/N). If so, for what areas?..

2. LARGE SCALE MAPS

The maps from the Inland Revenue Office have an embossed or rubber stamp, examples of which are on the accompanying sheet.

2a. Did you accept any sheets? (Yes/No)

2b. Approximately how many sheets do you have which are complete with property boundaries?

....

2c. Is it possible to estimate what proportion of the county (or similar area) is so covered?

....

2d. Do they appear to exist for particular geographical areas? Do they exist, for example, for urban rather than rural areas?

....

....

....

(Please continue on a separate sheet if necessary)

2e. Do you have any knowledge of the availability of the working copies at your local District Valuation Office of the Inland Revenue?

3. FORMS 37–LAND

3a. What policy did you adopt for these forms?
(i) Not to accept.
(ii) To accept only for parishes without 'Domesday Books'.
(iii) To accept all.

3b. If accepted, were some or all subsequently destroyed? (Yes/No)

3c. For how many parishes do you have a copy of Form 37–Land?

....

3d. What proportion of the total number of parishes does this represent?

....

3e. If they do exist for some parishes but not others, is there any particular area of your county etc. which is so covered?

....

....

....

4. OTHER RECORDS RELATING TO THE 1910 FINANCE ACT

Could you please tell us of any other material of relevance to our study. Could you please list the name and reference number of such material below.

....

....

....

5. OTHER RESEARCHERS

Could you please tell us whether you know of any researchers who are currently using this material, or who have published work which makes some use of the material. Their name and address would be very helpful.

....

....

....

6. OTHER ARCHIVES

Please could you tell us whether there are any other archive repositories in your area which hold 1910 Finance Act material, e.g. libraries

....

....

....

THANK YOU VERY MUCH FOR YOUR COOPERATION. WE HOPE THAT WHEN OUR RESULTS EMERGE, YOU WILL FIND THEM OF SOME USE.

Bibliography

Only those sources which have been cited directly in the text are included in this list, except those noted in Section 4. There are also very many other smaller references available in newspapers and magazines of the period. The literature on land taxation and the taxation of land values between 1880 and 1920 is enormous and much of it is ephemeral. Pamphlet collections at the Bodleian Library, Oxford, British Library etc. are detailed in S.Ward (1976), see below.

1 Unpublished primary sources

Public Record Office (PRO)

ADM 116/1279 (Form 4 holdings of the Admiralty)
CAB 37/96/161; 37/97/9 'The taxation of land values in New York'
F 6/16 (Form 4 holdings of Forestry Commission)
IR 9/62–64 (Various specimen forms)
IR 40/2502 Record of the Test Case of Lady Emily Frances Smyth v Commissioners of Inland Revenue (the Norton Malreward Case)
IR 58 (Field Books)
IR 63/32A, N21.6 notes the consequences of the Lumsden case 1911–13
IR 63/35, N22.1 E.J. Harper, 'Cost of Valuation' (2 May 1914) f.168
IR 74/146 M3.11 H. Thomas, 'Draft notes on the organisation of the land valuation department' (June 1910)
IR 74/148 Report by the Chief Valuer and Deputy Chief Valuer upon the Progress of the Original Valuation under the Finance (1909–10) Act 1910, together with copies of the Board's order thereon 1912–13
IR 74/218 Account of the History and Responsibilities of the Valuation Office (1920)
IR 83/54 Notes on the calculation of Increment Value Duty
IR 91 Valuation Books for the City of London and for Westminster (Paddington)
IR 121, 124–35 (Record Sheet Plans)
LAR 1/24 (File no. 005/9)
MAF 38/865 Correspondence re use of National Farm Survey to replace missing information caused by losses due to enemy air raids
MAF 68/2435 (Agricultural Returns 1910, Parish Summaries for West Yorkshire)

RAIL 1057/1714 (Forms 4 for Rhymney Railway)
T 171/40 Builders' deputation to Lloyd George 17 April 1913
PRO Circular 3/CCA/6 1968
PRO circular GEN 33/2/2 (19 September 1979)

Scottish Record Office (Edinburgh)

SRO IRS 88 Survey Books (Drafts)
IRS 66 Field Books
IRS 67 Valuation Office maps

National Archives of Ireland (Dublin)

VO Finance Act 1910 (various boxes) Field Books, correspondence, administration of Valuation Office 1910–20

Public Record Office Northern Ireland (PRONI)

VAL 1B71A–1B719A (1830 valuation records)
VAL 2B and 2A (Griffith's valuation and maps)
VAL 12B (Annual Revision lists)
VAL 3B and 3A (First General Valuation 1935)

National Library of Wales (Aberystwyth)

Lloyd George Papers

Chief Valuer's Library, Inland Revenue Head Office, Carey Street, London

'Duties on Land Values: Interim report of the Committee appointed to draw up Regulations for carrying into effect the provisions of the Finance bill, 1909, dealing with the duties on Land Values' (nd, np)
Box 3/4 Chief Valuer's memo on grouping of hereditaments 8 February 1912
Box 3/5 Edgar Harper letter to C. J. Howell-Thomas

County, local and other record repositories

Bath City Council Corporation Finance Committee files, ref. F (77)
Bedfordshire CRO W3500–7 (Whitbread Papers) and PM 2937/11 (Pym Papers)
Berkshire CRO P/DVO 6 Acc. 2487/55 Valuation Book
Brighton Urban Studies Centre: Photographs of Brighton streets in the 1930s prior to slum clearance
Cambridgeshire County Record Office R84/62; 470/0105 (Valuation Book for Over, including Rampton and Willingham) and 470/0106 (Valuation Book for Isleham)
Cleveland County Archives Department, Middlesbrough. Valuation Books

Cornwall Record Office AD 772/256–83; and Crantock DV(1)/136; DV(2)/21,77 (Vicar of Crantock's appeal)

Cumbria Record Office (Barrow-in-Furness) Bd/HJ Box 311; BT/V/1, Boxes 1 and 2 (Forms 37); Carlisle Office: TIR 4/4, 20, 22, 26, 28, 33, 44, 50, 53, 63, 78, 79, 81, 86, 88, 93, 105 (Valuation Books); DB 98/49–50; DB 66/11–15

Derbyshire County Record Office D595R/4/1/71–3; D595R/3/1/7a; 1114; Valuation Books D595R/4/1/73; 4/2/1–359

Devon Record Office (Exeter): Tithe apportionments for Aveton Gifford, Loddiswell, East Allington and Woodleigh; Valuation Books for each parish (uncat.)

Dudley Public Library: Valuation Books for Rowley Regis

East Sussex Record Office C/C42/1 (Forms 4); Battle Abbey, BAT 2326; Hastings area, SAY 2897; Glynde, Acc. 4262. East Sussex Record Office, Ashburnham Mss (uncatalogued), Williams to Collins, p.2; Valuation Books IRV 1/18, 23, 29, 110; IRV 1/3 Ashburnham Valuation Book; Ashburnham Mss, uncatalogued estate rental 1909

Glamorgan Archives Service: Valuation Books D/D PRO/VAL/1/134

Gloucestershire Record Office D2428 1/53: Register of land values of licensed premises, Gloucester Sub-District; D2428 (plan of the county's ITPs)

Greater London Council Record Office, Bedford Mss E/BER/C9/E12/8 'Brief account of the principal steps taken in the settlement of the Provisional Valuations on the Bedford (London) Estates, 1910–17'

Gwynedd Archives and Museums Service (Archifau ac Amgueddfeydd): Valuation Books, XLTD/11 (2 vols.)

Hampshire CRO 152 M 82/9/1 (Valuation Book for Ashley Walk ITP), Forms 4 (unlisted); Forms 37–Land (unlisted)

Herefordshire Record Office (Hereford and Worcs.) AG 9/91 Valuation Book for Almeley, Eardisley AG9/96, Kinnersley AG9/97, Lyonshall AG9/43 and Kington (Urban and Rural) AG9/37–8; K10/10 Newport Estate Catalogue

Hertfordshire Record Office IR3/33/1–2 Forms 37; Hertfordshire RO IR1/1 Working Sheet Maps; IR1/1–520

Hove Reference Library: Sussex Industrial Archaeology Society drawer: Duties on Land Values 1910, Form 4–Land. The parishes covered are Hailsham, Hellingly, St Thomas in the Cliffe (Lewes), Wartling, Jevington, Warbleton, Arlington, Pevensey and Alfriston

Leicestershire Record Office DE 2064/1a–1096 (Forms 37): Valuation Books for Leicestershire and Rutland

Lincolnshire Archives Office 6 Tax (Forms 37) Binbrook 6 Tax/42/53; 2BD 7/136 Sale particulars of Binbrook Estate 12 July 1907

Norfolk County Record Office P/DLV 1/114 Valuation Book (Cromer ITP)

Northumberland Record Office NRO.1952 (Forms 37); ZBL.62/11 (Forms 4, 7 and 36, Matfen and Westwater Estates)

North Yorkshire County Record Office: Valuation Books for Grassington and Buckden (uncat.)

Nottinghamshire Record Office: Valuation Book for Laxton

Powys Library, Llandrindod Wells M/LVR VB. 23 (uncat.Valuation Book); Forms 37 (uncat.)

St John's College, Cambridge library, Forms 4 and 36 (6 boxes)

Somerset Record Office DD/IR W 46/3; T 2/2; 8/4 (Valuation Books)

Southwark Local Studies Library: Valuation Books for Camden ITP

Surrey Record Office 2415 Finance (1909–10) Act 1910: 'Domesday Books' Public Records presented under the Public Records Act 1958, s.3(6), 2; Guildford Muniment Room 85/42/5; 85/29/9 (Bray family papers re 1910 legislation); 173/19/43–45 and 98 (Onslow collection)

Warwickshire County Record Office CR 1978/2/34

West Sussex Record Office Add. MSS. 32872–6 (Forms 4 relating to various parishes in West Sussex but also including Brighton, Guildford, Croydon, Datchet, Langley, Penarth, Hemel Hempstead and Highgate)

West Yorkshire Archive Service C243/1–546; N269, C275; L280; Wakefield Office C243/106 to 110; C243/121 to 122 Valuation Books

Wiltshire Record Office, Form 24–Land (Inside Valuation Book for Pewsey 1910, Inland Revenue register no. 57. Pewsey); Domesday Books, Register 126 Corsley and Upton Scudamore ITP

2 Printed primary sources

Commission on Local Taxation, Final Report (Cd. 638) 1901 – 'Separate Report on Urban Rating and Site Values', 165–6

Commissioners of HM Inland Revenue, 54th Report, Year ended 31 March 1911, BPP 1911 (Cmd. 5833), XXIX

Commissioners of HM Inland Revenue, 55th Report, Year ended 31 March 1912, BPP 1912–13 (Cd. 6344), XXIX

Commissioners of HM Inland Revenue, 64th Report, Year ended 31 March 1921, BPP 1921 (Cmd. 1436), XIV

Daily News 21 June 1912

Finance (1909–10) Act (10 Edward 7) 1910

Finance Act (10 and 11 Geo. 5) 1920

Financial Times 17 September 1910

House of Commons Debates, 1909, iv, cc.472–548; vi, c.1123; vii; 1911, xxii, 927, c.1912; cc.1538–9

Inland Revenue Valuation Department, *List of Valuation Districts* (HMSO, London, October 1910)

Inland Revenue Valuation Department, *List of Valuation Districts with the Income Tax Divisions in each District and of the District Valuers in Great Britain* (HMSO, London, October 1910)

Inland Revenue Valuation Department, *Return showing the names and qualifications of the Chief Valuer, Superintending Valuers . . . 31st Day of January 1911* (for Mr Austen Chamberlain, Return to an Order of the House of Commons, 14 March 1911, HMSO, London, March 1911)

Inland Revenue Valuation Department, *Instructions to Valuers* Vol. I (HMSO, London 1910)

Inland Revenue Valuation Department, *General Instructions to Land Valuation Officers* (HMSO, London 1910)

Kelly's Directory of Lincolnshire (1913)

Kelly's Directory for Norfolk (1912)

Land Values: Journal of the Movement for the Taxation of Land Values July 1910, 6; September 1910, 67; October 1910, 91, 100–1; November 1910, 115; January 1911, 182; May 1911, 268; November 1911, 136
Lloyd George Liberal Magazine 1 (October 1920)
Ordnance Survey Annual Report 1911–12, BPP 1912–13, XLII
Punch 24 August 1910, 128; 25 October 1911
Return on the Owners of Land, England and Wales (1872–73), BPP LXXII, 1874
Royal Commission on the Historical Monuments of England, *Village Farm, Swaton, Lincolnshire* (Farmsteads Survey, September 1993) and *Grange Farm, Little Hale, Lincolnshire* (September 1993)
Scottish Record Office, *Valuation Office Records: Records created under Finance (1909–10) Act* (SR509183.053) (nd)
Select Committee on the Land Values Taxation, etc. (Scotland) Bill, Report, 13 December 1906 (reprinted in *Land Values*, January 1907)
Select Committee on Land Values, Final Report, BPP 1920 (Cmd. 556) XIX
The Times 27 February 1906; 26 March 1908; 12 March 1910; 14 April 1910; 13 October 1913; 23 October 1913; 10 November 1913; 21 July 1914

3 Secondary sources (contemporary and recent)

Aalen, F. H., 'Approaches to the working class housing problem in late Victorian Dublin: the Dublin Artisans' Dwelling Company and the Guinness (later Iveagh) Trust', in R. J. Bender (ed.), *Neuere Forschungen zur Sozialgeographie von Irland*, Mannheimer Geographische Arbeiten 17 (Mannheim 1984)
Aalen, F. H., 'Health and housing in Dublin *c.*1850 to 1921', in F. H. Aalen and K.Whelan, (eds.), *Dublin – City and County: From Prehistory to Present* (Dublin 1992)
Adams, I. H., *Agrarian Landscape Terms: A Glossary for Historical Geography* (Institute of British Geographers Special Publication No. 9, London 1976)
Aggs, W. H., *The Finance (1909–10) Act 1910, with Full Notes on the Land Duties* (London 1910)
Ailesbury, Lord, 'The rural problem: some reflections on the Land Enquiry', *The Nineteenth Century and After* 74 (1913), 1087
Andrews, J. H., *A Paper Landscape: The Ordnance Survey in Nineteenth-Century Ireland* (Oxford 1975)
Andrews, J. H., *History in the Ordnance Map: An Introduction for Irish Readers* (2nd edn. 1993, Kerry)
Anon., 'Our old brown mother', *Land Values* (November 1910), 115
Anon., 'The Land Inquisition', *The Spectator* 105 (3 September 1910), 337
Anon., 'Inequalities in valuation', *Land Values* (January 1911), 182
Anon., 'The progress of the valuation', *Land Values* (November 1912), 270
Anon., 'The collapse of the Land Taxes', *The Spectator* 110 (19 April 1913), 646–7
Anon., 'The arch-depopulator of the Highlands', *The Nineteenth Century* (June 1914), 1413
Anon., *The Valuation Office 1910–85: Establishing a Tradition* (Inland Revenue 1985)
Arensberg, C. and Kimball, S., *Family and Community in Ireland* (Cambridge, Mass. 1940)

Armitage, G., 'Ordnance Survey Land Valuation Plans', *Sheetlines* 35 (1993), 8–9
Aspinall, P. J., *Building Applications and the Building Industry in Nineteenth-Century Towns: The Scope for Statistical Analysis* (Research Memorandum 68, Centre for Urban and Regional Studies, University of Birmingham, 1978)
Barnwell, P., 'The National Farm Survey, 1941–43', *Journal of the Historic Farm Buildings Group* 7 (1993), 12–19
Bateman, J., *The Great Landowners of Great Britain and Ireland* (London 1883)
Beckett, J. V., *The Aristocracy in England 1660–1914* (Oxford 1986)
Beckett, J. V., *The East Midlands from AD 1000* (London 1988)
Bew, P., *Land and the National Question in Ireland 1858–82* (Dublin 1978)
Bickerdike, C. F., 'The principle of land value taxation', *The Economic Journal* 22 (March 1912), 1–15
Blewett, N., *The Peers, the Parties and the People: The General Elections of 1910* (London 1972)
Booth, Charles, *Rates and the Housing Question in London: An Argument for the Rating of Site Values* (English League for the Taxation of Site Values 1904)
Bowles, Zara, 'Rights of way and the 1910 Finance Act', *Rights of Way Law Review* (September 1990), Section 9.3, 17–18
Brandon, P. F., 'A twentieth-century squire in his landscape', *Southern History* 2 (1980), 191–220
Brooks, E. M., 'Inland Revenue Valuation Records', *Hampshire Field Club Archaeological Society Newsletter* 3 (1985), 7–8
Brown, J., 'Scottish and English land legislation, 1905–1911', *Scottish Historical Review* 47 (1968), 72–85
Burrow, E. J. (ed.), *Cromer and Sheringham as Holiday and Health Resorts* (Cheltenham, nd, *c.* 1920)
Cameron, D. K., *The Ballad and the Plough: A Portrait of Life in the Old Scottish Farmtouns* (London 1978)
Campbell, R. H., *Owners and Occupiers: Changes in Rural Society in South-West Scotland before 1914* (Aberdeen 1991)
Cannadine, D., *The Decline and Fall of the British Aristocracy* (Yale 1990)
Carllew, Loris (pseudonym), *Alice in Plunderland* (London 1910)
Carter, H. and Lewis, C. Roy, *An Urban Geography of England and Wales in the Nineteenth Century* (London 1990)
Caudwell, W. G., 'Horsham: the development of a Wealden town in the early twentieth century', University of Sussex, unpubl. MA dissertation, 1986
Chamberlain, Mary, *Fenwomen: A Portrait of Women in an English Village* (London 1975)
Churchill, Winston, *The People's Rights* (London 1909, repr. 1970)
Cloke, P., Philo, C. and Sadler, D., *Approaching Human Geography* (London 1991)
Clout, H. D. and Sutton, K., 'The cadastre as a source for French rural studies', *Agricultural History* 43 (1969), 215–23
Constantine, S., *Lloyd George* (Lancaster Pamphlets 1992)
Coppock, J. T., 'The relationship of farm and parish boundaries – a study in the use of agricultural statistics', *Geographical Studies* 2(1) (1955), 12–26
Coppock, J. T., 'The parish as a geographical/statistical unit', *Tijdschrift v.Econ en Sociale Geog.* 51, (1960), 317–26

Coppock, J. T., 'Mapping the agricultural returns: a neglected tool of historical geography', in Michael Reed (ed.), *Discovering Past Landscapes* (London 1984), 14–36

Daly, M., 'Late nineteenth and early twentieth century Dublin', in D.Harkness and M.O'Dowd (eds.), *The Town in Ireland* (Historical Studies 13, Dublin 1981), 221–52

Daly, M., 'Social structure of the Dublin working class 1871–1911', *Irish Historical Studies* 23 (1982), 121–33

Dangerfield, G., *The Strange Death of Liberal England* (London 1935)

Darby, H. C., 'The economic geography of England, AD 1000–1250', in H.C.Darby (ed.), *An Historical Geography of England before AD 1800* (Cambridge 1963 edn)

Darby, H. C., Glasscock, R. E., Sheail, J. and Versey, G. R., 'The changing geographical distribution of wealth in England 1086–1334–1525', *Journal of Historical Geography* 5 (1979), 249–56

Daunton, M.J., *House and Home: Working Class Housing, 1850–1914* (London 1983)

David, E. (ed.), *Inside Asquith's Cabinet: From the Diaries of Charles Hobhouse* (London 1977)

Davies, C.E., *Land Valuation under the Finance (1909–10) Act 1910: The New Land Duties, Licensing Duties, Stamp Duties, and Alteration in Death Duties* (London, Estates Gazette, 1910)

Davies, C.E., *Reports on Land Valuation Appeals: Finance (1909–10) Act 1910,* 3 vols. (London 1913–14)

Davies, G.L. Herries and Mollan, R.C. (eds.), *Richard Griffith 1784–1878* (Dublin 1980)

Davies, M. F., *Life in an English Village: An Economic and Historical Survey of the Parish of Corsley in Wiltshire* (London 1909)

Davies, W. K. D., 'Toward an integrated study of central places', in H.Carter and W. K. D. Davies (eds.), *Urban Essays: Studies in the Geography of Wales* (London 1970)

Davies, W. Watkin, *Lloyd George 1863–1914* (London 1939)

Dictionary of National Biography (Twentieth-century supplement 1901–11, ed. Sir Sidney Lee, Oxford 1912)

Douglas, Roy, *The History of the Liberal Party* (London 1971)

Douglas, Roy, 'God gave the land to the people', in A.Morris (ed.), *Edwardian Radicalism 1900–1914: Some Aspects of British Radicalism* (London 1974)

Douglas, Roy, *Land, People and Politics: A History of the Land Question in the United Kingdom, 1878–1952* (London 1976)

Duckworth, G. H., 'The making, prevention and unmaking of a slum', *Journal of the Royal Institute of British Architects* 33 (1936), 327–37

Dunford, M. and Perrons, D., *The Arena of Capital* (London 1983)

Du Parcq, H., *Life of David Lloyd George,* 3 vols. (London 1911–13)

Dymond, D., *The Norfolk Landscape* (London 1985)

Dyos, H. J., *Victorian Suburb: A Study of the Growth of Camberwell* (Leicester 1961)

Emy, H. V. 'The Land Campaign: Lloyd George as a social reformer', in A. J. P. Taylor (ed.), *Lloyd George: Twelve Essays* (London 1971), 35–9

English, Barbara, *The Great Landowners of East Yorkshire 1530–1910* (Hassocks 1990)

Eve, C. Gerald, 'Systems of land valuation in the United Kingdom', in W. Broomhall

(ed.), *The Country Gentlemen's Estate Book* (The Country Gentlemen's Association, London 1914)
Farrant, S., with Fossey, K. and Peasgood, A. (eds.), *The Growth of Brighton and Hove 1840–1939* (Centre for Continuing Education, University of Sussex 1981)
Ferguson, P., 'The maps and records of the Valuation Office', University College Dublin, Unpubl. BA dissertation 1977
Fitzpatrick, D., 'Irish farming families before the First World War', *Comparative Studies in Society and History* 25 (1983), 339–74
Fitzrandolph, Helen E. and Hay, M.Doriel, *The Rural Industries of England and Wales*, 3 vols. (Oxford 1926–7, reprinted Wakefield 1977)
Foot, William, *Maps for Family History: A Guide to the Records of the Tithe, Valuation Office, and National Farm Surveys of England and Wales, 1836–1943* (Public Record Office, London 1994)
Fossey, K., 'Slums and tenements 1840–1900', in S.Farrant with K.Fossey and A.Peasgood (eds.), *The Growth of Brighton and Hove 1840–1939* (Centre for Continuing Education, University of Sussex 1981)
Foucault, M., *The Archaeology of Knowledge* (Tavistock 1972)
George, Henry, *Progress and Poverty* (London 1881)
George, William, *My Brother and I* (London 1958)
Gilbert, B., 'David Lloyd George: land, the budget, and social reform', *American Historical Review* 81(5) (1976), 10–59
Gilbert, B., 'David Lloyd George: the reform of British landholding and the Budget of 1914', *The Historical Journal* 21 (1978), 117–41
Gilbert, B., *David Lloyd George: A Political Life – the Architect of Change 1863–1912* (London 1987)
Girouard, M., 'The birth of a sea-side resort: Cromer, Norfolk', *Country Life* 150 (1971), 424–6, 502–5
Godfrey, J. D., 'A century of agriculture on the South Downs *c.*1840–1940', University of Sussex, unpubl. D.Phil. thesis (forthcoming)
Gradidge, R., *Dream Houses: The Edwardian Ideal* (London 1980)
Gregory, D., *Geographical Imaginations* (Oxford 1994)
Gribbon, S., 'An Irish city: Belfast 1911', in D.Harkness and M.O'Dowd (eds.), *The Town in Ireland* (Historical Studies 13, Dublin 1981), 203–20
Grigg, D., 'The changing agricultural geography of England: a commentary on the sources available for the reconstruction of the agricultural geography of England, 1770–1850', *Transactions of the Institute of British Geographers* 41 (1967), 73–96
Grigg, John, *Lloyd George: The People's Champion 1902–1911* (London 1978)
Guy, Robert, 'Scottish feuing and Increment Duty', *Glasgow Herald* 11 May 1910
Harley, J. B., 'Historical geography and its evidence: reflections on modelling sources', in Alan R.H.Baker and Mark Billinge (eds.), *Period and Place: Research Methods in Historical Geography* (Cambridge 1982), 261–73
Harley, J. B., 'Maps, knowledge and power', in D.Cosgrove and S.Daniels (eds.), *The Iconography of Landscape* (Cambridge 1988)
Harper, Sir Edgar, 'Lloyd George's land taxes' (Paper to International Conference on Land-Value Taxation, Edinburgh 1929, Reprinted in *Land and Liberty* (nd))
Havinden, M. A., *Estate Villages: A Study of the Berkshire Villages of Ardington and Lockinge* (University of Reading 1966)

Hey, David, *Yorkshire from AD 1000* (London 1986)
Hobhouse, L. T., *Liberalism* (reprinted Westport, Conn. 1980)
Hoppeen, K. T., 'Landownership and power in nineteenth-century Ireland: the decline of an elite', in R.Gibson and M.Blinkhorn (eds.), *Landownership and Power in Modern Europe* (London 1991), 164–80
Hunter, J., *The Making of the Crofting Community* (Edinburgh 1976)
Hyder, J. *The Land Song* (UCTLV Leaflet no.61, 1912)
Johnson, Sir Alexander, *The Inland Revenue* (The New Whitehall Series 13, London 1965)
Jordan, D., 'Merchants, "strong farmers" and Fenians: the post-famine political elite and the land war', in C.H.E.Philpin (ed.), *Nationalism and Popular Protest in Ireland* (Cambridge 1987), 320–48
Jowitt, A., 'Late Victorian and Edwardian Bradford', in J. A. Jowitt and R. K. S.Taylor (eds.), *Bradford 1890–1914: The Cradle of the Independent Labour Party* (Bradford Centre Occasional Papers 2, 1980)
Kain, R. J. P. and Baigent, E., *The Cadastral Map in the Service of the State* (Chicago 1992)
Kain, R. J. P., Fry, R. E. J. and Holt, H. M. E., *An Atlas and Index to the Tithe Files of Mid Nineteenth-Century England and Wales* (Cambridge 1986)
Kain, R. J. P. and Oliver, R., *The Tithe Maps of England and Wales* (Cambridge 1995)
Kain, R. J. P. and Prince, H. C., *The Tithe Surveys of England and Wales* (Cambridge 1985)
Keating, P. (ed.), *Into Unknown England, 1866–1913* (London 1976)
Lake, C. J., 'Finance (1909–10) Act, 1910: a list of the forms in common use and an abstract of exemptions', *Surveyors' Institution* (September 1910), 275–83
Land Enquiry Committee, Report of, *The Land* vol.I (1913) and vol.II (1914)
Land Union, *The Land Union's Reasons for Repeal of the New Land Taxes and Land Valuation* (London 1910)
Land Union, *The New Land Taxes and Mineral Rights Duty: The Land Union's Handbook on Provisional Valuations* (London, n.d. [1911])
Larsson, G., *Land Registration and Cadastral Systems* (London 1991)
Law, C. M., 'The growth of urban population in England and Wales, 1801–1911', *Transactions of the Institute of British Geographers* 41 (1967), 125–43
Lawrence, E. P., *Henry George in the British Isles* (E.Lancing, Michigan 1957)
Lintott, H., 'Mapping rural landownership. An evaluation of three national surveys as cartographic databases', University of Sussex, unpubl. MA term-paper, 1995
Lloyd George, D., *The People's Budget* (London 1909)
Lloyd George, D., 'The issues of the budget', *The Nation* (30 October 1909), 182
Lloyd George, D., *Better Times: Speeches by the Right Hon. D.Lloyd George, MP* (London 1910)
Lloyd George, D., *The People's Will* (London 1910)
Lloyd George, D., *Slings and Arrows: Sayings Chosen from the Speeches of the Rt. Hon. David Lloyd George, O.M., M.P.*, ed. P. Guedalla (London 1929)
London School of Economics, *The New Survey of London Life and Labour*, vol.VI (Survey of Social Conditions) no.2, *The Western Area* (London 1932)
Long, H. C., *The Edwardian House: The Middle Class Home in Britain 1880–1914* (Manchester 1993)

Mallet, B., *British Budgets 1887–88 to 1912–13* (London 1913)

Martins, P. Wade, *An Historical Atlas of Norfolk* (Norfolk Museums Service, Norwich 1993)

Masterman, Lucy, *C.F.G. Masterman: A Biography* (London 1939)

Mill, J.S., *Principles of Political Economy* (London 1891 edn)

Morgan, K. O., *Wales in British Politics 1868–1922* (Cardiff 1970)

Morgan, K. O., *Rebirth of a Nation: Wales 1880–1980* (Oxford 1981)

Morris, A. J. A., *C.P. Trevelyan 1870–1948: Portrait of a Radical* (Belfast 1977)

Morris, R. J., 'Clubs, societies and Associations', in F. M. L. Thompson (ed.), *The Cambridge Social History of Britain 1750–1950* vol.III (Cambridge 1990), 395–443

Mortimore, M. J., 'Landownership and urban growth in Bradford and its environs in the West Riding conurbation, 1850–1950', *Transactions of the Institute of British Geographers* 46 (1969), 105–19

Murray, B. K., 'The politics of the People's Budget', *The Historical Journal* 16 (1973), 555–60

Murray, B. K., 'Lloyd George and the land: the issue of Site-Value Rating', in J. A. Benyon *et al.* (eds.), *Studies in Local History: Essays in Honour of Professor Winifred Maxwell* (Cape Town 1976)

Murray, B. K., *The People's Budget 1909–10: Lloyd George and Liberal Politics* (Oxford 1980)

Newton-Robinson, Charles, 'The blight of the Land Taxes: why they must be repealed', *The Nineteenth Century and After* 68 (1910), 389–98

Newton-Robinson, Charles, 'The blight of the Land Taxes: a retrospect and a prospect', *The Nineteenth Century and After* 69 (1911), 1073–85

Newton-Robinson, Charles, 'The blight of the Land Taxes: why they must be repealed', *The Nineteenth Century and After* 72 (1912), 96–109

Northfield Report, The (HMSO, London 1979)

Offer, A., *Property and Politics 1870–1914: Landownership, Law, Ideology and Urban Development in England* (Cambridge 1981)

Oliver, Richard, *Ordnance Survey Maps: A Concise Guide for Historians* (Charles Close Society for the Study of Ordnance Survey maps, British Library, London 1993)

Olney, R. J. (ed.), *Labouring Life on the Lincolnshire Wolds: A Study of Binbrook in the Mid-Nineteenth Century* (Occasional Papers in Lincolnshire History and Archaeology 2, 1975)

Ottewill, D., *The Edwardian Garden* (Yale 1989)

Outhwaite, R. L., *The Rating of Land Values: The Case for Hastings, Harrogate, Glasgow* (London 1912)

Paton, M., 'Urban property and the slum', University of London, unpubl. MA dissertation, 1989

Peacock, A. J., 'Land Reform 1880–1919', University of Southampton, unpubl. MA thesis, 1961

Pease, J. A., *The Government's Record, 1906, 1907, 1908 and 1909: Four Years of Liberal Legislation and Administration* (London 1909)

Phillips, G. D., *The Diehards: Aristocratic Society and Politics in Edwardian England* (Cambridge, Mass. 1979)

Philo, C., 'Thoughts, words and "creative locational acts"', in F.W.Boal and D.N.Livingstone (eds.), *The Behavioural Environment* (London 1989), 205–34

Pike, W. T., *Nottinghamshire and Derbyshire at the Opening of the Twentieth Century: Contemporary Biographies* (London 1901)

Pomfret, J. E., *The Struggle for Land in Ireland 1800–1923* (New York 1930)

Public Record Office, *Valuation Office: Records Created under the Finance (1909–10) Act* (information leaflet no.68, London 1988)

Public Record Office, *How to Find and Use Valuation Office Maps and Field Books* (London 1992)

Raffan, P. Wilson, *The Policy of the Land Values Group in the House of Commons: An Address Delivered by P. Wilson Raffan at the National Liberal Club Political and Economic Circle* (UCTLV, London 1912)

Rawding, C. K. (ed.), *Keelby Parish and People, 1831–1881* (Keelby 1987)

Rawding, Charles (ed.), *Binbrook in the Nineteenth Century* (Binbrook WEA, Binbrook 1989)

Rawding, Charles, 'To the Glory of God? The building of Binbrook St. Mary and St. Gabriel', *Lincolnshire History and Archaeology* 25 (1990), 41–6

Rawding, Charles 'The iconography of churches: a case study of landownership and power in nineteenth-century Lincolnshire', *Journal of Historical Geography* 16 (2) (1990), 157–76

Rawding, Charles (ed.), *Binbrook 1900–1939* (Binbrook WEA, Binbrook 1991)

Rawding, Charles and Short, Brian, 'Binbrook in 1910: the use of the Finance (1909–10) Act Records', *Lincolnshire History and Archaeology* 28 (1993), 58–65

Rawlence, E. A., '"A perfectly impartial assessment": some reflections on the land valuations', *The Nineteenth Century* 75 (March 1914), 609–17

Read, D., 'Introduction: Crisis Age or Golden Age?', in D.Read (ed.), *Edwardian England* (Historical Association, London 1982), 14–39

Rees, A. D., *Life in a Welsh Countryside* (Cardiff 1950)

Rees, D. Morgan, *The Industrial Archaeology of Wales* (Newton Abbot 1975)

Riley, Bill, 'Highways and the 1909–10 Finance Act maps', *Byway and Bridleway: Journal of the Byways and Bridleways Trust* 1 (1992), 4

Rowland, P., *The Last Liberal Governments: The Promised Land, 1905–1910* (London 1968)

Rowland, P., *Lloyd George* (London 1975)

Rowlands, Marie, *The West Midlands from A.D. 1000* (London 1987)

Royle, S., 'Irish manuscript census records: a neglected source of information', *Irish Geography* 11 (1978), 110–25

Rubinstein, W. D., 'New men of wealth and the purchase of land in nineteenth-century England', *Past and Present* 92 (1981),125–47

Russell, A. K., *Liberal Landslide: The General Election of 1906* (Newton Abbot 1973)

Scottish Record Office, *Valuation Office Records: Records Created under Finance (1909–10) Act 1910* (Edinburgh nd [*c.*1993])

Scrutton, T. E., 'The work of the commercial courts', *Cambridge Law Journal* 1 (1921), 8

Searle, C., 'The odd corner of England: a case study of a rural social formation in transition, Cumbria, *c.*1700–*c.*1914', University of Essex, unpubl. Ph.D. thesis, 1983

Short, Brian, *The Geography of England and Wales in 1910: An Evaluation of Lloyd George's 'Domesday' of Landownership* (Historical Geography Research Series 22, 1989)

Short, Brian, 'Local demographic studies of Edwardian England and Wales: the use of the Lloyd George "Domesday" of landownership', *Local Population Studies* 51 (1993), 62–72

Short, Brian, *The Lloyd George Finance Act Material* (Short Guides to Records, 2nd Series) (Historical Association and British Association for Local History, London 1994)

Short, Brian, 'The twentieth century', in David Hey (ed.), *The Oxford Companion to Family and Local History* (Oxford 1996), 456–62

Short, Brian (ed.), *Scarpfoot Parish: Plumpton 1830–1880* (Centre for Continuing Education, University of Sussex 1981)

Short, Brian and Reed, Mick, 'The Finance (1909–10) Act 1910 Records', *Journal of the Society of Archivists* 8 (1) (1986), 82–3

Short, Brian and Reed, Mick, 'An Edwardian land survey: the Finance (1909–10) Act 1910 records', *Journal of the Society of Archivists* 8 (2) (1986), 95–103

Short, Brian and Reed, Mick, *Landownership and Society in Edwardian England: The Finance (1909–10) Act 1910 Records* (University of Sussex 1987)

Short, Brian, Reed, Mick and Caudwell, William, 'The county of Sussex in 1910: sources for a new analysis', *Sussex Archeological Collections* 125 (1987), 199–224

Short, Brian and Watkins, Charles, 'The National Farm Survey of England and Wales 1941–43', *Area* 26 (3) (1994), 288–93

Sinclair, C., *Tracing Scottish Local History: A Guide to Local History Research in the Scottish Record Office* (HMSO, Edinburgh 1994)

Smyth, W. J., 'Landholding changes, kinship networks and class transformation in rural Ireland: a case study from County Tipperary', *Irish Geography* 24 (1991), 1–9

Spring, D., 'Land and politics in Edwardian England', *Agricultural History* 58 (1984), 33

Spufford, Margaret, *Contrasting Communities: English Villagers in the Sixteenth and Seventeenth Centuries* (Cambridge 1974)

Taylor, W. J., 'Notes on sources: Valuation Records', *Lancashire* 8 (2) (1987), 26–8 (Lancashire Family History and Heraldry Society)

Thane, P., 'Government and society in England and Wales, 1750–1914', in F.M.L.Thompson (ed.), *The Cambridge Social History of Britain 1750–1950* vol.III (Cambridge 1990), 1–61

Thompson, F. M. L., *English Landed Society in the Nineteenth Century* (London 1963)

Thompson, F. M. L., *Chartered Surveyors: The Growth of a Profession* (London 1968)

Thompson, P., *The Edwardians: The Remaking of British Society* (2nd edn, London 1992)

Turner, M. E. and Mills, D. (eds.), *Land and Property: The English Land Tax 1692–1832* (London 1986)

Turnock, D., *The Historical Geography of Scotland since 1707: Geographical Aspects of Modernisation* (Cambridge 1982)

Turnor, Christopher, *Land Problems and National Welfare* (London 1911)

Verinder, F., *Form IV: What Next?* (United Committee for the Taxation of Land Values, London 1911)

Vincent, J. (ed.), *The Crawford Papers: The Journals of David Lindsay 27th Earl of*

Crawford and 10th Earl of Balcarres 1871–1940 during the Years 1892 to 1940 (Manchester 1984)
Wallace, A. R., *Land Nationalisation* (London 1896)
Ward, S., 'Land reform in England 1800–1940: a summary', *Agrartorteneti Szemle: Historia Rerum Rusticarum*, supplement 18 (1976), 37–44
Ward, S., 'Land reform in England 1880–1914', University of Reading, unpubl. Ph.D. thesis, 1976
Waymark, Janet, 'Landed estates in Dorset since 1870: their survival and influence', University of London unpubl. Ph.D. thesis, 1995
Wedgwood, J. C., *Land Values: Why and How They Should Be Taxed* (Land Values Publication Department, London 1911)
Wedgwood, J. C., *Memoirs of a Fighting Life* (London 1940)
White, J. D., *Land-Value Reform in Theory and Practice* (London 1948)
Whitehead, A., 'Social fields and social networks in an English rural area, with special reference to stratification', University of Wales, unpubl. Ph.D. thesis, 1971
Williams, M., *The Draining of the Somerset Levels* (Cambridge 1970)
Wilson, J. M., *Imperial Gazetteer of Scotland* vol.I (London, nd)
Wilson, O., 'Landownership and land use: continuity and change in the north Pennines' (University of Durham, Department of Geography Graduate Discussion Paper no.23, 1988)
Winterbotham, H., *The National Plans* (Ordnance Survey Professional Papers, new series no.16, HMSO, London 1934)
Wood, H., 'The tragedy of the Irish public records', *Irish Genealogist* 1 (1937), 67–71
Wright, Susan, 'Image and analysis: new directions in community studies', in B. Short (ed.), *The English Rural Community: Image and Analysis* (Cambridge 1992), 195–217
Wrigley, C., *David Lloyd George and the British Labour Movement: Peace and War* (Hassocks 1976)
Yardley, R. B., *Land Value Taxation and Rating: A Critical Survey of the Aims and Proposals with a History of the Movement* (printed for the Land Union by W. H. and L. Collingridge, London 1930)

4 Other relevant publications not otherwise referred to in the text

4a Unpublished theses

Bouquet, Mary Rose, 'The sexual division of labour: the farm household in a Devon parish', University of Cambridge, unpubl. Ph.D. thesis, 1981
Cowcher, W. B., 'A Dissertation upon the incidence of local rates and taxes upon the unearned increment of land', University of Oxford, B.Litt. thesis, 1914
Ho, Ping-Ti, 'Land and State in Great Britain 1873–1910: A Study of Land Reform Movements and Land Policies', University of Columbia, Ph.D. thesis, 1952

4b Official publications (including full details of all relevant Reports of Commissioners of Inland Revenue)

The Finance (1909–10) Act 1910
The Revenue Act 1911

The Finance Act 1920
BPP 1911 (*c.* 88) Return of Valuation Staff
BPP 1911 (Cd. 5833) XXIX The 54th Report of the Commissioners of HM Inland Revenue, Year ended 31 March 1911
BPP 1912–13 (Cd. 6344) XXIX The 55th Report of the Commissioners of HM Inland Revenue, Year ended 31 March 1912
BPP 1913 (Cd. 7000) XXVIII The 56th Report of the Commissioners of HM Inland Revenue, Year ended 31 March 1913
BPP 1914 (Cd. 7572) XXXVI The 57th Report of the Commissioners of HM Inland Revenue, Year ended 31 March 1914
BPP 1915–16 (Cd. 8116) XXIV The 58th Report of the Commissioners of HM Inland Revenue, Year ended 31 March 1915
BPP 1916 (Cd. 8425) XI The 59th Report of the Commissioners of HM Inland Revenue, Year ended 31 March 1916
BPP 1917–18 (Cd. 8887) XV The 60th Report of the Commissioners of HM Inland Revenue, Year ended 31 March 1917
BPP 1918 (Cd. 9151) X The 61st Report of the Commissioners of HM Inland Revenue, Year ended 31 March 1918
BPP 1919 (Cmd. 502) XXIV The 62nd Report of the Commissioners of HM Inland Revenue, Year ended 31 March 1919
BPP 1920 (Cmd. 1083) XVIII The 63rd Report of the Commissioners of HM Inland Revenue, Year ended 31 March 1920
BPP 1920 (Cmd. 918) Minute containing the Instructions given to the Board of the Inland Revenue by the Chancellor of the Exchequer in regard to the Recommendations of the Select Committee on National Expenditure concerning the Land Valuation Department
BPP 1921 (Cmd. 1436) XVI The 64th Report of the Commissioners of HM Inland Revenue, Year ended 31 March 1921

4c Contemporary sources

Chapman, S.J., 'The incidence of some Land Taxes and the dispersion of differential advantages', *Economic Journal* 22 (September 1912), 489–92
Devonshire, G.H. and Samuel, F., *Duties on Land Values* (London 1910)
Napier, T.B., *The New Land Taxes and their Practical Application: Being an Examination and Explanation from a Legal Point of View of the Land Clauses of the Finance (1909–10) Act 1910 and the Revenue Act 1911* (London 1912)
Scheftel, Y., *The Taxation of Land Value* (Boston 1916)
UCTLV, *Land Valuation: A Reply to the Land Union 'Guide'* (United Committee for the Taxation of Land Values, London 1910)

4d Secondary sources

Bouquet, Mary Rose, *Family, Servants and Visitors* (Norwich 1985)

Index

Page numbers in italics denote figures or tables.

Index entries related to the Finance (1909–10) Act 1910 Survey are gathered under that main heading and a series of major subheadings. Under the subheading 'administration', material on England and Wales, Ireland and Scotland is indexed separately, but the subheadings 'archives and documentation' and 'case studies' have entries to those subjects for all over Britain.

Cambridge Studies in Historical Geography

---

1 Period and place: research methods in historical geography. *Edited by* A. R. H. BAKER *and* M. BILLINGE

2 The historical geography of Scotland since 1707: geographical aspects of modernisation. DAVID TURNOCK

3 Historical understanding in geography: an idealist approach. LEONARD GUELKE

4 English industrial cities of the nineteenth century: a social geography. R. J. DENNIS*

5 Explorations in historical geography: interpretative essays. *Edited by* A. R. H. BAKER *and* DEREK GREGORY

6 The tithe surveys of England and Wales. R. J. P. KAIN *and* H. C. PRINCE

7 Human territoriality: its theory and history. ROBERT DAVID SACK

8 The West Indies: patterns of development, culture and environmental change since 1492. DAVID WATTS*

9 The iconography of landscape: essays in the symbolic representation, design and use of past environments. *Edited by* DENIS COSGROVE *and* STEPHEN DANIELS*

10 Urban historical geography: recent progress in Britain and Germany. *Edited by* DIETRICH DENECKE *and* GARETH SHAW

11 An historical geography of modern Australia: the restive fringe. J. M. POWELL*

12 The sugar-cane industry: an historical geography from its origins to 1914. J. H. GALLOWAY

13 Poverty, ethnicity and the American city, 1840–1925: changing conceptions of the slum and ghetto. DAVID WARD*

14 Peasants, politicians and producers: the organisation of agriculture in France since 1918. M. C. CLEARY

15 The underdraining of farmland in England during the nineteenth century. A. D. M. PHILLIPS

16 Migration in Colonial Spanish America. *Edited by* DAVID ROBINSON

17 Urbanising Britain: essays on class and community in the nineteenth century. *Edited by* GERRY KEARNS *and* CHARLES W. J. WITHERS

18 Ideology and landscape in historical perspective: essays on the meanings of some places in the past. *Edited by* ALAN R. H. BAKER *and* GIDEON BIGER

19 Power and pauperism: the workhouse system, 1834–1884. FELIX DRIVER

20 Trade and urban development in Poland: an economic geography of Cracow from its origins to 1795. F. W. CARTER

21 An historical geography of France. XAVIER DE PLANHOL

22 Peasantry to capitalism: Western Östergötland in the nineteenth century. GÖRAN HOPPE *and* JOHN LANGTON

23 Agricultural revolution in England: the transformation of the agrarian economy 1500–1850. MARK OVERTON*

24 Marc Bloch, sociology and geography: encountering changing disciplines. SUSAN W. FRIEDMAN

25 Land and society in Edwardian Britain. BRIAN SHORT

Titles marked with an asterisk* are available in paperback